ADVANCES IN

Applied Microbiology

VOLUME 26

CONTRIBUTORS TO THIS VOLUME

John R. Chipley

D. G. Cooper

Howard Dalton

R. S. Hanson

Ching-Tsang Hou

N. Kosaric

Allen I. Laskin

D. C. M. Ng

Ramesh N. Patel

Jerome J. Perry

I. Russell

G. C. Stewart

Joji Takahashi

J. E. Zajic

ADVANCES IN

Applied Microbiology

Edited by D. PERLMAN

School of Pharmacy
The University of Wisconsin
Madison, Wisconsin

VOLUME 26

 1980

ACADEMIC PRESS
A Subsidiary of Harcourt Brace Jovanovich, Publishers
New York   London   Toronto   Sydney   San Francisco

COPYRIGHT © 1980, BY ACADEMIC PRESS, INC.
ALL RIGHTS RESERVED.
NO PART OF THIS PUBLICATION MAY BE REPRODUCED OR
TRANSMITTED IN ANY FORM OR BY ANY MEANS, ELECTRONIC
OR MECHANICAL, INCLUDING PHOTOCOPY, RECORDING, OR ANY
INFORMATION STORAGE AND RETRIEVAL SYSTEM, WITHOUT
PERMISSION IN WRITING FROM THE PUBLISHER.

ACADEMIC PRESS, INC.
111 Fifth Avenue, New York, New York 10003

United Kingdom Edition published by
ACADEMIC PRESS, INC. (LONDON) LTD.
24/28 Oval Road, London NW1 7DX

LIBRARY OF CONGRESS CATALOG CARD NUMBER: 59–13823

ISBN 0–12–002626–0

PRINTED IN THE UNITED STATES OF AMERICA

80 81 82 83   9 8 7 6 5 4 3 2 1

CONTENTS

LIST OF CONTRIBUTORS ix
DAVID PERLMAN: 1920–1980 xi

Microbial Oxidation of Gaseous Hydrocarbons

CHING-TSANG HOU

Text 1

Ecology and Diversity of Methylotrophic Organisms

R. S. HANSON

I.	Introduction	3
II.	Occurrence and Activities of Aerobic Methane-Oxidizing Bacteria	4
III.	Isolation of Methane-Oxidizing Organisms	18
IV.	Diversity of Aerobic Methylotrophs	21
V.	Regulation of One-Carbon Metabolism in Facultative Methylotrophs	30
	References	35

Epoxidation and Ketone Formation by C_1-Utilizing Microbes

CHING-TSANG HOU, RAMESH N. PATEL, AND ALLEN I. LASKIN

I.	Microbial Epoxidation of Gaseous 1-Alkenes	41
II.	Microbial Methyl Ketone Formation	54
III.	Conclusion	67
	References	68

Oxidation of Hydrocarbons by Methane Monooxygenases from a Variety of Microbes

HOWARD DALTON

I.	Introduction	71
II.	Substrates of Methane Monooxygenases	73
III.	Methane Monooxygenase	78
IV.	Whole Cell Oxidation of Propene	85
	References	86

Propane Utilization by Microorganisms

JEROME J. PERRY

I.	Introduction	89
II.	Isolation and Types	90
III.	Induction of the Propane-Oxidizing System	95
IV.	Metabolism of Propane and Related Compounds	99
V.	Growth Yield for Microorganisms Utilizing Hydrocarbon Substrates	106
VI.	Products from Propane-Oxidizing Bacteria	108
VII.	Cooxidations Involving Propane	110
VIII.	Toxicity of Propane and Related Compounds	111
IX.	Prospecting with Propane-Oxidizing Microbes	111
	References	112

Production of Intracellular and Extracellular Protein from *n*-Butane by *Pseudomonas butanovora* sp. nov.

JOJI TAKAHASHI

I.	Introduction	117
II.	Properties of *Pseudomonas butanovora* sp. nov.	118
III.	Cellular Growth and Accumulation of Extracellular Protein	121
IV.	Properties of Intracellular and Extracellular Protein	124
V.	Concluding Remarks	126
	References	126

Effects of Microwave Irradiation on Microorganisms

JOHN R. CHIPLEY

I.	Introduction	129
II.	Effects of Microwave Irradiation on Microorganisms	130
III.	Effects of Microwave Irradiation on Other Biologic Systems	137
IV.	Thermal versus Nonthermal Effects of Microwave Irradiation	138
	References	143

Ethanol Production by Fermentation: An Alternative Liquid Fuel

N. Kosaric, D. C. M. Ng, I. Russell, and G. C. Stewart

I.	Introduction	148
II.	Alcohol as a Fuel	151
III.	Overview of Ethanol Production Processes	152
IV.	Ethanol from Sugars	154
V.	Ethanol from Starch	165
VI.	Ethanol from Cellulosic Materials	176
VII.	Ethanol from Other Wastes	199
VIII.	Economic Analysis of Ethanol Production Processes	204
IX.	Energy Considerations	218
X.	Summary and Conclusions	220
	References	224

Surface-Active Compounds from Microorganisms

D. G. Cooper and J. E. Zajic

I.	Introduction	229
II.	Carbohydrate-Containing Surfactants	231
III.	Amino Acid-Containing Surfactants	239
IV.	Phospholipids	242
V.	Fatty Acids and Neutral Lipids	245
VI.	Conclusions	250
	References	250

Index 255
Contents of Previous Volumes 261

LIST OF CONTRIBUTORS

Numbers in parentheses indicate the pages on which the authors' contributions begin.

JOHN R. CHIPLEY, *U. S. Tobacco Company, Nashville, Tennessee 37202* (129)

D. G. COOPER, *Chemical and Biochemical Engineering, Faculty of Engineering Science, The University of Western Ontario, London, Ontario, Canada N6A 5B9* (229)

HOWARD DALTON, *Department of Biological Sciences, University of Warwick, Coventry CV4 7AL, England* (71)

R. S. HANSON, *Department of Bacteriology, University of Wisconsin, Madison, Wisconsin 53706* (3)

CHING-TSANG HOU, *Corporate Research Science Laboratory, Exxon Research and Engineering Company, Linden, New Jersey 07036* (1, 41)

N. KOSARIC, *Chemical and Biochemical Engineering, Faculty of Engineering Science, The University of Western Ontario, London, Ontario, Canada N6A 5B9* (147)

ALLEN I. LASKIN, *Corporate Research Science Laboratory, Exxon Research and Engineering Company, Linden, New Jersey 07036* (41)

D. C. M. NG, *Chemical and Biochemical Engineering, Faculty of Engineering Science, The University of Western Ontario, London, Ontario, Canada N6A 5B9* (147)

RAMESH N. PATEL, *Corporate Research Science Laboratory, Exxon Research and Engineering Company, Linden, New Jersey 07036* (41)

JEROME J. PERRY, *Department of Microbiology, North Carolina State University, Raleigh, North Carolina 27650* (89)

I. RUSSELL, *Brewing Research and Development Department, The Labatt Brewing Company, Box 5050, London, Ontario, Canada N6A 4M3* (147)

G. C. STEWART, *Brewing Research and Development Department, The Labatt Brewing Company, Box 5050, London, Ontario, Canada, N6A 4M3* (147)

JOJI TAKAHASHI, *Institute of Applied Biochemistry, University of Tsukuba, Niihari-gun, Ibaraki, 300-31, Japan* (117)

J. E. ZAJIC,* *Chemical and Biochemical Engineering, Faculty of Engineering Science, The University of Western Ontario, London, Ontario, Canada N6A 5B9* (229)

*Present address: College of Science, University of Texas at El Paso, El Paso, Texas 79968.

David Perlman

DAVID PERLMAN
1920-1980

David Perlman, editor of *Advances in Applied Microbiology*, died January 29, 1980 following nearly three years of a valiant and courageous struggle with cancer.

Born in Madison, Wisconsin, David Perlman grew up and was educated in the academic environment of the University of Wisconsin, where his father, Selig Perlman, the distinguished labor historian, taught for 45 years. After completion of his graduate studies in microbial biochemistry under the tutelage of Professor Marvin J. Johnson and the late Professor William H. Peterson, Dr. Perlman worked for short periods at Hoffmann-La Roche, Inc. and Merck and Co. before joining the Squibb Institute for Medical Research, where he remained until his return to the University of Wisconsin School of Pharmacy as professor of pharmaceutical biochemistry in 1967. From 1968 to 1975 Dr. Perlman served as Dean of the School of Pharmacy while teaching undergraduate and graduate students and coordinating an expanded research program. During his administration, the pharmacy student enrollment doubled, the teaching staff increased 150%, and research space was enlarged 75%. Dr. Perlman subsequently held the Kremers Professorship of Biochemical Pharmacology at Wisconsin.

Over a professional career of some 34 years, David Perlman developed scientific interests of remarkable versatility and productivity, as researcher, educator, scholar, book editor, essayist, historian of applied microbiology, leader of scientific societies, and organizer of more than 30 symposia, conferences, short courses, workshops, and annual programs.

Much of his research focused on the development and/or improvement of fermentation processes for citric acid, 2,3-butylene glycol, penicillin, streptomycin, neomycin, vitamin B_{12}, tetracycline, ascorbic acid, riboflavin, and some aspects of mammalian cell culture. He discovered the enzyme mannosidostreptomycinase which converts mannosidostreptomycin to the clinically more useful streptomycin. With his co-workers, Dr. Perlman pioneered microbial hydroxylation of steroids, a technique which led to the development of biotransformation systems for sterols, antibiotics, alkaloids, and organic acids. Twenty-eight patents and more than 350 papers document his manifold research and scholarly works.

Dr. Perlman edited or coedited 28 books, the most recent being *Microbial Technology* (Second Edition), *Advances in Applied Microbiology* (since 1967), and *Annual Reports on Fermentation Processes* (since 1977).

His leadership and achievements in microbial biochemistry and fermentation technology, both in the United States and overseas, have been acknowledged with many honors: a fellowship from the John Simon Guggenheim Memorial Foundation; fellowships in The New York Academy of Sciences,

The American Academy of Microbiology, and The Academy of Pharmaceutical Sciences; The James M. VanLanen Distinguished Service Award, and the first Marvin J. Johnson Research Award, both presented by the Division of Microbial and Biochemical Technology, American Chemical Society; the Fisher Scientific Co. Award for Applied and Environmental Microbiology, given by the American Society for Microbiology; the Charles Thom Award for Research in Applied Microbiology, awarded by the Society of Industrial Microbiology; and the Pasteur Award for Applied Microbiology, from the Illinois Section, American Society for Microbiology.

Dr. Káto (Lenard) Perlman, Dave's wife, will continue the joint research study they shared.

The David Perlman Lectureship has been established by the University of Wisconsin Foundation to receive memorial contributions. It is sponsored by the School of Pharmacy, Department of Biochemistry, Department of Bacteriology, and Mrs. David Perlman (address: The David Perlman Lectureship, c/o Mr. Robert B. Rennebohm, University of Wisconsin Foundation, 702 Langdon Street, Madison, Wisconsin 53706).

H. J. Peppler

ADVANCES IN

Applied Microbiology

VOLUME 26

Microbial Oxidation of Gaseous Hydrocarbons

CHING-TSANG HOU

Corporate Research Science Laboratory,
Exxon Research and Engineering Company,
Linden, New Jersey

Gaseous (C_1–C_4) hydrocarbons are abundant in nature and in the fractions of oil refineries. They have recently become attractive as raw materials for chemical syntheses as well as for single-cell protein production.

Since the first methane-oxidizing bacterium was isolated by Söhngen in 1905, the progress of research in this area has been relatively slow. Prior to 1970, most of the activity on microbial oxidation of gaseous hydrocarbons was concentrated in Foster's laboratory in Texas and in Quayle's laboratory in Sheffield, England. In 1970, Whittenbury and his co-workers isolated many methane-utilizing bacteria and classified these bacteria into several groups on the basis of morphology, fine structure, and type of resting stage formed. In 1974, Hanson and his co-workers isolated facultative methane utilizers. Currently, the methylotrophic bacteria can be classified into two major classes, depending on whether they are obligate or facultative. The obligate methylotrophs have three subgroups: (1) Those with a type I membrane structure can utilize methane and have the ribulose monophosphate pathway of carbon assimilation; (2) those with a type II membrane structure can grow on methane and have so-called Icl^--serine pathway; and (3) those with no internal membrane structure are unable to use methane and have the ribulose monophosphate pathway. The facultative methylotrophs have two main subgroups. The first group utilizes methane and has the Icl^--serine pathway. The other group cannot utilize methane. Within this group, some have the ribulose phosphate pathway, others have the Icl^--serine pathway, and others have the so-called Icl^+-serine pathway.

The other gaseous alkanes are also utilized predominantly by bacteria. Most frequently cited are *Mycobacterium* and *Pseudomonas* species, but *Nocardia, Streptomyces, Flavobacter, Alkaligenes, Brevibacter, Corynebacteria, Bacillus,* and other genera are represented.

A seminar, "Microbial Oxidation of Gaseous Hydrocarbons," which included alkanes from methane to butane, was held at the 79th meeting of the American Society for Microbiology at Los Angeles. The authors of the following five chapters were the invited speakers of that seminar. These authors have agreed to publish their talks here; they include these topics: growth of microbes, genetic manipulation of these microbes, oxidation reactions, and discussion of the enzymes involved in the initial attack of gaseous hydrocarbons. It is hoped that this presentation may stimulate progress in the area of microbial oxidation of gaseous hydrocarbons.

Ecology and Diversity of Methylotrophic Organisms

R. S. Hanson

*Department of Bacteriology,
University of Wisconsin,
Madison, Wisconsin*

I.	Introduction	3
II.	Occurrence and Activities of Aerobic Methane-Oxidizing Bacteria	4
	A. Technique Used to Detect Methylotrophs in Nature	4
	B. Examples of the Distribution and Activities of Aerobic Methane-Oxidizing Organisms in Some Aquatic Environments	7
	C. Anaerobic Methane Oxidation in Aquatic Environments	12
III.	Isolation of Methane-Oxidizing Organisms	18
IV.	Diversity of Aerobic Methylotrophs	21
	A. Methane-Oxidizing Bacteria	21
	B. Yeasts That Oxidize Methane	26
	C. Bacteria and Fungi That Grow on Methanol and Methylamines but Do Not Oxidize Methane	27
V.	Regulation of One-Carbon Metabolism in Facultative Methylotrophs	30
	Is Methane Oxidation Coded for by Plasmids in Facultative Methane Oxidizers?	32
	References	35

I. Introduction

Methylotrophs are microorganisms recognized by their ability to grow on compounds that contain no carbon carbon bonds and to assimilate carbon as formaldehyde or a mixture of formaldehyde and carbon dioxide (Anthony, 1975; Colby *et al.*, 1979; Quayle, 1972; Ribbons *et al.*, 1970). Methane, methanol, and N-methyl compounds (methylamines) are the substrates most often utilized. Of these, methane, produced in anaerobic environments containing decomposable organic matter (Wolfe, 1971), is the most abundant in nature. It has been estimated that $1-4 \times 10^{15}$ gm of atmospheric methane are produced annually by biologic processes (Ehalt, 1976). The amount available from nonbiologic sources has been estimated as between 20% and 100% of that from biologic sources (Gold, 1979). Microbes that utilize methane have been detected in nearly every type of natural environment (Colby *et al.*, 1979; Dworkin and Foster, 1956; Hutton and Zobell, 1949; Quayle, 1972; Leadbetter and Foster, 1958; Heyer, 1977; Silverman, 1964; Whittenbury *et al.*, 1970a, 1976) and they are known to play a significant role in the carbon cycle of aquatic ecosystems (Cappenberg, 1972; Harrits and Hanson, 1980; Rudd and Hamilton, 1978).

The isolation and characterization of many methane-oxidizing bacteria by Whittenbury et al. (1970a) and the studies of Quayle's laboratory (Quayle, 1972; Quayle and Ferenci, 1978) have led the way to an understanding of the biology and biochemistry of the growth of microorganisms on one-carbon compounds. However, few quantitative studies on the ecology of methane-oxidizing bacteria have been published. The importance of specific methylotrophic organisms in nature has not been established. Indeed, we cannot be certain that the dominant methylotrophs in nature have been isolated and characterized. It seems apparent from recent biogeochemical and ecological studies that some remain to be discovered.

The purpose of this contribution is to review information on the ecology and diversity of microorganisms that oxidize reduced one-carbon compounds. Reviews by Anthony (1975), Colby et al. (1979), Quayle (1972), and Quayle and Ferenci (1978) were often used as sources of information during the preparation of this article. Readers are referred to these sources, the proceedings of two recent conferences on C_1 metabolism (Schlegel et al., 1976; Skryabin et al., 1977), and the other contributions in this volume for more detailed information on the microbiology and biochemistry of one-carbon metabolism.

II. Occurrence and Activities of Aerobic Methane-Oxidizing Bacteria

Hutton and Zobell (1949) pointed out that the widespread occurrence of methane-oxidizing bacteria in soil was generally appreciated by 1920. Their studies confirmed that these bacteria are abundant in soils and marine sediments where both methane and oxygen are present. Heyer (1977) examined 250 samples from a variety of natural sources by enrichment techniques and found that 90% of them contained previously known methane-oxidizing bacteria. Only the acid soils of coniferous woods and heath failed to yield methylotrophs.

Several studies, some of which are described in this review, indicate that these organisms play a more important role in carbon cycling in aquatic ecosystems.

A. Techniques Used to Detect Methylotrophs in Nature

The presence of methane-oxidizing bacteria in nature has been determined by colony counts (Cappenberg, 1972; Sorokin, 1961) and by measurements of methane oxidation (Harrits and Hanson, 1980; Howard et al., 1971; Panganiban et al., 1979; Patt et al., 1974; Rudd et al., 1974). The first procedure detects only those bacteria capable of growth on the medium chosen, which in most cases has been a mineral-salts medium that does not contain organic carbon. Organisms that form colonies on agar media, incubated under an atmosphere containing methane, include those that grow on organic and inorganic contaminants in the agar and agar-digesting bacteria, as well as methylotrophs (Whittenbury et al., 1970a). More precise colony counts of methylotrophs have been obtained by filtering dilutions of samples onto sterile bacteriological filters (Cappenberg, 1972). The filters were placed onto porous filter pads and incubated in petri dishes with a small amount of mineral-salts medium. Replicate filters were incubated under air, methane plus air, and methane alone in the dark. Microcolonies were counted after staining with methylene blue. The development of colonies in the presence of methane and air indicated that the procedure specifically detected methylotrophs. This was further confirmed in one case by incubating filters with [^{14}C]methane. It was observed that nearly all of the colonies that grew on nitrocellulose filters incorporated [^{14}C]methane into cell material precipitable by cold trichloroacetic acid (R. S. Hanson, unpublished results). However, not all colonies were pure cultures when examined microscopically. Colony counts and most probable number estimations are expected to underestimate the methane-oxidizing potential of natural samples. Whittenbury et al. (1976) estimated that they could detect only 10% of the methylotrophs present in natural samples and pure cultures by colony counts.

Enrichment culture techniques have shown that the growth of different bacteria is favored by small changes in the enrichment conditions. Whittenbury et al. (1976) pointed out the organisms isolated from a source often reflect the enrichment and plating procedures used rather than the dominant organisms in the sample. Several different methylotrophs, including mesophiles, thermophiles, and thermoduric bacteria, have been obtained from a single source (Malashenko, 1976; Whittenbury et al., 1976). Enrichments from most natural samples yield methanotrophic bacteria even when the methane-oxidizing potential of the samples is undetectable by the most sensitive methods available. Therefore, enrichments are not indicators of the relative abundance of organisms or of the relative importance of a specific organism in an ecosystem.

The detection of methane-oxidizing organisms by enrichment from natural samples can be accomplished simply by use of a modified Söhngen apparatus (Hutton and Zobell, 1949). Two bottles fitted with two-hole rubber stoppers were connected by glass tubing. The first bottle, closed to the air, was partially filled with the elective culture medium, inoculated with material from a natural source, and equilibrated with a methane–air mixture. The second bottle was filled with sterile media and was open to the air. As growth occurred, methane and oxygen were consumed and media flowed from the second bottle into the first. The rate and amount of methane oxidation was calculated from the amount of media transferred. These measurements should be confirmed by gas chromatographic estimations of methane consumption. Growth and gas consumption at the expense of contaminants in natural gas or impurities in methane or medium components can give false-positive results (Naguib and Overbeck, 1970).

The methane oxidizing potential of natural samples is most conveniently and sensitively measured by adding ^{14}C-labeled methane to samples in sealed containers. After a suitable period of incubation, the amount of labeled carbon dioxide and radioactive cell material were estimated (Rudd et al., 1974; Patt et al., 1974). In the procedure described by Rudd and his colleagues sodium hydroxide was added to the samples after an appropriate incubation period in order to terminate methane oxidation. Nitrogen was used to purge the samples of radioactive methane and an aliquot of the samples was counted in a scintillation counter. The samples were then acidified and the radioactive carbon dioxide was removed by bubbling nitrogen through the containers. The radioactivity remaining in the mixture was that incorporated into cell material. The radioactivity in carbon dioxide, lost from the acidified samples, was estimated by difference. The amounts of methane carbon converted to carbon dioxide and incorporated into cell material were calculated from the specific activity of the radioactive methane. The latter can be measured by using a gas chromatograph–gas proportion scintillation counter (Harrits and Hanson, 1980) or after combustion to carbon dioxide (Rudd et al., 1974).

Patt et al. (1974) and Panganiban et al. (1975) measured the methane-oxidizing potential of lakewater samples exposed to radioactive methane by acidifying the samples and collecting the radioactive carbon dioxide in an alkaline trapping solution. The amount of radioactive carbon assimilated was estimated by treating a portion of each sample that had been exposed to [^{14}C]methane with cold trichloroacetic acid. The precipitate was collected on glass fiber filters. The filters were dried and the radioactivity of the precipitate was determined in a scintillation counter.

The radiochemical purity of the methane used in ecological studies is very important. Several nonmethylotrophs are capable of utilizing labeled gaseous impurities (CO_2, ethane, propane, etc.). Daniels (1977) has described a convenient method for the production of labeled methane from [^{14}C]carbon dioxide by cultures of a methanogenic bacterium. The radiochemical purity of the methane produced by this method is excellent and the cost of the [^{14}C]methane is a fraction of that available from commercial sources.

B. Examples of the Distribution and Activities of Aerobic Methane-Oxidizing Organisms in Some Aquatic Environments

Studies of dimictic freshwater lakes have shown similar distributions of methane-oxidizing bacteria in several lakes and the data from these lakes can be used to assess the factors that influence the activities of methylotrophs in nature. Dimictic lakes stratify in summer to form a warm aerobic upper layer, the epilimnion, over a cool, oxygen-depleted bottom layer, the hypolimnion (Fig. 1). Rapid decreases in temperature and oxygen with increasing depth occur within the metalimnion, a narrow part of the water column separating the hypolimnion and epilimnion (Figs. 1, 2, and 3).

Methane is produced in anoxic sediments of these lakes where the redox potential is low and is distributed to the water column by diffusion (Rudd and Hamilton, 1978), by turbulence (Harrits and Hanson, 1980), and during seasonal turnovers of dimictic lakes (Rudd and Hamilton, 1978).

In the fall the epilimnion cools and turnover occurs. Mixing of the water column results, the whole lake is oxygenated, and there is a uniform distribution of all chemical and physical parameters throughout the water column.

The rates of production and oxidation of methane have been measured in several ecosystems. Data from two lakes are presented below. These lakes

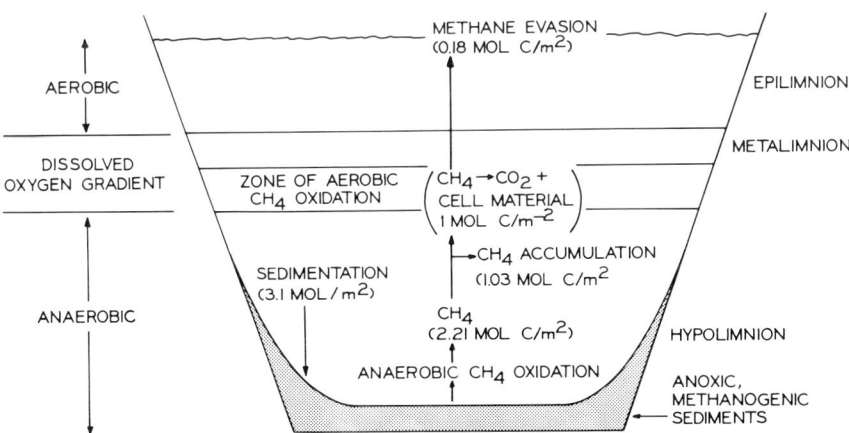

FIG. 1. Carbon sedimentation rates, methane production, and methane utilization in Lake Mendota, Madison, Wisconsin from 6/14/77 to 9/14/77. The numbers in parentheses indicate the amount of carbon produced or consumed over the 3-month period by each process. This figure represents a summary of data published by Fallon *et al.* (1980) and Harrits and Hanson (1980). Partial carbon budgets for Lake 227, Ontario, Canada, have been published by Rudd and Hamilton (1978).

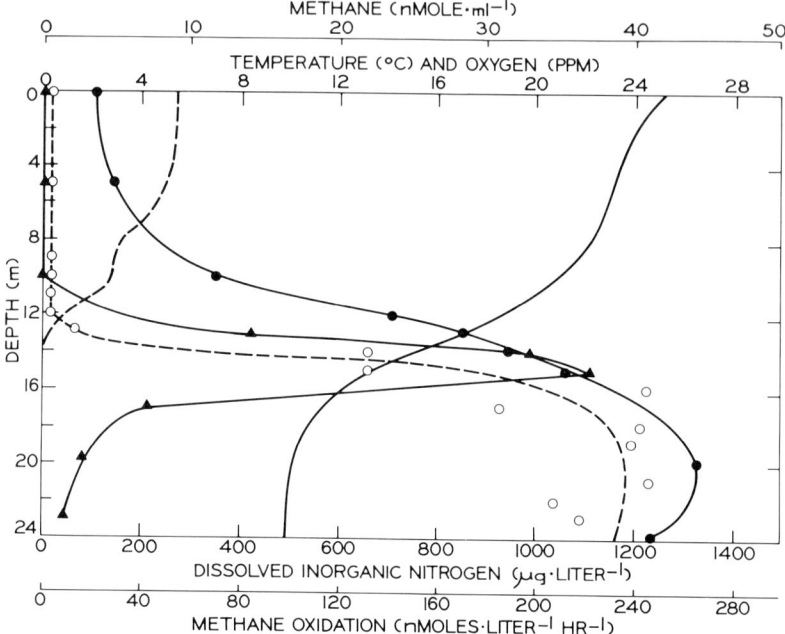

FIG. 2. Methane concentrations (○), methane oxidation rates (▲), dissolved inorganic nitrogen concentrations (●), temperature (———), and oxygen concentration (---) in the lake water column of Lake Mendota, Madison, Wisconsin during stratification (7/14/77). (From Harrits and Hanson, 1980, reproduced with permission.)

have been chosen because the factors that affect methane oxidation are best understood in these well-characterized ecosystems and because they illustrate that different factors determine the localization and activities of methylotrophs in different environments. Lake 227 is a small (5×10^4 m^2), relatively shallow (10 m deep), artificially eutrophied, shield lake in the Experimental Lakes Area of Northwestern Ontario, Canada. Lake Mendota, Madison, Wisconsin, is a larger (3.9×10^7 m^2), deeper (22 m deep), naturally eutrophic lake.

Fifty-five percent of the annual carbon input into Lake 227 was regenerated as methane (Rudd and Hamilton, 1978). Sixty percent of the methane available was oxidized to carbon dioxide and cell material and 40% escaped to the atmosphere. The rate of methane production in Lake 227 in the summer of 1974 was 10.8 mmol C/m^2/day. Because vertical diffusion into the oxygenated part of the water column was very slow, methane oxidation consumed only 11% of the methane produced during the summer and the remainder of the methane accumulated in the hypolimnion of the stratified lake. Nearly all methane oxidation by microorganisms was confined to a narrow part of the 10-m water column at the bottom of the metalimnion where the dissolved oxygen content was 0.1–0.4 mg O_2/liter (Rudd and Hamilton, 1975; Rudd et al., 1974, 1976). Methane oxidation did not occur in the epilimnion although sufficient methane was present there to support growth of methylotrophs. Subsequent experiments demonstrated that the rate of methane oxidation in lake water samples decreased when the oxygen concentration exceeded 1 mg O_2/liter.

The dissolved inorganic nitrogen (DIN) content ($NH_3 + NO_3^- + NO_2^-$) of the metalimnion of Lake 227 was less than 3 μM. Methane oxidation became insensitive to oxygen when ammonia was added to samples from this location. The data of Rudd and Hamilton (1975) and Rudd et al. (1976) support the conclusion that methylotrophs are forced to fix nitrogen in order to grow in the stratified lake. The sensitivity of nitrogenase to oxygen restricted the microorganisms to that part of the water where the oxygen concentration was very low. Methane oxidation in seawater samples from Cape Lookout Blight was also found to be sensitive to oxygen until the DIN level exceeded 14 μM (Sansone and Martens, 1978). The downward diffu-

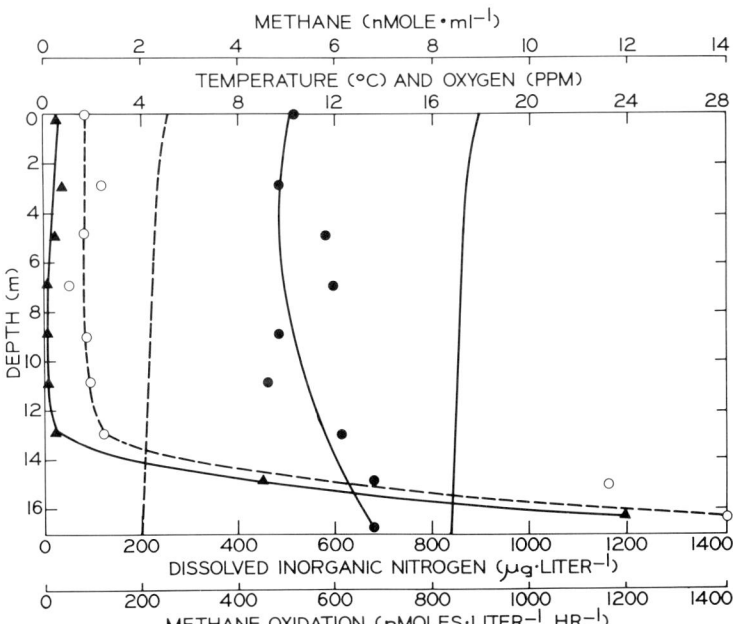

FIG. 3. Methane concentrations, methane oxidation rates, dissolved inorganic nitrogen concentrations, temperature, and oxygen concentration in the lake water column of Lake Mendota after fall turnover (9/29/77). The symbols used are the same as in Fig. 2. (From Harrits and Hanson, 1980, reproduced with permission.)

sion of oxygen is slow in Lake 227 and it is consumed by methylotrophs, thus causing the hypolimnion to become anoxic. Therefore, methane is stable in the hypolimnion.

When fall turnover occurred, the oxygen in the epilimnion mixed with the methane and dissolved inorganic nitrogen compounds of the hypolimnion. The microorganisms that oxidized methane were no longer oxygen sensitive, methane oxidation was observed throughout the water column, and 95% of the annual methane oxidation occurred after mixing. The oxidation of methane caused rapid utilization of the dissolved oxygen in the lake and in 1973-1974, when an early freezeover occurred, the lake became totally anoxic (Rudd and Hamilton, 1978).

An analysis of the carbon budget for Lake Mendota showed that 54% of the carbon sedimented (68 mmol/m^2/day) was returned to the water column of the stratified lake during the summer of 1977 (Fig. 1). Nearly half of the methane produced (36 mmol/m^2/day) was oxidized by aerobic microorganisms during this period. Most of the remaining methane accumulated in the hypolimnion until fall turnover. High rates of oxidation, determined by measurements of the conversion of labeled methane to carbon dioxide and cell material (Fig. 2), and the highest populations of methylotrophs were confined to the bottom of the metalimnion where the oxygen concentration was 0.4 mg O_2/liter or less (Harrits and Hanson, 1980). In contrast to Lake 227, the dissolved inorganic nitrogen content of the metalimnion exceeded 30 μM during stratification. The rate of methane oxidation in lake water samples was insensitive to oxygen concentrations over the range 0.2-10 mg O_2/liter and the fixation of $^{15}N_2$ could not be detected. The absence of methane in the epilimnion and the absence of oxygen in the hypolimnion preclude their activities in these locations. The relatively high rates of aerobic methane oxidation during stratification in Lake Mendota compared to Lake 227 can be explained by the more rapid transfer of oxygen and methane into the metalimnion. Because Lake Mendota is much larger and less protected from wind than Lake 227, wind turbulence caused the water column to be less stable and greater mixing occurred at the metalimnion.

After fall turnover the remaining 50% of the methane produced during stratification of Lake Mendota was rapidly oxidized and the methylotrophs were confined to the bottom of the water column where methane was consumed as rapidly as it diffused upward from the sediments (Fig. 3).

These studies indicate that methylotrophs are capable of fixing nitrogen in nature. When methane was continuously bubbled through lake water samples, the methane-oxidizing microbes in Lake Mendota water increased in number as well as in activity after the combined inorganic nitrogen was exhausted (Harrits and Hanson, 1980). In the absence of fixed nitrogen they were oxygen sensitive, like the methylotrophs present during stratification of

Lake 227. It is interesting that these microbes possess this ability, although their nitrogen-fixing potential is not expressed *in situ* in Lake Mendota (Harrits and Hanson, 1980).

Methylotrophs have been shown to be localized in the water columns of other aquatic environments. In Lake Vechten, the highest populations of methane-oxidizing bacteria were found in the hypolimnion and the populations paralleled the concentration of methane (Cappenberg, 1972). The confinement of methanotrophs to the metalimnion and hypolimnion of freshwater lakes was also observed by Anagostidis and Overbeck (1966). Jannasch (1975), and Sorokin (1961). In these cases the availability of methane in the water columns rather than oxygen sensitivity appeared to dictate the location of methylotrophs. In each case methane was almost completely oxidized as rapidly as it diffused to a point in the water column where oxygen was available. Losses of methane by emanation in deep lakes, such as Lake Kivu, Africa (Jannasch, 1975), appear to be negligible.

The methane oxidized is converted to carbon dioxide and cell material in ratios of 3.6:1 in Lake Mendota (Harrits and Hanson, 1980) and 2:1 in Lake 227 (Rudd and Hamilton, 1975). The methylotrophs in nature therefore are less efficient in the assimilation of methane carbon than pure cultures of methylotrophs (Quayle and Ferenci, 1978). The carbon dioxide diffuses upward where it is utilized by CO_2-fixing phototrophs. The cell material can be a source of seston for secondary grazers in the ecosystem, then recycled by aerobic heterotrophs or by sedimentation and methanogenesis (Rudd and Hamilton, 1978). The carbon recycled by aerobic methylotrophs in Lake Mendota during one summer was approximately 3.7×10^7 mol (Fallon *et al.*, 1980). Perhaps the greatest ecological impact of these bacteria is on the oxygen budget of freshwater lakes (Overbeck and Ohle, 1964; Cappenberg, 1972; Rudd and Hamilton, 1978; Harrits and Hanson, 1980). They are believed to be primarily responsible for oxygen consumption at the metalimnion of several dimictic and meromictic lakes and after fall turnover of dimictic lakes. Their activities result in the production of a favorable environment for sulfate-reducing bacteria and other anaerobes in the hypolimnion (Cappenberg, 1972). Rudd and Hamilton (1978) have suggested that a nontoxic inhibitor of methane oxidation, such as acetylene produced from calcium carbide, may prevent oxygen utilization and kills of invertebrates and fishes that are a consequence of anoxia in frozen lakes.

Whittenbury *et al.* (1976) have pointed out that the environmentally important activities of methylotrophs need not be confined to one-carbon metabolism. These bacteria are capable of cooxidizing a large number of organic compounds (Stirling and Dalton, 1979), carbon monoxide (Ferenci, 1974), and ammonia (Whittenbury *et al.*, 1970a). These cooxidations are not necessarily useless activities. Ethanol, which would not support growth,

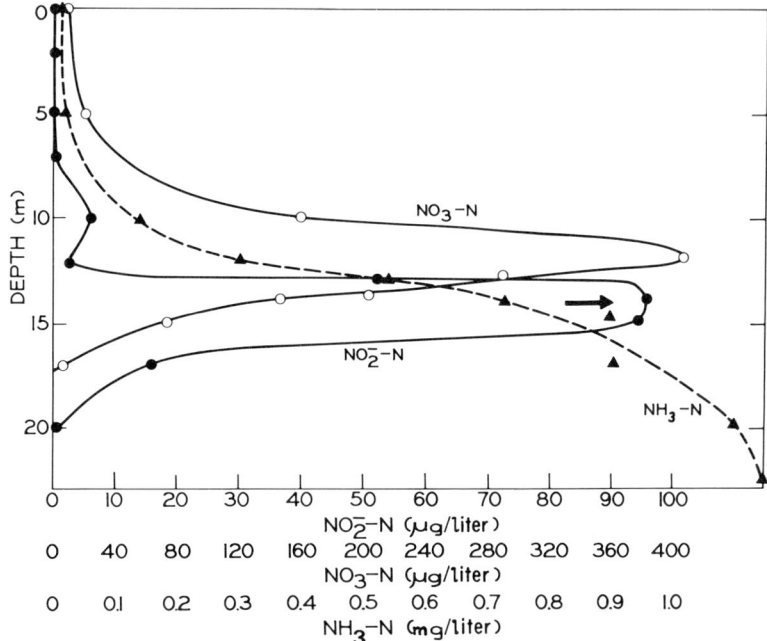

Fig. 4. The concentration of nitrogenous compounds as a function of depth in Lake Mendota, Madison, Wisconsin. The concentrations of nitrite, nitrate, and ammonia were measured as described by Harrits and Hanson (1980) in water samples collected on 6/14/77. The peak methane oxidation rate is indicated by the arrow. The shape of the methane oxidation rate versus depth curve on the same date (Fig. 2) closely parallels the NO_2^- concentration.

served as a hydrogen donor for nitrogen fixation (Dalton and Whittenbury, 1975). The coincidence of nitrite and methane oxidation in the lake water column of Lake Mendota (Fig. 4) suggests that methanotrophs may play a significant role in nitrification in this lake. Several methylotrophs are known to cooxidize ammonia to nitrite (Whittenbury et al., 1970a).

C. Anaerobic Methane Oxidation in Aquatic Environments

Methane has been thought to be a refractory substrate for anaerobic organisms. The oxidation of methane by aerobic bacteria involves molecular oxygen as a substrate in the initial hydroxylation reaction (Higgins and Quayle, 1970). Anaerobic oxidation requires a unique and undiscovered biochemical transformation of this compound (Quayle, 1972).

Measurements of methane distributions in marine sediments have indicated that methane is not conserved in the absence of oxygen in marine sediments (Reeburgh and Heggie, 1977). It is nearly absent in high-sulfate zones between the sediment surface and 20 cm below the surface and increases dramatically in the deeper sediments that are nearly devoid of sulfate (Claypool and Kaplan, 1974; Martens and Brenner, 1977; Reeburgh, 1976). For example, methane concentrations were 5–50 μmol/liter between the sediment water interface to 30–35 cm into the sediment of the Cariaco Trench and increased to more than 100 times this concentration deeper into the sediment (Reeburgh, 1976). The use of diffusion models led Barnes and Goldberg (1976) and Reeburgh (1976) to the conclusion that the concave upward methane versus depth profiles of several sediments could be explained by methane consumption in the sulfate-reducing zones. Martens and Brenner (1977) ruled out alternative explanations of their data on methane concentration versus depth on cores from sediments of Long Island Sound. A summary of methane and sulfate distributions in several anoxic marine sediments were used by Reeburgh and Heggie (1977) to support the hypothesis that the process governing the distributions of methane and sulfate in anoxic sediments was anaerobic methane oxidation or cooxidation by microorganisms. It was considered unlikely that oxidants other than sulfate could serve as electron acceptors for this process in marine sediments (Reeburgh, 1976), and Reeburgh and Heggie (1977) concluded that anaerobic methane oxidation did not occur in the absence of sulfate or in the presence of oxygen.

Kosiur and Warford (1979) added labeled acetate and lactate to Santa Barbara Basin sediments. The measured changes in the specific activities of $^{14}CH_4$ and $^{14}CO_2$ with time indicated to them that anaerobic methane oxidation occurred. Recent radiotracer experiments by Reeburgh (1979) provide more direct evidence that $^{14}CH_4$ is oxidized to $^{14}CO_2$ in sediments from Skan Bay, near Unalaska Island. The rate of oxidation was found to be highest in a narrow part of the sediment column near the base of the sulfate reduction zone where the product $(CH_4) \times (SO_4^{2-})$ was highest. Above this zone methane concentrations were low and below it sulfate was depleted. Estimations of $^{13}C/^{12}C$ ratios in carbon dioxide show a minimum of $\delta^{13}CO_2$ in approximately the same part of the anoxic sediment column where methane oxidation occurs (Reeburgh, 1979). They have interpreted the data on the distribution of the isotopes to indicate that methane oxidation is a significant source of carbon dioxide in the sulfate reduction zone of the sediment column.

The rate of anaerobic methane oxidation has been estimated to be between 68 μmol/liter/year and 0.232 mmol/liter/year (Barnes and Goldberg, 1976; Kosiur and Warford, 1979; Reeburgh and Heggie, 1977).

These studies support the hypothesis that methane diffusing upward from anoxic marine sediments is oxidized when it contacts sulfate that diffuses downward from the water column. Reeburgh and Heggie (1977) estimated that most of the methane produced in Cariaco Trench sediments is consumed by anaerobic oxidation in some sediments of the methane and little (1-10%) escapes to the water column. About half of the sediment sulfate flux was consumed by these reactions in Cariaco Trench sediments according to Reeburgh (1976). An extension of these results to other marine environments may show that anaerobic methane oxidation is a larger sink for methane on a global basis than aerobic oxidation.

Estimates of global methane production by biologic processes usually are based on the amount of methane that is evaded to the atmosphere. The recent work on the oxidation of methane in marine and freshwater environments indicates that only a fraction of the methane produced in the deep ocean reaches the water column and in deep lakes little reaches the water surface. Therefore, the amount of methane produced by biologic processes may be much greater than previously calculated.

Whelan (1974) noted that freshwater lakes with high sulfate concentrations have low methane concentrations and vice versa. Because sulfate concentrations in eutrophic freshwater range between 1 and 10 μM, the anaerobic oxidation of methane coupled to sulfate reduction would be expected to be less significant than in marine systems where sulfate concentrations usually exceed 10 mM. Most of the methane escapes to the water column in freshwater systems and aerobic methane oxidation would be expected to be the major methane sink. The reaction of methane with sulfate to produce CO_2 and H_2S ($CH_4 + SO_4 + 2 H^+ \rightarrow H_2 + CO_2 + H_2O$) is favorable under standard conditions (-4.5 kcal/mol) but does not occur at appreciable rates under 500°C in the absence of microbial activity (Freely and Kulp, 1957). The free energy change, when the sulfate concentration is 10 mM, (HCO_3^-) is 30 mM, (HS^-) is 0.5 mM, and methane is 0.1 mM, is -6.0 kcal/mol. These concentrations are similar to those found in marine environments (Martens and Brenner, 1977). Even more favorable conditions exist in some marine sediments (Reeburgh, 1976).

Davies and Yarbrough (1966) observed that the anaerobic sulfate-reducing bacterium, *Desulfovibrio desulfuricans*, oxidized ^{14}C-labeled methane to [^{14}C]carbon dioxide in the presence of lactate as the major carbon source. Sorokin (1957) and Postgate (1969), however, failed to detect methane oxidation by sulfate-reducing bacteria. Wertlieb and Vishniac (1967) reported that the phototrophic bacterium *Rhodopseudomonas gelatinosa* oxidized $^{14}CH_4$ to $^{14}CO_2$ and assimilated methane in the absence of air. This experiment has not been successfully repeated. It remains uncertain whether methane is cometabolized by heterotrophic sulfate-reducing bacteria, as suggested by Mechalas (1974) and Reeburgh (1976), or is metabolized by a unique group of anaerobic methylotrophs, as suggested by the work of Panganiban (1976).

Panganiban *et al.* (1979) have shown that anaerobic oxidation of ^{14}C-

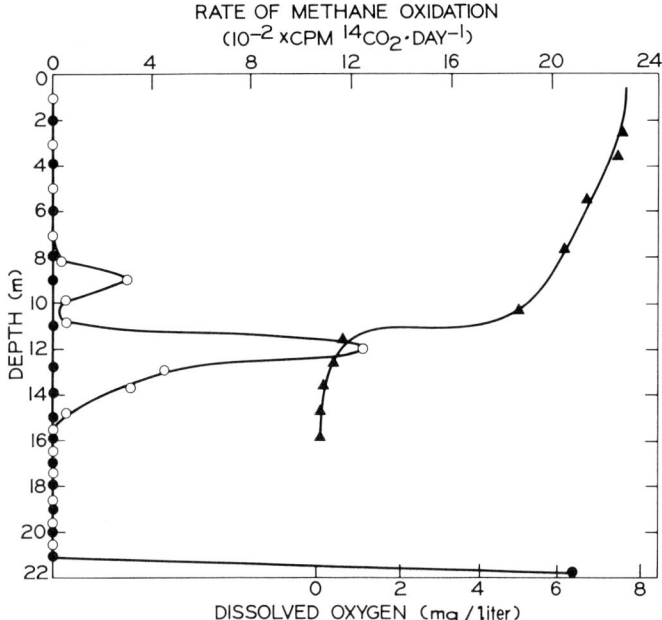

FIG. 5. Aerobic and anaerobic oxidation of methane as a function of depth in lake water samples from Lake Mendota, Madison, Wisconsin (7/12/75). The procedures employed for measurements of methane oxidation rates have been described by Panganiban et al. (1979). The specific activity of $^{14}CH_4$ used was 7.7 μCi/mmol. The rate of oxidation of labeled methane to $^{14}CO_2$ in the absence of air (•) occurs only at the sediment surface. The most rapid oxidation of methane (○) occurs at the metalimnion, which is indicated by a rapid decrease in dissolved oxygen content of the water (▲). (Adapted from Panganiban et al., 1979, with permission.)

labeled methane is converted to $^{14}CO_2$ in samples taken from the sediment surface of Lake Mendota, Madison, Wisconsin. Anaerobic oxidation occurred only at the sediment surface. No oxidation occurred in the presence of air in these samples (Fig. 5; Panganiban et al., 1979). Enrichments continued to oxidize methane only when sulfate (50 mM) was present as an electron acceptor and acetate or lactate served as carbon sources. The growth of the organisms in liquid media and on solid media containing sulfate as an electron acceptor and acetate as a carbon source was dependent on the presence of methane (Fig. 6). The production of [^{35}S]HS$^-$ from $^{35}SO_4^{2-}$ (Fig. 7) and the assimilation of [^{14}C]acetate (Panganiban et al., 1979) by enrichment cultures were also dependent on methane. Acetate was not oxidized to carbon dioxide by anaerobic lakewater samples or enrichment cultures (Panganiban et al., 1979). Methane carbon was not assimilated into cell material and in this respect this process differed from methane metabolism by all known aerobic methylotrophs.

FIG. 6. Anaerobic growth of an enrichment culture from Lake Mendota on methane. The enrichment cultures were obtained as described in Fig. 7. When turbidity developed in the culture after the second transfer, 0.5-ml samples were transferred to 15 × 150 mm tubes containing 4–5 ml of media (Panganiban et al., 1979). The tubes were sealed with rubber stoppers. The tube on the left contained an atmosphere of oxygen-free N_2. The tube on the right contained an oxygen-free atmosphere of CH_4 and N_2 (1:4 v/v). Growth, evidenced by turbidity, occurred only in tubes containing methane in the atmosphere.

Zehnder and Brock (1979) have shown that cultures of several methanogenic bacteria oxidize $^{14}CH_4$ to $^{14}CO_2$ while producing methane. The amount of methane simultaneously oxidized varied between 0.3 and 0.001% of that produced. A small amount of methane carbon was assimilated into cell material. It is conceivable that a consortium of methanogens and

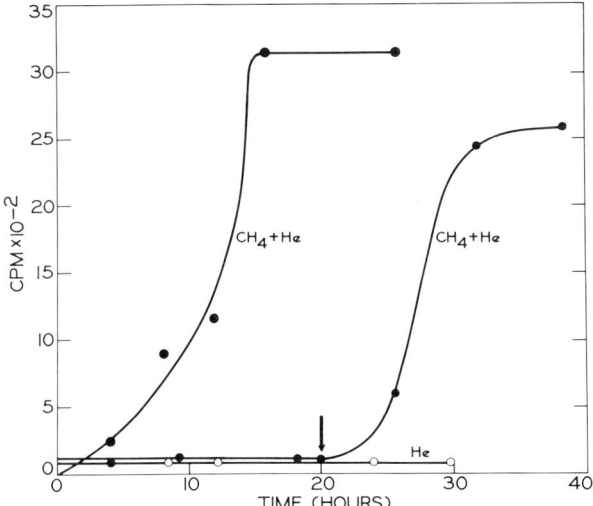

FIG. 7. The production of $^{35}H_2S$ from $^{35}SO_4^{2-}$ by enrichment cultures. The cultures used were obtained by inoculation of a mineral-salts medium containing 50 mM SO_4^{2-} and 0.05% sodium acetate (Panganiban et al., 1979) with sediment surface samples from Lake Mendota, Madison, Wisconsin. The culture was incubated at 37°C under an atmosphere containing methane and helium (1:4 v/v). When turbidity developed, 10 ml of this culture was transferred to 90 ml of the same medium in a 1-liter flask. This culture was incubated under identical conditions. After turbidity developed (48 hours), 5-ml aliquots were transferred to tubes sealed with rubber stoppers. One millicurie of $^{35}SO_4$ was added and the tubes were continuously purged with the gases indicated in the figure. The gas mixtures, methane:helium (1:4 v/v) and helium, were passed over hot copper filings to remove traces of oxygen. The effluent gases from the cultures were passed through an alkaline trapping solution (Panganiban et al., 1979). The radioactive hydrogen sulfide collected in the trapping solution was estimated in a scintillation counter (Panganiban et al., 1979). One culture was incubated without methane for 20 hours. Methane and helium were then bubbled through the culture from the time indicated by the arrow. Radioactive hydrogen sulfide was not produced by this culture until methane was present. Hydrogen sulfide was not formed in the control culture that did not receive methane.

sulfate reducers could accomplish a reversal of methanogenesis in environments with high methane partial pressures. Hydrogen produced by the methanogens can be utilized by sulfate reducers and the partial pressure of hydrogen could be reduced to the point where the reversal of methanogenesis is thermodynamically favorable (Zehnder and Brock, 1979).

Because bacterial enrichments grow and oxidize methane to carbon dioxide in the presence of sulfate and acetate (Panganiban et al., 1979) it appears that anaerobic sulfate-reducing bacteria play the major role in methane oxidation in anoxic environments of one freshwater lake. It is now obvious in light of the recent descriptions of methane consumption in marine systems that these experiments should be repeated in the more favorable marine environments, particularly those where methane and sulfate concentrations are both high and the free energy changes are more favorable. It is important to determine whether this reaction can be accomplished by a single species or requires mixed populations of microorganisms.

III. Isolation of Methane-Oxidizing Organisms

The principles, media, gas mixtures, and other conditions employed for the isolation of methylotrophic bacteria have been reviewed by Dworkin and Foster (1956), Foster and Davis (1966), Hutton and Zobell (1949), Quayle (1972), Leadbetter and Foster (1958), Naquib and Overbeck (1970), Silverman (1964), and Whittenbury et al. (1970a).

The inoculation of small amounts of soil, water, or other natural material into these liquid media followed by incubation under a methane–air atmosphere results in the appearance of turbid cultures or pellicles and gas utilization in a few days to 2 weeks. The incubation of a liquid-salts medium, inoculated with material from aquatic environments at 30°C without shaking, favored the growth of *Pseudomonas methanica* as a pink pellicle (Dworkin and Foster, 1956; Leadbetter and Foster, 1958). When loopfuls of the pellicle were streaked onto agar and incubated under methane and air at 30°C, pink colonies appeared in 5–7 days. This technique yielded pure cultures of *P. methanica* (*Methylocystis* of Whittenbury et al., 1970b) from several aquatic sources. Modifications of this technique, including shaking of enrichments or changes in incubation temperatures, promoted the growth of other organisms. *Pseudomonas methanitrificans* (believed to be a strain of *Methylosinus*, Whittenbury et al., 1970b) was isolated by Davis et al. (1964) from soils exposed to natural gas by successive transfer in liquid media devoid of combined nitrogen. The cultures were incubated with shaking at 30°C under an atmosphere of methane and air. *Methanomonas methanooxidans* was isolated from several sources by Brown et al. (1964) and Stocks and McCleskey (1964). The enrichments were shaken at 30°C under methane and air. After several transfers a sample was diluted and microcolonies were picked with the aid of a microscope. The colonies did not grow to a size easily visible without the use of a microscope.

Methylococcus capsulatus was isolated by Foster and Davis (1966) by elective culture in a liquid medium at 50–55°C. After streaking onto mineral-salts agar, visible colonies formed in 10–14 days at 37°C under a methane–air atmosphere.

In spite of vigorous attempts to isolate methane-oxidizing bacteria in pure culture, only three well-defined species were described before 1970.

Whittenbury et al. (1970a) obtained pure cultures of more than 100 gram-negative, aerobic, methane-utilizing bacteria. They inoculated 25 ml of a mineral-salts media (with NO_3 or NH_3 as N sources) in 250-ml bottles with approximately 1 gm of soil or water from a variety of sources. Mud, soil, and water from several countries were used as inocula. The bottles were sealed with SUBASEAL® caps and 20 ml of methane were injected. The bottles were incubated at 30, 45, and 55°C without shaking. After turbidity developed in the cultures (3–4 days) they were serially diluted in sterile tap water and spread onto plates of a mineral-salts agar medium. The plates were incubated in a methane–air atmosphere and examined at 3-day intervals for 3–4 days.

Small, usually transparent colonies of non-methane-utilizing bacteria appeared rapidly and reached their maximum size in 3–4 days. Opaque colonies of methylotrophs were 10- to 100-fold fewer in number and continued to increase in size for 2–3 weeks. These colonies were picked with the aid of a microscope when small (0.2 mm) and were transferred to agar slopes. The slopes were incubated with methane–air for 2 weeks.

The success of Whittenbury et al. (1970a) in isolating new species as well as in reisolating some previously described bacteria was apparently because of the short enrichment period employed, which limited losses of methylotrophs by predation and prevented overgrowth of bacteria that grew on substrates other than methane. The isolation of cultures from microcolonies as soon as they were visible under a plate microscope and frequent observations of the plates seem to have been necessary for the isolation of large numbers of methylotrophs in pure culture. Experience with several successful isolation attempts obviously played a role in subsequent isolations.

Trotsenko (1975) was similarly successful in isolating 100 colonies of obligate methylotrophs using a modification of the technique of Whittenbury et al. (1970a). Soil and water samples were inoculated into 0.5 ml of a mineral-salts medium and incubated standing with a 1:1 mixture of methane and air at 30°C. Dilutions of enrichments were used to inoculate media solidified with silica gel (Galchenko, 1975). Microcolonies were picked from silica gel plates with the aid of a microscope and transferred to silica gel slopes. The use of silica gel rather than agar has been reported to be more effective because of its reduced toxicity for methylotrophs (Aiyer, 1920; Naguib and Overbeck, 1970) and because the number of colonies of non-methane-utilizing bacteria on the plates was reduced (Galchenko, 1975).

Another effective technique involved dilution and filtration of enrichment cultures onto nitrocellulose filters (Patt et al., 1974). The filters, wetted with a mineral-salts medium, were incubated under a methane–air atmosphere. The filters were also used to purify cultures by streaking (Patt et al., 1974). This technique also reduced the amount of contamination by non-methane-utilizing bacteria and permitted growth of strains that were sensitive to substances in agar.

Nearly all enrichments contain more than one kind of bacterium. Stocks and McClesky (1964) observed that the methanol-oxidizing bacterium *Vibrio extorquens* was almost a permanent escort of methane oxidizers. In other enrichments bacteria of the genus *Hyphomicrobium* are present (Wilkinson *et al.*, 1974).

Stable mixed cultures have been obtained by enrichment. Some mixed cultures grow with shorter generation times and give higher growth yields than pure cultures of methane-utilizing bacteria (Vary and Johnson, 1967; Wilkinson *et al.*, 1974). When these cultures were diluted and spread on a mineral-salts agar, colonies often contained two or three morphologically distinct bacteria (Vary and Johnson, 1967). The methane-oxidizing bacteria are difficult to obtain in pure cultures. The non-methane-utilizing members of the consortia can usually be isolated in pure culture by streaking onto a mineral-salts agar without methanol or more complex organic substrates (Lamb, 1974; Schilling, 1978; Wilkinson *et al.*, 1974).

The interactions between members of two mixed cultures have been described. The mixed culture described by Wilkinson *et al.* (1974) contained a *Pseudomonas* sp. that grew on minimal-medium agar plates when methane was supplied, a *Hyphomicrobium* sp., an *Acinetobacter* sp., and a *Flavobacter* sp. Low concentrations of methanol reversibly inhibited growth of the *Pseudomonas* sp. This methane-utilizing organism would not grow in liquid cultures because it produced sufficient methanol during the oxidation of methane to inhibit its own growth. The *Hyphomicrobium* sp. utilized the small concentrations of methanol; therefore, inhibition of growth by methanol was prevented in the mixed cultures. The *Acinetobacter* sp. and *Flavobacter* sp. apparently grew on excreted organic materials or products of cell lysis that may also have been toxic to the methane-oxidizing *Pseudomonas* sp. The poor affinity of the *Pseudomonas* sp. for methanol coupled with the toxicity of low concentrations of methanol and organic substances caused this organism to be dependent on others in liquid culture.

Similar, but more complex interactions, were observed in a mixed culture studied by Lamb (1974). This consortium, originally isolated and studied by Sheehan and Johnson (1971), contained three organisms. The methylotrophic bacterium grew on agar media with methane or methanol when vitamin B_{12} was supplied. A second citrate-utilizing organism grew on organic products of the metabolism of the methylotroph and excreted vitamin B_{12}. These two organisms formed a stable methane-utilizing mixed culture. Addition of the third organism, a methanol-utilizing bacterium, increased the rate of biomass production of the mixed culture.

It has been our experience that these consortia that are difficult to purify are commonly isolated when enrichments are prolonged for several transfers or when the largest colonies are selected from agar plates after 2–3 weeks of incubation with methane. This has been particularly true when nitrate and ammonia are omitted from the enrichment and purification media in attempts to enrich for nitrogen-fixing species (Schilling, 1978). These consortia might be selected after several transfers because they outgrow the more independent methylotrophs. These observations raise questions about the organisms that dominate in nature. The bacteria in the consortia have not been extensively characterized.

Heating of samples at 55–80°C prior to elective culture eliminates predators and other bacteria. This method would be expected to select for species of methylotrophs that are thermoduric or thermotolerant or those that form thermotolerant resting stages. *Methylobacterium organophilum* XX, a facultative methylotroph, was isolated after heating lakewater samples (Patt et al., 1974). The use of antibiotics may also assist in the reisolation of some species. Mary L. O'Connor (personal communication) observed, and we have confirmed, that *M. organophilum* R6 (Patel et al., 1978) is resistant to 50 μg/ml of tetracycline. *Methylobacterium organophilum* XX has exhibited a high degree of resistance to rifamycin (M. L. O'Connor, personal communication).

Elective culture techniques with low-pH media have been used by Wolf (1979) to isolate four strains of methylotrophic yeast.

One of the persistent problems at our laboratory has been the growth of fungi on agar plates containing colonies of methylotrophs. Once established, they are very difficult to eliminate. The addition of 100 μg/ml of cyclohexamide to agar media has helped reduce the frequency of these troublesome contaminants.

Elective culture techniques used in the past have employed mineral-salts media to prevent overgrowth of heterotrophs. Therefore, methylotrophs that require additional growth factors have not been isolated except as members of consortia. Rudd and Hamilton (1975) observed that methylotrophs in enrichments from Lake 227 required unidentified factors present in lake water for growth and continued methane oxidation.

IV. Diversity of Aerobic Methylotrophs

A. Methane-Oxidizing Bacteria

Kaserer (1905) was the first to report the presence of methane-oxidizing organisms in soil and Söhngen (1906) was the first to isolate an organism able to grow on methane as the carbon and energy source. He named the bacterium *Bacillus methanicus*. The name was changed to *Methanomonas methanica* by Orla-Jensen (1909), to *Pseudomonas methanica* by Dworkin and Foster (1956), and finally to *Methylomonas methanica* (Foster and Daivs, 1966). Hutton and Zobell (1949) and Strawinski and Tortorich (1955) isolated several other cultures of obligate methylotrophs that were not completely characterized. The only well-characterized isolates prior to 1969 were *Methanomonas methanica* (Dworkin and Foster, 1956), *Methanomonas methanica* (Dworkin and Foster, 1956), *Methanomonas methanooxidans* (Brown and Strawinski, 1956), and *Methylococcus capsulatus* (Foster and Davis, 1966). The now classical studies of Whittenbury *et al.* (1970a,b) on methane-oxidizing bacteria provided a basis for their classification. All of over 100 strains isolated were gram-negative, catalase-positive, aerobic bacteria that utilized compounds without carbon–carbon bonds as substrates. All were obligate methylotrophs. Only methane, methanol, and dimethylether served as carbon and energy sources. Many of the isolates formed heat- or desiccation-resistant resting stages and cells of all strains contained complex intracytoplasmic membrane structures (Davies and Whittenbury, 1970; Proctor *et al.*, 1969; Quayle, 1972; Smith and Ribbons, 1970).

The types of intracytoplasmic membrane structures found in methylotrophs have been subdivided into two basic types (Whittenbury *et al.*, 1970b). Type I bacteria contain bundles of disk-shaped membrane vesicles distributed throughout the cell (Davies and Whittenbury, 1970). Type II bacteria contain a system of paired membranes throughout the cell or concentrated at the periphery (Davies and Whittenbury, 1970).

All the methane-oxidizing bacteria oxidize methane to carbon dioxide. Methanol, formaldehyde, and formate are intermediates in the catabolic pathway (Anthony, 1975; Colby *et al.*, 1979; Quayle, 1972; Ribbons *et al.*, 1970). Formaldehyde or formaldehyde plus carbon dioxide are assimilated for the synthesis of cell material (Quayle, 1972; Quayle and Ferenci, 1978; Colby *et al.*, 1979).

All type I methane-oxidizing bacteria assimilate carbon via the ribulose monophosphate (RMP) pathway, described by Kemp (1974), Kemp and Quayle (1967), and Lawrence *et al.* (1970), or by one of its variations (Quayle and Ferenci, 1978). The initial reactions in this pathway, shown in Eq. (1), are catalyzed by 3-hexulose phosphate synthase (HPS) and phospho-3-hexulose isomerase (PHI).

$$\text{HCHO} + \begin{array}{c} CH_2OH \\ | \\ CO \\ | \\ HCOH \\ | \\ HCON \\ | \\ CH_2OPO_3^{2-} \end{array} \xrightarrow{\text{HPS}} \begin{array}{c} CH_2OH \\ | \\ HOCH \\ | \\ C=O \\ | \\ HCOH \\ | \\ HCOH \\ | \\ CH_2OPO_3^{2-} \end{array} \xrightarrow{\text{PHI}} \begin{array}{c} CH_2OH \\ | \\ C=O \\ | \\ HOCH \\ | \\ HCOH \\ | \\ HCOH \\ | \\ CH_2OPO_3^{2-} \end{array} \quad (1)$$

Ribulose 6-phosphate
D-Erythro-L-glycero-3-hexulose 6-phosphate
Fructose 6-phosphate

In the first stage of the pathway, three molecules of formaldehyde are condensed with three molecules of ribulose 6-phosphate to yield three molecules of fructose 6-phosphate. One molecule of fructose 6-phosphate is cleaved to yield two molecules of three-carbon compounds in stage 2. Finally, three molecules of the C_1 acceptor, ribulose 6-phosphate, are regenerated from two molecules of fructose 6-phosphate and one molecule of glyceraldehyde 3-phosphate to complete the cycle in stage 3. The three-carbon compound generated from three molecules of formaldehyde is used for cell carbon synthesis via pathways present in heterotrophic and autotrophic bacteria. Detailed descriptions of the enzymes and the variations in stages 2 and 3 of the pathway found in methylotrophic bacteria have been presented by Quayle and Ferenci (1978).

The use of formaldehyde as the sole carbon source for biosynthesis by the ribulose monophosphate pathway requires very little or no ATP input depending on the enzymes involved in phases II or III (Quayle and Ferenci, 1978). It is theoretically possible to synthesize pyruvate from 3 mol of formaldehyde in an ATP-yielding sequence of reactions by a variant of this pathway.

Type II methane-oxidizing bacteria all utilize the serine pathway for formaldehyde fixation. The reactions of this pathway are shown in Fig. 8. The synthesis of cell carbon by this pathway consumes more ATP than the RMP pathways, but less than the Calvin cycle for CO_2 fixation.

The methane-oxidizing bacteria were divided into the five different groups by Whittenbury *et al.* (1970b), as shown in the accompanying tabulation.

Group	Morphology	Intracytoplasmid Membrane type	Resting stage description
Methylosinus	Rod or pear shaped	II	Exospore
Methylocystis	Rod or vibroid	II	"Lipid" cyst
Methylomonas	Rod	I	Immature *Azotobacter*-type
Methylobacter	Rod	I	*Azotobacter* type cyst
Methylococcus	Coccus	I	Immature *Azotobacter*-type cyst

Two reports by Patt *et al.* (1974) and Patel *et al.* (1978) have described the isolation of facultative methane-oxidizing bacteria; bacteria capable of growth on methane as well as more complex substrates including succinate, glucose, and nutrient broth. These organisms have been named strains of *Methylobacterium organophilum*. They contain type II membranes and have several other features in common with the type II obligate methane-oxidizing bacteria.

Colby *et al.* (1979) have presented a tentative classification scheme that includes obligate and facultative methane-oxidizing bacteria and incorpo-

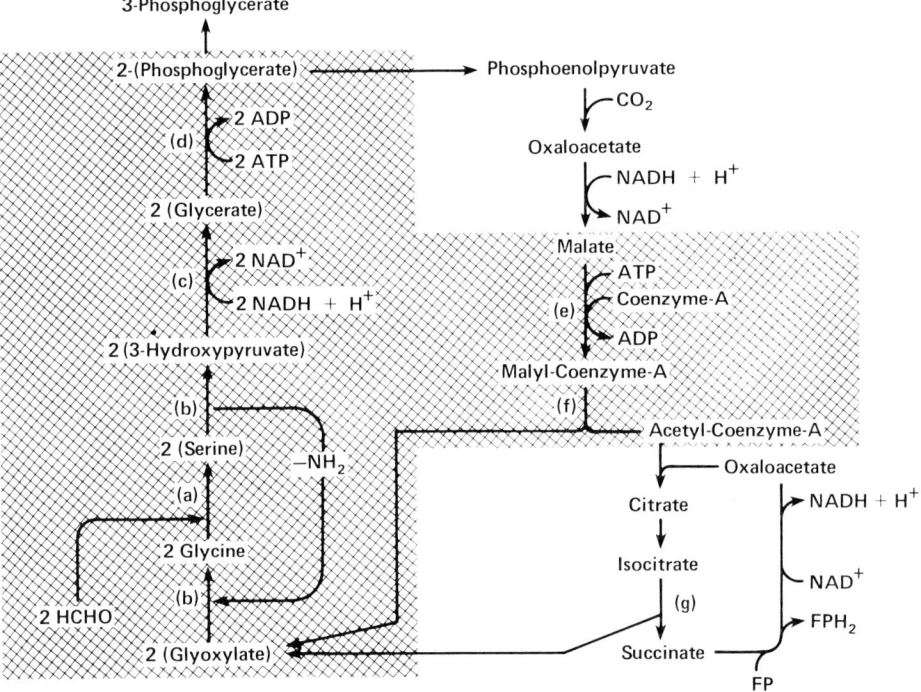

FIG. 8. The Icl⁺ serine pathway of formaldehyde fixation. The net reaction is:

2 HCHO + CO$_2$ + FAD + 2 NADH + H$^+$ + 3 ATP → phosphoglycerate + 3 ADP + 2 NAD + FADH$_2$

The reactions unique to formaldehyde assimilation (shaded area) are catalyzed by the following enzymes: (a) serine transhydroxymethylase (EC 2.1.2.1); (b) serine-glycine aminotransferase; (c) hydroxypyruvate reductase (EC 1.1.1.29); (d) glycerate kinase (EC 2.7.1.31); (e) malate thiokinase (EC 6.2.1-); (f) malyl-CoA glyoxylate-lyase (EC 4.1.3.24). Reactions catalyzed by phosphoenolpyruvate carboxylase (EC 4.1.1.32) and malate dehydrogenase (EC 1.1.1.37) complete the cycle. The one-carbon acceptor, glyoxalate, is regenerated from acetyl-CoA via the glyoxalate cycle (bottom right portion of the figure) in some type II methylotrophs and by the oxidation of acetyl-CoA to glyoxalate via an unknown reaction sequence in those methylotrophs that lack enzyme g, isocitrate lyase (EC 4.1.3.1).

rates more recent information. The scheme still divides the bacteria into the two major groups (type I and type II) and subgroups.

Type I organisms are all obligate methylotrophs, form *Azotobacter*-like cysts, assimilate formaldehyde via the ribulose monophosphate pathway (RMP), lack a complete tricarboxylic acid cycle (they are devoid of α-ketoglutarate dehydrogenase), contain type I intracytoplasmic membranes, and the predominant fatty acids of the lipids contain 16 carbon atoms.

Subgroup A of type I methylotrophs includes those bacteria with a DNA base ratio of 50-54 mol % G + C, are rods or cocci, and do not fix CO_2 autotrophically. *Methylomonas methanica* and *Methylomonas albus* are examples of organisms in this subgroup.

Subgroup B includes those type I organisms with a DNA base ratio of 62.5 mol % G + C that are cocci, grow at 45°C, and contain phosphoribulokinase plus ribulose diphosphate carboxylase, key enzymes of the Bensen-Calvin cycle for autotrophic CO_2 fixation, as well as enzymes of the RMP pathway for formaldehyde fixation. Isocitrate dehydrogenase is NAD^+ dependent. *Methylococcus capsulatus* possesses the properties of this subgroup.

Type II methane-oxidizing bacteria form exospores or lipid cysts, utilize the serine pathway for formaldehyde fixation, have a complete tricarboxylic acid cycle, and contain type II intracytoplasmic membranes and the major fatty acid in lipids has a chain length of 18 carbon atoms. Type II bacteria that have been examined have a DNA base ratio of 62.5 mol % G + C and an $NADP^+$-dependent isocitrate dehydrogenase.

The first subgroup of type II methylotrophs includes the obligate methylotrophs, *Methylomonas methanooxidans* and *Methylosinus trichosporium*. The facultative methane-oxidizing bacteria are assigned to the second subgroup.

The validity of this classification scheme has not been adequately tested. Not all methane-oxidizing bacteria are known to possess all the characteristics of a proposed group or subgroup. Romanovskaya *et al.* (1978) have reviewed the descriptions of obligate methylotrophic bacteria and have published detailed descriptions of species, species considered *incertae sedis*, and a diagnostic key.

Some type I methane-oxidizing bacteria that assimilate most of their carbon by the ribulose monophosphate pathway also contain low levels of enzymes of the serine pathway for formaldehyde fixation (Romanovskaya *et al.*, 1978; Colby *et al.*, 1979). Hydroxypyruvate reductase has been found in extracts of *M. capsulatus* and *P. methanica* (Whittenbury *et al.*, 1976; Colby *et al.*, 1979). Trotsenko (1975) and Shishkina *et al.* (1976) found hydroxypyruvate reductase in 11 of 12 strains of type II methylotrophs but the activity of this enzyme increased markedly when formate was added as a supplementary carbon source for the growth of *Methylobacter bovis* (Trotsenko, 1975). Serine-glyoxalate aminotransferase was found in several type I methane oxidizers as well (Colby *et al.*, 1979; Romanovskaya *et al.*, 1978; Trotsenko, 1975). The metabolic function of these enzymes in type I methylotrophs is unknown. Bamford and Quayle (1977) have pointed out that the presence of one or two enzymes of a pathway in an organism does not indicate that the whole pathway functions. There is no published evidence for the simultaneous functioning of both pathways for formaldehyde fixation in a methane-oxidizing bacterium (Colby et al., 1979). The presence of ribulose diphosphate carboxylase and phosphoribulokinase have been demonstrated in extracts of *M. capsulatus* (Taylor, 1977) and fixation of small amounts of CO_2 by the Benson–Calvin cycle occurs during growth of this organism on methane (Colby et al., 1979).

Although small amounts of carbon can be assimilated by a secondary route in this one type I methane-oxidizing bacterium, the major division of methane oxidizers into two groups based on the major pathways utilized for formaldehyde assimilation as well as other properties remains valid.

Dalton and Whittenbury (1975) have shown that *Methylococcus capsulatus* (Bath) contained glutamate dehydrogenase as well as enzymes of the glutamate cycle (glutamine synthase and glutamine–oxaloglutarate aminotransferase) for ammonia assimilation. Glutamate dehydrogenase levels were high when ammonia was used as a nitrogen source, whereas glutamate synthase levels were much higher and glutamate dehydrogenase was repressed during growth on nitrogen. Shishkina and Trotsenko (1979) investigated nitrogen metabolism in several methane-oxidizing bacteria and found that three type I methylotrophs (*Methylomonas methanica*, *Methylobacterium bovis*, and *Methylobacterium vinelandii*) contained glutamate dehydrogenase plus glutamine synthase. The type II methylotrophs tested, *Methylosinus trichosporium*, *Methylocystis fuscus*, and *Methylocystis methanolicus*, contained enzymes of the glutamate cycle and lacked glutamate dehydrogenase.

Most type II and some type I methane oxidizers are capable of fixing nitrogen (Colby et al., 1979; Dalton and Whittenbury, 1975; deBont and Mulder, 1974).

Different sterols have been found in *Methylococcus capsulatus* (Bird et al., 1971) and *Methylobacterium organophilum* XX (Patt and Hanson, 1978). It is not known whether these differences extend to other type I and type II methane-oxidizing bacteria.

B. Yeasts That Oxidize Methane

Rudd and Hamilton (1975) noted that large yeastlike organisms dominated in lakewater samples from Lake 227 that were exposed to methane in the laboratory.

Five strains of yeast capable of oxidizing methane have been isolated in this laboratory by H. Wolf (Wolf and Hanson, 1978). Pure cultures were obtained after enrichments, using soil and water samples, in a rich medium at pH 3.5 and a temperature of 20°C. When turbidity developed, the suspensions were diluted and spread onto agar plates of a mineral-salts medium supplemented with vitamins and amino acids; the pH was 5.5. The plates were incubated under an atmosphere of 70% CH_4, 20% air, and 10% CO_2. High CO_2 concentrations appear to be essential for growth on methane. Colonies were transferred to a rich agar medium for final purification and were finally streaked onto the surface of sterile filters that were wetted with the mineral-salts medium plus vitamins and amino acids and incubated with the methane air–CO_2 mixture indicated above. All colonies were pink and contained large budding cells, and when thin sections were examined in the electron microscope the cells were shown to contain a nucleus, mitochondria, and microbodies.

Growth in liquid cultures was shown to be dependent on methane. Resting-cell suspensions exhibited methane-dependent oxygen consumption and oxidized $^{14}CH_4$ to $^{14}CO_2$. All were facultative methylotrophs (Wolf and Hanson, 1978).

Four strains differed from each other in nutritional requirements, sugar and hydrocarbon utilization, the ability to use nitrate as a nitrogen source, optimum growth temperature, and the formation of ballistospores. The four different isolates were identified as *Rhodotorula glutinus*, *Rhodotorula rubra*, *Sporobolomyces roseus*, and *Sporobolomyces gracilis* (Wolf, 1979).

Microbodies previously shown to contain methanol oxidase and catalase in methanol-oxidizing yeasts (Fukui *et al.*, 1975; Roggenkamp *et al.*, 1975; van Dijken *et al.*, 1976; Veenhuis *et al.*, 1976) were detected in *S. roseus* and *R. glutinus* when they were grown on methane but were absent in cells grown on glucose. None of the isolates grew with methanol as a carbon and energy source. Methane-grown cells of *S. roseus* oxidized methane at higher rates than cells grown on glucose as a carbon and energy source. Enzymes associated with the oxidation of one-carbon intermediates were also present at higher levels in cells grown on methane (Wolf, 1979).

These facultative methylotrophic prokaryotes offer interesting possibilities for comparative studies of the biochemistry of methane oxidation. Unfortunately, the slow growth rates (generation times on methane generally were 1 week or longer depending on the strain) and low cell yields have made it difficult to purify or further characterize the enzymes involved in one-carbon metabolism.

C. Bacteria and Fungi That Grow on Methanol and Methylamines but Do Not Oxidize Methane

Although few cultures of facultative methane-oxidizing bacteria have been described, most of the many bacteria isolated from enrichments using methylamine and methanol as substrates also grow on heterotrophic substrates (Anthony, 1975; Colby et al., 1979; Quayle, 1972). *Bacterium* 4B6 and *Bacterium* C2A1 described by Colby and Zatman (1975), *Pseudomonas* W6 (Muller and Babel, 1977), *Methylobacterium methylovora* strain M8-5 (Kuono et al., 1973), *Methylophilus methylotrophicum* (Bvrom and Ousby, 1975), and *Methylomonas* M15 (Sahm and Wagner, 1975a) are exceptions.

A comparison of assimilatory pathways utilized by some methylotrophs that also grow on heterotrophic substrates is provided in Table I. More complete lists of facultative methylotrophs and their properties are available in reviews by Anthony (1975), Colby et al. (1979), and Quayle (1972). From the partial list presented here, it is obvious that there is a greater diversity in morphology, physiological capabilities, and one-carbon assimilatory pathways among these bacteria than among the methane oxidizers. Whereas all well-characterized methane oxidizers are gram-negative bacteria, both gram-positive and gram-negative bacteria are found to grow on methanol. The morphological diversity encompasses the stalked, budding bacteria of

TABLE I
PATHWAYS OF CARBON ASSIMILATION BY SOME FACULTATIVE METHYLOTROPHS[a]

Organism	One-carbon substrates used for growth	C_1 assimilation pathway
Methane-oxidizing bacteria		
Methylobacterium organophilum XX	Methane, methanol	Icl$^-$ serine
Methylobacterium organophilum R6	Methane	Serine
Bacteria unable to grow on methane		
Pseudomonas AM1	Methanol, methylamines	Icl$^-$ serine
Organism JB1	Methanol	Icl$^-$ serine
Hyphomicrobium spp.	Methanol, methylamines	Icl$^-$ serine
Organism JB1	Methanol	Icl$^-$ serine
Bacterium 5B1	Methylamines	Icl$^-$ serine
Pseudomonas MA	Methylamines	Icl$^+$ serine
Pseudomonas aminovorans	Methylamines	Icl$^+$ serine
Pseudomonas MS	Methylamines	Icl$^+$ serine
Bacillus spp. PM6		Ribulose monophosphate
Pseudomonas C	Methanol	Ribulose monophosphate
Pseudomonas oleovorans	Methanol, methylamines	Ribulose monophosphate
Anthrobacter globiformis	Methylamines	Ribulose monophosphate
Streptomyces spp.	Methanol	Ribulose monophosphate and serine
Brevibacterium sp. 24	Methylamines	Ribulose monophosphate
Mycobacterium sp. 10	Methylamines	Ribulose monophosphate

[a] This information has been gathered from reviews by Anthony (1975) and Colby et al. (1979), and manuscripts by Bellion and Spain (1976) and Lignova and Trotsenko (1977). The many references to the original descriptions of the organisms are available from these sources.

the genus *Hyphomicrobium*, pseudomonads, streptomyces, mycobacteria, and spore-forming bacilli.

In addition to the bacteria listed, some chemolithotrophs and phototrophs are also able to use methanol as an alternate substrate although they are not methylotrophs in the strictest sense of the definition because they assimilate carbon via the Calvin cycle (Colby et al., 1979). Several other organisms that are capable of using formate and CO as carbon sources also assimilate carbon as CO_2. This is to be expected because the reduction of formate to formaldehyde is endergonic.

Rhodopseudomonas acidophila, a phototrophic bacterium, is capable of utilizing methanol as a carbon source when it is supplied in place of, or in addition to, bicarbonate during anaerobic growth with light (Quayle and Pfennig, 1975). Aerobic growth of this organism in the dark on methanol has been achieved by using reduced oxygen tensions (Pfennig and Siefert, 1977). Methanol is oxidized to CO_2 and carbon is assimilated as CO_2 by the ribulose diphosphate pathway under both conditions.

Paracoccus denitrificans (Cox and Quayle, 1975), *Microcyclus aquaticus* (Lignova et al., 1978), *Thiobacillus novellus* (Chandra and Shetna, 1976), and *Micrococcus denitrificans* (Cox and Quayle, 1975) are chemolithotrophs that assimilate carbon from methanol. Like *Rhodopseudomonas acidophila*, they first oxidize methanol to CO_2 and use the autotrophic route for biosynthesis of cell material. It is curious that these bacteria that have increased their metabolic diversity by evolution of the ability to oxidize methanol did not also acquire a more energy-conserving pathway for assimilation of carbon. "These organisms are probably autotrophs by design and methylotrophs by accident" (Quayle and Ferenci, 1978, p. 269).

Three facultative methanol-utilizing organisms, *Achromobacter* 1L, *Pseudomonas* 8, and *Mycobacterium* 50, isolated from activated sludge by Lignova and Trotsenko (1979) apparently utilize the ribulose diphosphate pathway during growth on methanol.

Pseudomonas gazotropha is another interesting example of an organism that has both methylotrophic and autotrophic capabilities. During growth on CO, the ribulose diphosphate pathway for CO_2 fixation is used, whereas the more efficient serine pathway for formaldehyde fixation functions during growth on methanol (Romanova and Nozhevinikova, 1977).

Unlike the methane-oxidizing bacteria that are not known to contain both pathways for formaldehyde assimilation, *Streptomyces* sp. 239 apparently can use both the RMP and serine pathways for formaldehyde fixation during growth on methanol (Kato et al., 1977). The serine pathway is present in young cultures, whereas the ribulose monophosphate pathway is dominant during midexponential growth.

Methanol-oxidizing yeast belonging to the genera *Pichia*, *Saccharomyces*,

Hansenula, Rhodotorula, Kloeckera, Candida, and *Torulopsis* have been isolated (Ogata *et al.*, 1969; Tani *et al.*, 1978). One mycelial fungus, *Trichoderma lignorium,* is also capable of utilizing methanol (Tye and Willets, 1973). The incubation of yeast cultures with $^{14}CH_3OH$ resulted in the early labeling of sugar phosphates (Fujii and Tonomura, 1974). However, hexulose phosphate synthase activity could not be detected in extracts (Fujii and Tonomura, 1974; Sahm and Wagner, 1975b). Van Dijken and his colleagues (1979; Quayle and Ferenci, 1978) have proposed that a new pathway involving the condensation of formaldehyde with pentose phosphate to yield dihydroxyacetone and glyceraldehyde 3-phosphate as the initial reaction in one-carbon assimilation by these organisms. This pathway was named the dihydroxyacetone pathway for formaldehyde fixation. Further support for the existence of this pathway in *Hansenula polymorpha* and *Candida boidinii* has been provided by studies of mutants unable to grow on methanol. These mutants lacked key enzymes believed to function in the dihydroxyacetone pathway (O'Connor and Quayle, 1979).

V. Regulation of One-Carbon Metabolism in Facultative Methylotrophs

Facultative methylotrophs that use methanol as a carbon and energy source and assimilate formaldehyde via the serine pathway can be divided into those that use the glyoxylate cycle to regenerate the one-carbon acceptor molecule from acetyl-CoA (Fig. 8) and those that lack isocitrate lyase (Icl) (Bellion and Spain, 1976). The pink methylotrophs, such as *Pseudomonas* AM1, that use the Icl$^-$ serine pathway appear to oxidize acetyl-CoA directly to glyoxylate (Anthony, 1975). The biochemistry of this reaction sequence remains to be described.

Methylobacterium organophilum XX is a type II facultative methylotroph that contains all the enzymes of the Icl$^-$ serine pathway formaldehyde assimilation (Bellion and Spain, 1976; O'Connor and Hanson, 1977), methane monooxygenase, a methanol dehydrogenase, which also catalyzes the oxidation of formaldehyde (Wolf and Hanson, 1979) and formate dehydrogenase (O'Connor and Hanson, 1978). The bacterium utilizes several sugars, succinate, malate, acetate, methanol, and methane as sole sources of carbon and energy (Patt *et al.*, 1974) and grows well on rich media, such as nutrient broth. Succinate and methanol are the preferred carbon and energy sources. Growth on methane is slow (generation time $\cong$ 5 hours). The oxidation of methane by resting cells is restricted to cells grown on methane as a sole source of carbon and energy source. Methane oxidation is induced in 4–5 hours after glucose-grown cells are transferred to a mineral-salts medium and incubated under an atmosphere of methane (Patt *et al.*, 1975).

Methanol-grown cells oxidize methanol but not methane and the one-carbon substrates are not oxidized by cells grown on glucose. The oxidation of glucose is reduced in cells grown on one-carbon substrates because of repression of glucose-6-phosphate dehydrogenase (Patt et al., 1975).

Methane monooxygenase (Patt et al., 1975) has been detected only in cells grown with methane, whereas methanol induces the remaining enzymes uniquely involved in one-carbon metabolism; methanol dehydrogenase, formaldehyde dehydrogenase, serine-glyoxalate aminotransferase, hydroxypyruvate reductase, glycerate kinase and a glyoxalate activated serine transhydroxymethylase (O'Connor and Hanson, 1977). Succinate partially repressed the synthesis of these two enzymes when methanol was present as an inducer (O'Connor and Hanson, 1977). The formation of intracytoplasmic membranes in cells of *M. organophilum* occurs only in cells grown on methane (Patt et al., 1978).

Coordinate induction or repression of C_1-specific enzymes has also been observed in the facultative methanol-oxidizing bacteria, *Pseudomonas* AM1 (Dunstan et al., 1972) and *P. aminovorans* (Boulton and Large, 1977).

Two serine transhydroxymethylase isoenzymes have been shown to be present in *M. organophilum* XX (O'Connor and Hanson, 1975). The isoenzyme found in cells grown on succinate has a molecular weight of 10^5 and the active enzyme is believed to contain a single subunit. An additional isoenzyme is induced during growth on methanol. This isoenzyme has a subunit molecular weight of 5×10^4 and the active enzyme appears to a tetramer. The methanol-induced serine transhydroxymethylase is activated by glyoxalate.

The isoenzyme that functions during growth on succinate is believed to catalyze the synthesis of glycine from serine (Eq. 2).

$$CH_2\text{—}CH\text{—}COOH + \text{tetrahydrofolate} \rightleftharpoons H_2NCH_2\text{—}COOH + N^5,N^{10}\text{-methylene-tetrahydrofolate} \quad (2)$$
$$\underset{NH_2}{|}$$

The methylene tetrahydrofolate produced in this reaction serves as a one-carbon donor in other biosynthetic reactions.

The isoenzyme that is induced by growth on methanol is believed to participate (Eq. 4) in the assimilation of formaldehyde that is produced from methanol in Eqs. (3) and (4):

$$HCHO + \text{tetrahydrofolate} \rightleftharpoons N^5, N^{10}\text{-tetrahydrofolate} \quad (3)$$

$$N^5, N^{10}\text{-methylene tetrahydrofolate} + \text{glycine} \rightleftharpoons \text{serine} + \text{tetrahydrofolate} \quad (4)$$

Serine then enters the pathway described in Fig. 8. Activation of the serine transhydroxymethylase by glyoxylate, the immediate precursor of glycine, apparently represents "feedforward regulation" of the entry of carbon into this pathway.

Isocitrate lyase, a key enzyme of the anaplerotic glyoxylate cycle, functions with malate synthase to replenish intermediates of the tricarboxylic acid cycle removed for biosynthesis during growth of bacteria on acetate or substrates converted to acetate (Kornberg, 1966). It also plays a role in the regeneration of glyoxalate in the Icl$^+$ serine pathway for formaldehyde fixation (see Fig. 8) in type II methylotrophs. Bellion and Woodson (1975) observed that the isocitrate lyase present during growth of *Pseudomonas* MA on methylamine was different by several criteria from the isofunctional enzyme present in acetate-grown cells. Thus, there are two examples of the evolution of independently regulated, isofunctional enzymes serving distinct purposes in the metabolism of facultative methylotrophs.

The isolation of revertable pleiotropic mutants of *M. organophilum* that are deficient in methanol dehydrogenase, the glyoxalate-activated serine transhydroxymethylase, hydroxypyruvate reductase, serine-glyoxalate aminotransferase, and glycerate kinase suggests that the synthesis of these enzymes is coordinately regulated (O'Connor and Hanson, 1977). Similar mutants of *P. aminovorans* lacking several C_1 specific enzymes were isolated by Banford and O'Connor (1979).

The development of a transformation system in *M. organophilum* (O'Connor *et al.*, 1977) provided an opportunity for mapping mutants that were shown to be deficient in enzymes uniquely involved in one-carbon metabolism. The data obtained by O'Connor and Hanson (1978) indicated that the genetic determinants for these enzymes were physically linked and shared a common regulatory gene.

Is Methane Oxidation Coded for by Plasmids in Facultative Methane Oxidizers?

It is surprising that it has been difficult to isolate facultative methane-oxidizing bacteria with a range of diversity equivalent to that found in the facultative methanol- and methylamine-oxidizing bacteria. Only the evolution of a methane monooxygenase system is required to accomplish the oxidation of methane to methanol. Methane is abundant and ubiquitous in nature, while methanol and methylamines are less available. Silverman (1964) comments that obligate methane oxidation could be appreciated in pseudomonads if they occupied a unique ecological niche. However, methane-oxidizing bacteria are cohabitants with facultative methanol-oxidizing bacteria and aerobic heterotrophs and there is nothing obviously restrictive or unusual about their habitats.

It is possible that adenosine triphosphate (ATP) synthesis from methane requires the synthesis of complex intracytoplasmic membranes with specialized functions and that their presence is incompatable with ATP generation from heterotrophic substrates. These factors may have restricted their evolution in bacteria that have an alternate means of existence.

Alternatively, it is possible that facultative methane-oxidizing bacteria are more difficult to cultivate on methane in the laboratory and so have escaped detection. In several searches for methylotrophs, colonies that grew on nutrient agar or other media with heterotrophic substrates were discarded. Cappenberg (1972) claimed that nearly all the methane-oxidizing bacteria isolated as primary colonies from Lake Vechten, Netherlands, grew on rich media. However, these cultures were not characterized further. Perry (1968) reported that *Brevibacterium* strain JOB5 grew on n-alkanes between C_1 and C_2 as well as methane. This was the only well-documented claim of a facultative methylotroph before 1974. Other workers have claimed that facultative methane-oxidizing bacteria lost the ability to oxidize methane or hydrocarbons in laboratory culture (Aiyer, 1920). These observations were attributed to impure cultures or the growth of cultures on compounds other than methane that contaminated the gas supply. Every facultative methane-oxidizing isolate that oxidized $^{14}CH_4$ and appeared to be in pure culture that we have worked with (Patt *et al.*, 1974), except *M. organophilum* R6 (Patel *et al.*, 1978), has exhibited this instability. *Methylobacterium organophilum* XX, the only isolate of ours that has retained the ability to oxidize methane, appeared relatively stable for 2 years. However, our latest experience is that it is very difficult to revive from stock cultures by restreaking on a minimal salts medium with methane. Methane-oxidizing cultures are maintained by frequent transfer of young cultures or small colonies onto media with methane as the only carbon and energy source. After a few transfers in media containing methanol or succinate the ability of some cultures to oxidize methane is permanently lost. The cultures that are unable to grow on methane retain the ability to grow well on methanol and other substrates.

The instability of the methane oxidizing phenotype in *M. organophilum* XX and the independent regulation of methane and methanol metabolism in this bacterium suggested to us that the genetic determinants for the monooxygenase reaction and perhaps other functions associated with methane metabolism, but not methanol oxidation or formaldehyde assimilation, are located on a plasmid. A search for covalently closed circular DNA (CCC DNA) in *M. organophilum* XX and *M. organophilum* R6 was undertaken. Strains of *M. organophilum* XX, capable of oxidizing methane, were shown to contain CCC DNA by examining DNA preparations on alkaline sucrose gradients. These data, as well as an examination of CCC DNA preparations by agarose gel electrophoresis and by electron microscopy (Schilling, 1978), indicate the presence of a single plasmid in this organism. Plasmid DNA was isolated from methane-grown cells by the procedure of Currier and Nester (1976) and has been shown to have a molecular weight of approximately 8×10^7 by agarose gel electrophoresis (Schilling, 1978). A larger plasmid with an approximate molecular weight of 1.4×10^8 has been isolated from *M. organophilum* R6 (K. W. Wruck, W. A. Hackett, and R. S. Hanson, unpublished results).

We have been unable to detect CCC DNA in strains of *M. organophilum* XX that are unable to grow on methane (Schilling, 1978). The uncertainties involved in the detection of CCC DNA when it is known to be present in a bacterium weakens the hypothesis that a plasmid codes for methane oxidation in this bacterium. We have not yet been able to prove transformation of methane$^-$ strains to a methane$^+$ phenotype using plasmid DNA. Warner and Higgins (1977) found three plasmids in *Pseudomonas* AM1, a facultative methanol-oxidizing bacterium. They failed to detect CCC DNA in obligate methylotrophs.

Methylobacterium organophilum XX resembles the pink facultative pseudomonads possessing the Icl$^-$ serine pathway (i.e., *Pseudomonas* AM1 and organism JB1) except for its ability to oxidize methane. These bacteria may have oxidized methane in nature and the loss of a plasmid would have converted them to their present phenotype. We therefore isolated several pink methylotrophic bacteria from Lake Mendota, the source of *M. organophilum*. These isolates grow on methanol but not methane. DNA was isolated from each culture and *M. organophilum* XX. The DNA from *M. organophilum* XX was labeled *in vitro* by nick translation (Kawai *et al.*, 1973) and tested for homology with DNA from new isolates as well as *Pseudomonas* AM1 and organism JB1 (Table II). A nonpigmented type II obligate methylotroph, *Methylosinus trichosporium* OB3B, and a type I obligate methylotroph isolated from Lake Mendota were tested for relatedness in the same way. The pink facultative methylotrophs and *M. trichosporium*, which assimilates carbon by the Icl$^+$ serine pathway, had similar DNA base ratios (62.5 to 66 mol % G + C), whereas the type I isolate tested had a DNA base ratio of 52.5 mol % G + C. *Pseudomonas* AM1 and organism JB1 show greater relatedness to *M. organophilum* XX than facultative methanol oxidizers isolated from Lake Mendota (Table II). The DNA preparations from *M. organophilum* grown on methane and glucose were also indistinguishable in hybridization assays.

Thus, *Methylobacterium organophilum* appears to be different from any other isolate but more closely related to the pink facultative methanol oxidizers than to the nonpigmented type II obligate methylotroph and the type I methylotroph tested. We were not able to isolate a plasmidless strain of *M. organophilum* from nature or to find further support for the hypothesis that this organism differs from a known facultative methylotroph by the presence of a plasmid that codes for methane oxidation.

TABLE II
Homology between *Methylobacterium organophilum* XX and DNA Preparations from Other Methylotrophic Bacteria[a]

Organisms	Growth substrate	% Homology with [^{3}H]*M. organophilum* DNA
M. organophilum XX	Methane	95–100
M. organophilum XX	Succinate	95–100
Pseudomonas AM1	Methanol	45–48
Organism JB1	Methanol	43–46
Isolate 100	Methanol	30
Isolate f.s.5	Methanol	20–24
Isolate 8P	Methanol	22
Isolate f.s.7	Methanol	24
Isolate NP	Methanol	18
Methylosinus trichosporium OB3B	Methane	2.0
Isolate 4P	Methanol	2.0

[a] Isolates 100, f.s.5, 8P, s.7, and NP were obtained as pure cultures from enrichments inoculated with lake water from Lake Mendota, Madison, Wisconsin. All were pink facultative methylotrophs that contained hydroxypyruvate reductase and lacked hexulose phosphate synthase. Isolate 4P is a type I methylotroph that lacks hydroxypyruvate reductase.

DNA was isolated from *M. organophilum* XX cells grown on methane (O'Connor and Hanson, 1978) and labeled by nick translation (Kawai *et al.*, 1973). The specific activity was approximately 180,000 cpm/μg. Sheared radioactive *M. organophilum* XX DNA (2,000 cpm/0.011 μg) was mixed with unlabeled DNA from the organisms indicated. The DNA preparations were dissolved in 0.14 M phosphate buffer. The mixtures were heated at 100°C for 10 min and then at 65°C until a $C_0 t_{\frac{1}{2}}$ value of 100 (Britten and Kohne, 1968) was reached (24–48 hours). Single-stranded DNA was separated from double-stranded DNA by chromatography on hydroxylapetite columns (Brenner *et al.*, 1969).

Acknowledgments

The author is indebted to Dr. W. S. Reeburgh, University of Alaska, Fairbanks, for his permission to cite unpublished results. The research from the author's laboratory that is reported in this chapter was supported by the College of Agricultural and Life Sciences, University of Wisconsin, Madison, and by a grant from the National Science Foundation (PCM-7809744).

References

Aiyer, P. A. S. (1920). *Mem. Dep. Agric. India, Chem. Ser.* **5**, 173.
Anagostidis, K., and Overbeck, J. (1966). *Ber. Dtsch. Bot. Ges.* **79**, 163.
Anthony, C. (1975). *Sci. Prog. (Oxford)* **62**, 167–206.
Bamford, C. W., and O'Connor, M. L. (1979). *J. Gen. Microbiol.* **110**, 143.
Bamford, C. W., and Quayle, J. R. (1977). *J. Gen. Microbiol.* **101**, 259.
Barnes, R. O., and Goldberg, E. D. (1976). *Geology* **5**, 297.

Bellion, E., and Spain, J. C. (1976). *Can. J. Microbiol.* **22**, 404.
Bellion, E., and Woodson, J. (1975). *J. Bacteriol.* **122**, 557.
Bird, C. W., Lynch, C. M., Pirt, F. J., Reid, W. W., Brooks, C. J. W., and Middleditch, B. C. (1971). *Nature (London)* **230**, 473.
Boulton, C. A., and Large, P. J. (1977). *J. Gen. Microbiol.* **101**, 151.
Brenner, D. J., Fanning, G. R., Rake, A. V., and Johnson, K. E. (1969). *Anal. Biochem.* **28**, 447.
Britten, R. J., and Kohne, D. E. (1968). *Science* **161**, 529.
Brown, L. R., and Strawinski, R. J. (1956). *Bacteriol. Proc.* **56**, 18.
Brown, L. R., Strawinski, R. J., and McCleskey, C. S. (1964). *Can. J. Microbiol.* **10**, 791.
Byrom, D., and Ousby, J. C. (1975). *Microb. Growth C_1-Compd., Proc. Int. Symp.*, 1974 p. 23.
Cappenberg, T. E. (1972). *Hydrobiologia* **40**, 471.
Chandra, T. S., and Shetna, Y. I. (1976). *J. Bacteriol.* **131**, 289.
Claypool, G. E., and Kaplan, I. R. (1974). *In* "Natural Gases in Marine Sediments" (I. R. Kaplan, ed.), pp. 99–140. Plenum, New York.
Colby, J., and Zatman, L. J. (1975). *Biochem. J.* **148**, 513.
Colby, J., Dalton, H., and Whittenbury, R. (1979). *Annu. Rev. Microbiol.* **33**, 481.
Cox, R. B., and Quayle, J. R. (1975). *Biochem. J.* **150**, 569.
Currier, T. C., and Nester, E. W. (1976). *J. Bacteriol.* **126**, 157.
Dalton, H., and Whittenbury, R. (1975). *In* "Microbial Production and Utilization of Gases" (H. G. Schlegel, G. Gottschalk, and N. Pfennig, eds.), pp. 379–388. Goltze, Göttingen.
Daniels, L. (1977). Ph.D. Thesis, University of Wisconsin, Madison.
Davies, J. B., and Yarborough, H. E. (1966). *Chem. Geol.* **1**, 137.
Davies, S. L., and Whittenbury, R. (1970). *J. Gen. Microbiol.* **61**, 227.
Davis, J. B., Coty, V. G., and Stanley, J. P. (1964). *J. Bacteriol.* **90**, 102.
deBont, J. A. M., and Mulder, E. G. (1974). *J. Gen. Microbiol.* **83**, 113.
Dunstan, P. M., Anthony, C., and Drabble, W. T. (1972). *Biochem. J.* **128**, 107.
Dworkin, M., and Foster, J. W. (1956). *J. Bacteriol.* **72**, 646–659.
Ehalt, D. H. (1976). *In* "Microbial Production and Utilization of Gases" (H. G. Schlegel, G. Gottschalk, and N. Pfennig, eds.), pp. 13–22. Goltze, Göttingen.
Fallon, R., Harrits, S., Brock, T. D., and Hanson, R. S. (1980). *Limnol. Oceanogr.* (in press).
Ferenci, T. (1974). *FEBS Lett.* **41**, 94.
Foster, J. W., and Davis, R. H. (1966). *J. Bacteriol.* **91**, 1924.
Freely, H. W., and Kulp, J. L. (1957). *Am. Assoc. Pet. Geol. Bull.* **41**, 1802.
Fujii, T., and Tonomura, K. (1974). *Agric. Biol. Chem.* **38**, 1763.
Fukui, S., Kawamoto, S., Yashuhara, S., Tanaka, A., Osumi, M., and Imaizumi, F. (1975). *Eur. J. Biochem.* **59**, 561.
Galchenko, V. F. (1975). *Appl. Biochem. Microbiol.* **11**, 447–450.
Gold, T. (1979). *J. Pet. Geol.* **1**, 3.
Harrits, S. (1979). M.S. Thesis, University of Wisconsin, Madison.
Harrits, S., and Hanson R. (1980). *Limnol. Oceanogr.* **26**, 213.
Heyer, J. (1977). *In* "Microbial Growth on C_1-Compounds" (G. K. Skryabin, M. V. Ivanov, E. N. Kondratjeva, G. A. Zavarzin, Yu, A. Trotsenko, and A. I. Nesterov, eds.), pp. 19–21. USSR Acad. Sci., Puschino.
Higgins, I. J., and Quayle, J. R. (1970). *Biochem. J.* **118**, 201.
Howard, D. L., Frea, J. I., and Pfister, R. M. (1971). *Proc., Conf. Great Lakes Res.* **14**, 236–240.
Hutton, W. E., and Zobell, C. E. (1949). *J. Bacteriol.* **58**, 463.
Jannasch, H. W. (1975). *Limnol. Oceanogr.* **20**, 860.

Kaserer, H. (1905). *Zentralbl. Bakteriol. Parasitend.* **16**, 681.
Kato, N., Tsuji, K., Ohashi, H., Tani, Y., and Ogata, K. (1977). *Agric. Biol. Chem.* **41**, 29.
Kawai, Y., Nonoyama, M., and Pagano, J. S. (1973). *J. Virol.* **12**, 1006.
Kemp, M. B. (1974). *Biochem. J.* **139**, 129.
Kemp, M. B., and Quayle, J. R. (1967). *Biochem. J.* **102**, 94.
Kornberg, H. L. (1966). *Essays Biochem.* **2**, 1–32.
Kosiur, D. R., and Warford, A. L. (1979). *Estuarine Coastal Mar. Sci.* **8**, 379.
Kuono, K., Oki, T., Nomura, H., and Ozaki, A. (1973). *J. Gen. Appl. Microbiol.* **19**, 11.
Lamb, S. C. (1974). Ph.D. Thesis, University of Wisconsin, Madison.
Lawrence, A. J., Kemp, M. B., and Quayle, J. R. (1970). *Biochem. J.* **116**, 631.
Leadbetter, E. R., and Foster, J. W. (1958). *Ark. Mikrobiol.* **30**, 91.
Lignova, N. V., and Trotsenko, Yu. A. (1977). *In* "Microbial Growth on C_1-Compounds" (G. K. Skryabin, M. V. Ivanov, E. N. Kondratjeva, G. A. Zavarzin, Yu. A. Trotsenko, and A. I. Nesterov, eds.), p. 37. USSR Acad. Sci., Puschino.
Lignova, N. V., and Trotsenko, Yu. A. (1979). *FEMS Microbiol. Lett.* **5**, 239.
Lignova, N. V., Namsaraev, B. B., and Trotsenko, Yu. A. (1978). *Microbiology* **47**, 168.
Malashenko, Yu. R. (1976). *In* "Microbial Production and Utilization of Gases" (H. G. Schlegel, G. Gottschalk, and N. Pfennig, eds.), pp. 293–300. Goltze, Göttingen.
Martens, C. S., and Brenner, R. A. (1977). *Limnol. Oceanogr.* **22**, 10.
Mechalas, B. J. (1974). *In* "Natural Gases in Marine Sediments" (I. R. Kaplan, ed.), pp. 11–25. Plenum, New York.
Muller, R., and Babel, W. (1977). *In* "Microbial Growth on C_1-Compounds" (G. K. Skryabin, M. V. Ivanov, E. N. Kondratjeva, G. A. Zavarzin, Yu. A. Trotsenko, and A. I. Nesterov, eds.), pp. 46–47. USSR Acad. Sci., Puschino.
Naguib, M., and Overbeck, J. (1970). *Z. Mikrobiol.* **10**, 17.
O'Connor, M. L., and Hanson, R. S. (1975). *J. Bacteriol.* **124**, 985.
O'Connor, M. L., and Hanson, R. S. (1977). *J. Gen. Microbiol.* **101**, 327.
O'Connor, M. L., and Hanson, R. S. (1978). *J. Gen. Microbiol.* **104**, 105.
O'Connor, M., and Quayle, J. R. (1979). *J. Gen. Microbiol.* **113**, 203.
O'Connor, M. L., Wopat, A., and Hanson, R. S. (1977). *J. Gen. Microbiol.* **98**, 265.
Ogata, K., Nishikawa, H., and Ohsugi, M. (1969). *Agric. Biol. Chem.* **33**, 1519.
Orla-Jensen, S. (1909). *Zentralbl. Bakteriol., Parasitend. Infektionskr., Abt. 1* **22**, 311.
Overbeck, J., and Ohle, W. (1964). *Verh., Int. Ver. Theor. Angew. Limnol.* **15**, 535.
Panganiban, A. T. (1976). *Abstr. Annu. Meet. Am. Soc. Microbiol.* p. 121.
Panganiban, A. T., Patt, T. E., Hart, W., and Hanson, R. S. (1979). *Appl. Environ. Microbiol.* **37**, 303.
Patel, R. N., Hou, C. T., Felix, A. (1978). *J. Bacteriol.* **133**, 352.
Patt, T. E., and Hanson, R. S. (1978). *J. Bacteriol.* **134**, 636.
Patt, T. E., Cole, G. C., Bland, J., and Hanson, R. S. (1974). *J. Bacteriol.* **120**, 955.
Patt, T. E., O'Conner, M., Cole, G. C., and Hanson, R. S. (1975). *In* "Microbial Production and Utilization of Gases" (H. G. Schlegel, G. Gottschalk, and N. Pfennig, eds.), p. 317. Goltze, Göttingen.
Perry, J. J. (1968). *Antonie van Leeuwenhoek* **34**, 27.
Pfennig, N., and Seifert, E. (1977). *In* "Microbial Growth on C_1-Compounds" (G. K. Skryabin, M. V. Ivanov, E. N. Kondratjeva, G. A. Zavarzin, Yu. A. Trotsenko, and A. I. Nesterov, eds.), pp. 146–147. USSR Acad. Sci., Puschino.
Postgate, J. R. (1969). *J. Gen. Microbiol.* **57**, 293.
Proctor, H. M., Harris, J. R., and Ribbons, D. W. (1969). *J. Appl. Bacteriol.* **32**, 118.
Quayle, J. R. (1972). *Adv. Microb. Physiol.* **7**, 119.
Quayle, J. R., and Ferenci, T. (1978). *Microbiol. Rev.* **42**, 251.

Quayle, J. R., and Pfennig, N. (1975). *Arch. Microbiol.* **102**, 193.
Reeburgh, W. S. (1976). *Earth Planet. Sci. Lett.* **28**, 337.
Reeburgh, W. S. (1980). In "The Dynamic Environment of the Ocean Floor" (K. Fanning and F. T. Manheim, eds.), Heath, Indianapolis, Indiana (in press).
Reeburgh, W. S., and Heggie, D. T. (1977). *Limnol. Oceanogr.* **22**, 1.
Ribbons, D. W., Harrison, J. E., and Wadzinski, A. M. (1970). *Annu. Rev. Microbiol.* **24**, 135.
Roggenkamp, R., Sahm, H., Hinkelman, W., and Wagner, F. (1975). *Eur. J. Biochem.* **59**, 231.
Romanova, A. K., and Nozhevnikova, A. N. (1977). In "Microbial Growth on C_1-Compounds" (G. K. Skryabin, M. V. Ivanov, E. N. Kondratjeva, G. A. Zavarzin, Yu. A. Trotsenko, and A. I. Nesterov, eds.), pp. 109–111. USSR Acad. Sci., Puschino.
Romanovskaya, V. A., Malashenko, Yu. R., and Bogachenko, V. N. (1978). *Mikrobiologiya* **47**, 96.
Rudd, J. W., and Hamilton, R. D. (1975). *Arch. Hydrobiol.* **75**, 522.
Rudd, J. W., and Hamilton, R. D. (1978). *Limnol. Oceanogr.* **23**, 337.
Rudd, J. W., Hamilton, R. D., and Campbell, N. E. R. (1974). *Limnol. Oceanogr.* **19**, 519.
Rudd, J. W., Furutani, A., Flett, R. J., and Hamilton, R. D. (1976). *Limnol. Oceanogr.* **21**, 357.
Sahm, H., and Wagner, F. (1975a). *Arch. Microbiol.* **97**, 163.
Sahm, H., and Wagner, F. (1975b). *Eur. J. Microbiol.* **2**, 147.
Sansone, F. J., and Martens, C. S. (1978). *Limnol. Oceanogr.* **23**, 349.
Schilling, C. (1978). M.S. Thesis, University of Wisconsin, Madison.
Schlegel, H. G., Gottschalk, G., and Pfennig, N., eds. (1976). "Microbial Production and Utilization of Gases." Goltze, Göttingen.
Sheehan, B. T., and Johnson, M. J. (1971). *Appl. Microbiol.* **21**, 511.
Shishkina, V. N., and Trotsenko, Yu. A. (1979). *FEMS Microbiol. Lett.* **5**, 187.
Shishkina, V. N., Yurchenko, V. V., Romanovskaya, V. A., Malashenko, Yu, R., and Trotsenko, Yu. A. (1976). *Mikrobiologiya* **45**, 359.
Silverman, M. P. (1964). *U.S. Bur. Mines, Inf. Circ.* **8246**, 1–37.
Skryabin, G. K., Ivanov, M. V., Kondratjeva, E. N., Zavarzin, G. A., Trotsenko, Yu. A., and Nesterov, A. I., eds. (1977). "Microbial Growth on C_1-Compounds." USSR Acad. Sci., Puschino.
Smith, U., and Ribbons, D. W. (1970). *Arch. Mikrobiol.* **74**, 116.
Söhngen, N. L. (1906). *Zentralbl. Bakteriol., Parasitend. Infektionskr.* **15**, 513.
Sorokin, Yu. I. (1957). *Dokl. Akad. Nauk. SSSR* **115**, 816.
Sorokin, Yu. S. (1961). *Mikrobiologiya* **30**, 928.
Stirling, D. I., and Dalton, H. (1979). *FEMS Microbiol. Lett.* **5**, 315.
Stocks, P. K., and McCleskey, C. S. (1964). *J. Bacteriol.* **110**, 890.
Strawinski, R. J., and Tortovich, J. A. (1955). *Bacteriol. Proc.* p. 27.
Tani, Y., Kato, N., and Yamada, H. (1978). *Adv. Appl. Microbiol.* **24**, 165–186.
Taylor, S. (1977). *FEMS Microbiol. Lett.* **2**, 305.
Trotsenko, Yu. A. (1975). In "Microbial Production and Utilization of Gases" (H. G. Schlegel, G. Gottschalk, and N. Pfennig, eds.), p. 329. Goltze, Göttingen.
Tye, R., and Willets, A. (1973). *J. Gen. Microbiol.* **77**, 1P.
van Dijken, J. P., Otto, R., and Harder, W. (1976). *Arch. Microbiol.* **111**, 137.
van Dijken, J. P., Veenhuis, M., Zwart, K., and Harder, W. (1979). *Proc. Soc. Gen. Microbiol.* **115**, 223.
Vary, P. S., and Johnson, M. S. (1967). *Appl. Microbiol.* **15**, 1473.
Veenhuis, M., van Dijken, J. P., and Harder, W. (1976). *Arch. Microbiol.* **111**, 123.
Warner, P. J., and Higgins, I. J. (1977). *FEMS Microbiol. Lett.* **1**, 339.
Wertlieb, D., and Vishniac, W. (1967). *J. Bacteriol.* **93**, 1722.

Whelan, T. (1974). *Estuarine Coastal Mar. Sci.* **2**, 407.
Whittenbury, R., Phillips, K. C., and Wilkinson, J. F. (1970a). *J. Gen. Microbiol.* **61**, 205.
Whittenbury, R., Davies, S. L., and Davey, J. F. (1970b). *J. Gen. Microbiol.* **61**, 219.
Whittenbury, R., Colby, J., Dalton, H., and Reed, H. L. (1976). *In* "Microbial Production and Utilization of Gases" (H. G. Schlegel, G. Gottschalk, and N. Pfennig, eds.), pp. 281–292. Goltze, Göttingen.
Wilkinson, T. G., Topiwola, H. H., and Hamer, G. (1974). *Biotechnol. Bioeng.* **16**, 41.
Wolf, H. J. (1979). Ph.D. Thesis, University of Wisconsin, Madison.
Wolf, H. J., and Hanson, R. S. (1978). *Appl. Environ. Microbiol.* **36**, 105.
Wolf, H. J., and Hanson, R. S. (1979). *J. Gen. Microbiol.* **114**, 187.
Wolfe, R. S. (1971). *Adv. Microb. Physiol.* **6**, 107.
Zehnder, A. J., and Brock, T. D. (1979). *J. Bacteriol.* **137**, 420.

Epoxidation and Ketone Formation by C_1-Utilizing Microbes

CHING-TSANG HOU, RAMESH N. PATEL, AND ALLEN I. LASKIN

*Corporate Research Science Laboratory,
Exxon Research and Engineering Company,
Linden, New Jersey*

I.	Microbial Epoxidation of Gaseous 1-Alkenes	41
	A. Introduction	41
	B. Current Industrial Processes for 1,2-Epoxides	42
	C. Bacterial Strains	43
	D. Assay Method	44
	E. Microbial Epoxidation of Propylene	45
II.	Microbial Methyl Ketone Formation	54
	A. Introduction	54
	B. Assay Method	55
	C. Oxidation of 2-Butanol to 2-Butanone	56
	D. Purification and Properties of a Secondary Alcohol-Specific Alcohol Dehydrogenase	62
III.	Conclusion	67
	References	68

I. Microbial Epoxidation of Gaseous 1-Alkenes

A. INTRODUCTION

The involvement of an oxygenase in the initial oxidative attack on methane by obligate methane-utilizing bacteria was suggested by Leadbetter and Foster (1959), based on the incorporation of ^{18}O from $^{18}O_2$ into the cellular constituents of *Pseudomonas methanica*. Strong evidence for the involvement of an oxygenase in the oxidation of methane also came from Higgins and Quayle (1970), who isolated $CH_3^{18}OH$ as the product of methane oxidation when suspensions of *Pseudomonas methanica* or *Methanomonas methanooxidans* were allowed to oxidize methane in $^{18}O_2$-enriched atmospheres. The subsequent observation of methane-stimulated reduced nicotine adenine dinucleotide (NADH) oxidation catalyzed by cell-free extracts of *Methylococcus capsulatus* (Ribbons and Michalover, 1970; Ribbons, 1975) or *Methylomonas methanica* (Ferenci, 1974) suggested that the enzyme responsible for this oxygenation was a monooxygenase. These workers relied on indirect enzyme assays, measuring methane-stimulated NADH disappearance spectrophotometrically or methane-stimulated O_2 disappearance polarographically. Tonge *et al.* (1975, 1977) purified the methane monooxygenase enzyme system from the cell-free particulate fraction of *Methylosinus trichosporium* into three components. Colby *et al.* (1977;

Colby and Dalton, 1978) purified a methane–monooxygenase enzyme system from the cell-free soluble fraction of *M. capsulatus* Bath and found that this system also consisted of three protein components.

The epoxidation of simple olefins was first demonstrated by Van der Linden (1963), who reported the production of 1 mg of 1,2-epoxyoctane in 30 minutes at 30°C from 1-octene by 52 mg of heptane-grown cells of *Pseudomonas aeruginosa* resuspended in 40 ml of buffer. He postulated that epoxides were formed by enzymes that were closely related, or identical, to the alkane-oxidizing (hydroxylation) system. The product epoxide was found to be not further oxidized enzymatically. Further studies on the epoxidation of 1-octene were provided by Cardini and Jurtshuk (1970), who demonstrated that a cell-free extract of a *Corynebacterium* sp. oxidized 1-octene to 1,2-epoxyoctane in addition to hydroxylation octane to octanol. The first pure alkane monooxygenase system was reported by Coon and his coworkers (1972). They resolved the ω-hydroxylation system of *Pseudomonas oleovorans* into three protein components: rubredoxin, NADH-rubredoxin reductase, and an ω-hydroxylase. Subsequently, we at Exxon (Abbott and Hou, 1973) demonstrated the epoxidation of 1-octene by whole cells of *P. oleovorans*. We also found that the methyl group of the product 1,2-epoxyoctane was susceptible to the hydroxylation reaction. The epoxidation of 1-octene was then studied at the enzyme level using the ω-hydroxylation enzyme system, obtained from Professor Coon. May and Abbott (1972, 1973) demonstrated that this well-characterized hydroxylation system indeed catalyzed the epoxidation of 1-octene. Their observations suggested mechanistic similarity between epoxidation and hydroxylation, with a single epoxidase/hydroxylase catalyzing both reactions. All of the epoxidation systems mentioned so far, however, were found to be not active on gaseous alkenes. We have found that resting-cell suspensions of methylotrophic bacteria epoxidized gaseous alkenes and hydroxidized gaseous alkanes (Hou *et al.*, 1979a). In addition, the activity of this monooxygenase, which catalyzed both the epoxidation and hydroxylation reactions, was found in the cell-free particulate fraction (Patel *et al.*, 1979a).

B. CURRENT INDUSTRIAL PROCESSES FOR 1,2-EPOXIDES

Epoxides have become extremely valuable products because of their ability to undergo a variety of chemical reactions. The products of epoxidation are industrially important because of their ability to polymerize under thermal, ionic, and free radical catalysis to form epoxy homopolymers and copolymers. Ethylene oxide and propylene oxide constitute the two important commercial epoxides. Over 90% of the current industrial output of propylene oxide is produced by either the chlorohydrin or the oxirane pro-

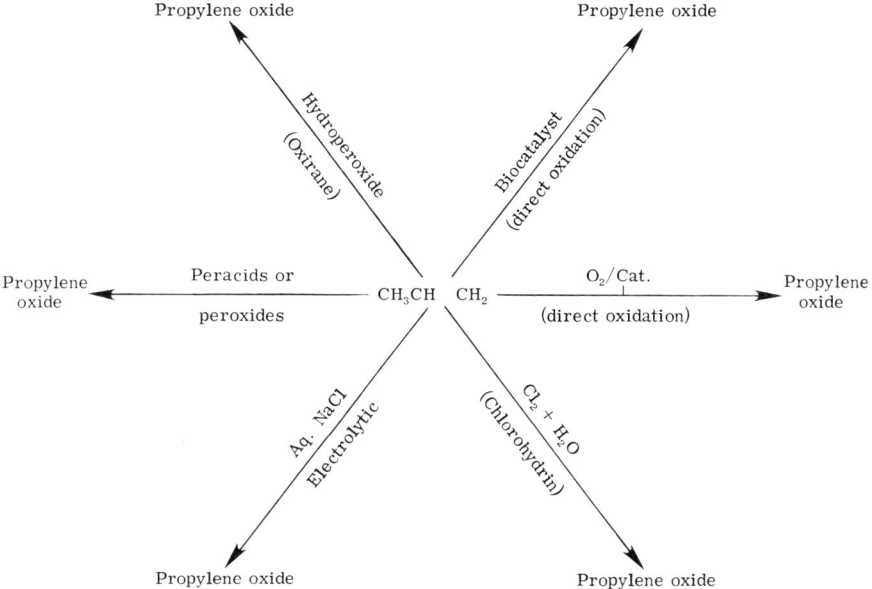

FIG. 1. Processes for propylene oxide.

cess (Fig. 1). Both of these processes require multiple steps and depend heavily on the market price of their by-products, such as styrene. The only direct oxidation process known is still not far beyond the laboratory curiosity stage. It requires a metal oxide as the catalyst.

C. Bacterial Strains

Methane-utilizing cultures were isolated from lake water from Warinanco Park, Linden, New Jersey and from lake and soil samples from Bayway Refinery, Linden, New Jersey. These methylotrophs were identified according to the classification given by Whittenbury et al. (1970) and Patt et al. (1976). They were: *Methylomonas* sp., eight strains; *Methylobacter* sp., five strains; *Methylococcus* sp., three strains; *Methylosinus* sp., two strains; *Methylocystis* sp., two strains; and *Methylobacterium organophilum*, one strain. *Methylosinus trichosporium* OB3b, *Methylosinus sporium* 5, *Methylocystis parvus* OBBP, *Methylomonas methanica* S_1, *Methylomonas* BG8, and *Methylobacter capsulatus* Y were kindly provided by R. Whittenbury (School of Biological Sciences, University of Warwick, Coventry, U.K.). *Methylobacterium organophilum* XX ATCC 24886 was kindly provided by R. S. Hanson (University of Wisconsin, Madison, Wisconsin). The organisms were maintained on mineral-salt plates in a desiccator jar under an atmosphere of methane and air (1:1 v/v) at 30°C. Organisms were grown at 30°C in 300-ml flasks containing 50 ml of mineral-salt medium (Foster and Davis, 1966) with methane and air (1:1 v/v) as the sole carbon and energy source. When methanol (at 0.3%) was used, the gaseous phase in the flask was air.

D. Assay Method

The harvested cells were washed twice with 0.05 M phosphate buffer, pH 7.0. The final pellet was resuspended in fresh buffer to obtain an optical density (OD) at 660 nm of 0.5. A 0.5-ml aliquot of this washed cell suspension (containing a given amount of dry cell mass) was placed in a 10-ml vial. The vial was sealed with a rubber cap to minimize evaporation. The gaseous phase of the vial was replaced with a gas mixture containing 50% gaseous alkene and 50% pure oxygen. The reaction mixture was incubated at 30°C in a water bath rotary shaker at 300 rpm. A 3-μl sample was removed with a syringe and was assayed by using a stainless steel column (20 ft by ⅛ in.) packed with 10% Carbowax 20M on 80/100 Chromosorb W (Supelco, Inc., Bellefonte, Pennsylvania). The column temperature was maintained isothermally at 100°C, and the carrier gas flow was 35 ml of helium per minute. The various epoxide products were identified by retention time comparisons and cochromatography with authentic standards. This identification was supplemented by observing the presence and absence of product peaks before and after bromination and acid hydrolysis. The amount of epoxide was determined from the peak area using a standard curve that had been constructed with authentic epoxides. Duplicate measurements were performed for each assay. A typical gas chromatogram with authentic C_3 hydrocarbons is shown in Fig. 2.

The conditions used in Fig. 2 could not separate propylene oxide from propanal. In order to rule out the possibility that propanal was an oxidation product of propylene, the reaction mixture was assayed using a glass column (6 ft long, 2 mm i.d.) 60/80 Tenax G.C. (Supelco, Inc., Bellefonte, Pennsylvania). The column temperature was maintained isothermally at 180°C; carrier gas flow was 35 ml of helium per minute. The retention times for propylene oxide and propanal were 13.5 and 15.7 minutes, respectively. Propylene glycol was assayed using this same Tenax G.C. glass column with column temperature maintained at 200°C.

For the cell-free system, the methane hydroxylation and propylene epoxidation activities were measured by determining methanol and propylene oxide formation by gas–liquid chromatography (GLC). A 0.2–ml sample of the particulate fraction (1 mg protein) was placed in a 5-ml vial. The vial was sealed with a rubber cap to minimize evaporation. The gaseous phase of the

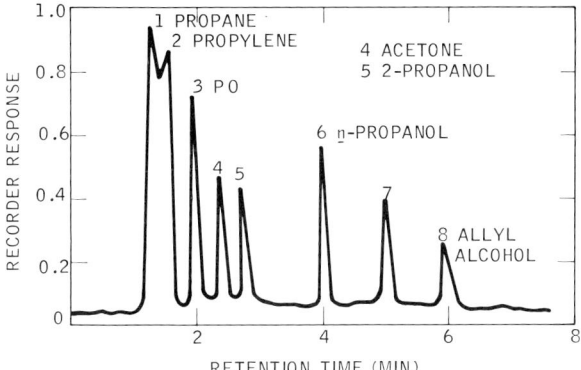

FIG. 2. A typical gas chromatogram for authentic C_3 hydrocarbons. A stainless steel column (20 ft by ⅛ in.) packed with 10% Carbowax 20M on 80/100 Chromosorb W was maintained isothermally at 100°C, and the carrier gas flow was 35 ml of helium per minute. PO, propylene oxide. From Hou *et al.* (1979a).

vial was replaced with a gas mixture containing 50% methane or propylene and 50% pure oxygen. In the case of methane hydroxylation, the buffer concentration of the reaction mixture was raised to 0.15 M sodium phosphate to inhibit further oxidation of methanol (Tonge *et al.*, 1975). The reaction mixture was incubated at 30°C on a water bath rotary shaker as described above. A 3-μl sample was removed with a syringe and was assayed by GLC.

E. MICROBIAL EPOXIDATION OF PROPYLENE

Resting cell suspensions of methane-grown cells of all newly isolated and known methane-utilizing cultures oxidized gaseous 1-alkenes to their corresponding 1,2-epoxides, which accumulated. Control experiments with heat-killed cells indicated that the epoxide was produced enzymatically. Methanol-grown cells showed no activity for either epoxidation of propylene or hydroxylation of methane. After the reaction, the cell suspensions were centrifuged to remove cells. The product propylene oxide was found totally in the supernatant fraction, i.e., the product epoxide accumulated extracellularly. No product peak other than propylene oxide from propylene was detected. The possibility of propanal as an oxidation product of propylene was ruled out on the basis of GLC analysis. A comparison of the activity for epoxidation of propylene by methane-grown microbes is listed in Table I. Epoxidation activity was found in all of the strains tested.

Independently of our studies, Colby *et al.* (1977) reported the ability of a nonspecific, soluble methane monooxygenase from *Methylococcus cap-*

TABLE I
EPOXIDATION OF PROPYLENE TO PROPYLENE OXIDE
BY METHANE-GROWN MICROBES[a]

Methylotrophs	Epoxidation rate (μmol/hour/mg of protein)
Obligate, type II	
Methylosinus sp. CRL 15	2.2
Methylosinus trichosporium OB3b	1.8
Methylosinus sp. CRL 16	1.6
Methylosinus sporium 5	1.0
Methylocystis sp. CRL 18	0.7
Methylocystis parvus OBBP	0.8
Obligate, type I	
Methylomonas sp. CRL 4	1.0
Methylomonas sp. CRL 21	1.2
Methylomonas methanica S$_1$	1.3
Methylomonas sp. CRL 8	0.6
Methylomonas albus BG 8	0.7
Methylomonas sp. CRL 17	0.5
Methylomonas sp. CRL 22	0.8
Methylomonas sp. CRL M6P	0.5
Methylomonas sp. CRL 20	1.0
Methylomonas sp. CRL 10	0.7
Methylobacter sp. CRL M6	0.6
Methylobacter sp. CRL 23	1.2
Methylobacter sp. CRL M1Y	0.7
Methylobacter sp. CRL 19	1.1
Methylobacter sp. CRL 5	0.8
Methylococcus capsulatus CRL M1	2.5
Methylococcus capsulatus Y	0.7
Methylococcus sp. CRL 25	0.6
Methylococcus sp. CRL 24	0.9
Facultative	
Methylobacterium organophilum CRL 26	1.3
Methylobacterium organophilum XX	0.9

[a] The reactions were ιducted as described in the text. Strains with CRL numbers are ne ιsolated cultures in our lab. From Hou et al. (1979a).

sulatus Bath to oxygena' -alkanes, n-alkenes, ethers, and alicyclic, aromatic, and heter· pounds. Stirling et al. (1979) have compared the methane mor e of the Bath strain with cell-free crude extracts obtained from ot ιs.

Representatives of three distinct groups of methylotrophic bacteria, Methylosinus trichosporium OB3b (an obligate methylotroph with type I membrane structure), Methylococcus capsulatus CRL M1 (an obligate methylotroph with type II membrane structure), and *Methylobacterium organophilum* CRL 26 (a facultative methylotroph), were chosen for further studies of microbial epoxidation. Because of the instability of the monooxygenase system, the optimal conditions for the epoxidation were studied in cell suspensions for methylotrophs, with particular reference to the epoxidation of propylene oxide.

In time course studies of epoxide formation, it was found that the rate of propylene oxide production was linear for the first 120 minutes for strains CRL M1 and OB3b, and for the first 60 minutes for strain CRL 26 (Fig. 3). Whether the apparent slower reaction rate after 1–2 hours incubation was caused by enzymatic or nonenzymatic degradation of propylene oxide or was because of product inhibition was explored. Propylene oxide (4.8 μmol) was added to viable or heat-killed cell suspensions in the presence or absence of substrate (propylene), and the mixture was incubated under standard assay conditions. No disappearance of propylene oxide or formation of propylene glycol was detected during 6 hours of incubation with either viable or heat-killed cell suspensions. However, after a prolonged incubation (24 hours), a slight disappearance of propylene oxide and the formation of propylene glycol was observed. The amount of the further oxidation of propylene oxide was identical with both viable and heat-killed cell suspensions. Data shown in Fig. 4 indicate that there was no enzymatic degradation of propylene oxide, and the nonenzymatic degradation of propylene oxide was negligible under our assay conditions. The viable cells incubated with both propylene and propylene oxide showed further production of propylene oxide. However, the rate of epoxidation was slower than in the standard assay system in which no external propylene oxide was added. Therefore, product inhibition might be counted as one of the reasons for the slower reaction rate after 2 hours of incubation.

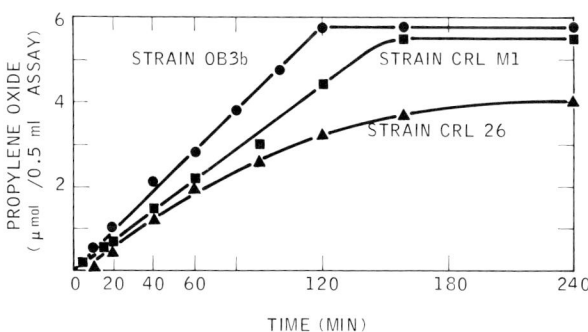

FIG. 3. Time course of propylene oxide production by resting cell suspensions of methylotrophic bacteria. From Hou *et al.* (1979a).

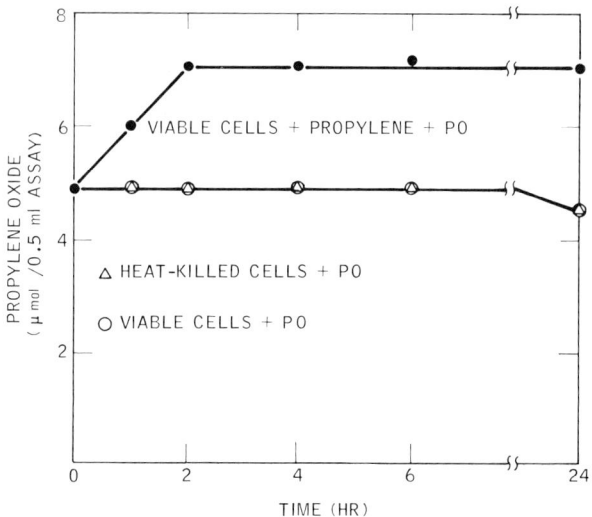

FIG. 4. Degradation of propylene oxide by resting-cell suspensions of *Methylococcus capsulatus* CRL M1. A 4.8- μmol amount of propylene oxide (PO) was added to each reaction mixture before incubation. Four milligrams of cells were used in each experiment.

Van der Linden (1963) demonstrated the production of 1,2-epoxyoctane from 1-octene by heptane-grown cells of *Pseudomonas* sp. and also stated that the epoxide was not further oxidized enzymatically. However, May and Abbott (1972, 1973) reported that when 1-octene was supplied as a substrate to the ω-hydroxylation enzyme system of *P. oleovorans*, both 8-hydroxy-1-octene and 1,2-epoxyoctane were formed. In addition, Abbott and Hou (1973) found that the methyl group of the latter compound was also susceptible to hydroxylation. Our present results, obtained from studies of viable cell suspensions of the methane-utilizing bacteria, however, indicated that propylene oxide was not further metabolized enzymatically.

Epoxide accumulation from 1-octene by *P. aeruginosa* was accompanied by the metabolism of a large quantity of 1-octene via methyl group oxidation (Van der Linden, 1963). In the epoxidation of propylene by cell suspensions of methane-utilizing bacteria, however, no formation of 3-hydroxy-1-propene was detected.

1. pH and Temperature

The optimum pH and temperature for epoxide production for all three of the methylotrophs tested was between 6.0 and 7.0 and around 35°C, respectively. At 40°C, there was an apparent decrease in the amount of epoxide accumulated. This is possibly a result both of the instability of the monooxygenase system at higher temperature and of the volatility of the product propylene oxide (bp 35°C).

2. Propylene Concentration

The production of propylene oxide by resting-cell suspensions of *M. capsulatus* CRL M1 was examined with various amounts of the substrate, propylene. The initial oxygen partial pressure in the gaseous phase was kept constant (50% v/v). Helium gas was used to balance the rest of the gaseous phase. A propylene concentration in the gaseous phase of approximately 15% (66 μmol) supported maximum propylene oxide production (Fig. 5). Higher propylene concentrations did not stimulate or inhibit the production of propylene oxide. The reaction rate appears to be dependent upon the solubility of both substrates, propylene and oxygen.

3. Substrate Specificity

The substrate specificity for the epoxidation of 1-alkenes by the three strains of methylotrophs was determined. In contrast to the *P. oleovorans* monooxygenase system (Abbott and Hou, 1973) which epoxidized only liquid 1-alkenes (from C_6 to C_{12}), the methane monooxygenase system epoxidized gaseous alkenes only (Table II). Some branched chain gaseous hydrocarbons, such as 2-methyl propane and 2-methyl propene, were also epoxidized. The highest production of epoxides was for propylene in all three strains tested.

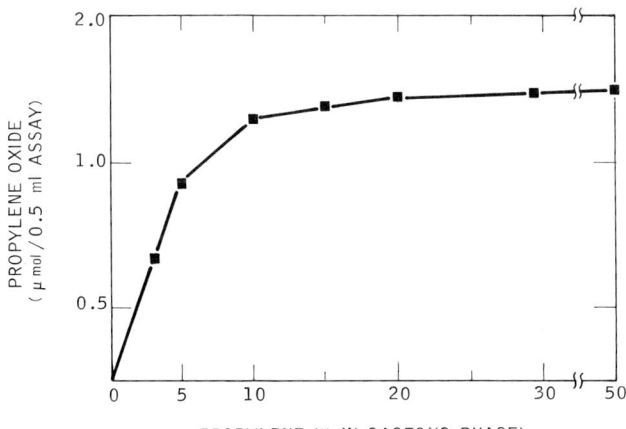

FIG. 5. Effect of propylene partial pressure on the production of propylene oxide by cell suspensions *Methylococcus capsulatus* CRL M1. Oxygen partial pressure was kept constant at 50%, v/v, of the gaseous phase. Propylene was diluted with helium to obtain various partial pressures of propylene. Propylene oxide was estimated after 30 minutes of incubation (3 mg of cells per vial). From Hou *et al.* (1979a).

TABLE II
Oxidation of Gaseous Alkenes and Methane by
Resting-Cell Suspensions of Methane-Grown Bacteria

Strains	Rate of oxidation (μmol/hour/assay[a])					
	Methane to methanol	Ethylene to ethylene oxide	Propylene to propylene oxide	1-Butene to 1,2-epoxy butane	Butadiene 1,2-epoxy butene	1-Pentene to 1,2-epoxy pentane
Obligate, type II membrane structure						
Methylosinus trichosporium OB3b	1.6	1.9	3.6	0.45	2.5	0
Obligate, type I membrane structure						
Methylococcus capsulatus CRL M1 NRRL B11219	2.5	5.5	5.5	1.3	4.4	0
Facultative						
Methylobacterium organophilum CRL 26 NRRL B 11222	0.7	0.9	2.5	0.9	2.8	0

[a] Two milligrams of cell (dry weight) per assay were used. From Hou *et al.* (1979a).

4. Inhibition Studies

The epoxidation of propylene to propylene oxide and the hydroxylation of methane to methanol by the three types of methane-utilizing bacteria were inhibited by metal-binding and metal-chelating agents, such as potassium cyanide, 1,10-phenanthroline, α,α-bipyridyl, thiourea, and imidazol (Table III). This suggests the involvement of metal ion(s) in both the epoxidation of propylene and the hydroxylation of methane. In addition, the extent of inhibition (percentage inhibition) for propylene and methane oxidation was similar, indicating that the epoxidation and hydroxylation reaction may be catalyzed by the same (or a similar) enzyme system.

The effect of methane on the oxidation of propylene to propylene oxide was studied in order to determine further whether a single enzyme system was responsible for both the oxidation of propylene and methane. The production of propylene oxide from propylene by a cell suspensions of *M. trichosporium* OB3b was assayed in the presence of a given amount of methane. The initial partial pressure of propylene in the gaseous phase was

TABLE III
EFFECT OF INHIBITORS ON THE EPOXIDATION OF PROPYLENE
AND THE HYDROXYLATION OF METHANE[a]

	Percentage inhibition					
	Methylosinus trichosporium OB3b		*Methylococcus capsulatus* CRL M1		*Methylobacterium organophilum* CRL 26	
Inhibitor	Epoxidation	Hydroxylation	Epoxidation	Hydroxylation	Epoxidation	Hydroxylation
Thiourea	100	100	100	100	100	100
1,10-Phenanthroline	90	92	95	95	90	90
α,α-Bipyridyl	100	90	100	100	100	100
Imidazole	95	90	95	95	100	100
Potassium cyanide	100	100	100	100	95	95

[a] The reactions were conducted as described in the text. The products were estimated by gas chromatography after 1 hour of incubation at 30°C. Each inhibitor was added at a final concentration of 1 mM. From Hou *et al.* (1979a).

kept constant in all of the experiments. As shown in Table IV, methane competes with propylene in the oxidation reaction by the whole cell system of strain OB3b. It is likely that the methane monooxygenase enzyme system catalyzes both the epoxidation of alkenes and the hydroxylation of methane. May and Abbott (1972, 1973) have reported that the ω-hydroxylation system from *P. oleovorans* catalyzed both the epoxidation of 1-octene and the hydroxylation of *n*-octane.

TABLE IV
EFFECT OF METHANE ON THE EPOXIDATION OF PROPYLENE[a]

Composition of gaseous phase	Propylene oxide formed (μmol)	Percentage inhibition
Propylene + helium + O_2 (25:25:50, v/v)	1.6	0
Propylene + methane + O_2 (25:25:50, v/v)	0.8	50

[a] Reactions were conducted as described in the text except that various gaseous compositions were used to maintain a constant propylene partial pressure. Cell suspensions of methane-grown *M. trichosporium* OB3b (3.6 mg) were used. Propylene oxide was estimated by gas chromatography after 15 minutes of the incubation. From Hou *et al.* (1979a).

5. Cell-Free System

We selected representatives of three distinct groups of methane-utilizing organisms to examine oxidation of n-alkanes (C_1 to C_4) and n-alkenes (C_2 to C_4) in cell-free systems. Cellular fractions were prepared from type I obligate methane-utilizing organisms, *Methylomonas* sp. CRL 17 and *Methylococcus capsulatus* (Texas); type II obligate methane-utilizing organisms, *Methylosinus trichosporium* (OB3b), and *Methylosinus* sp. (CRL-15); and a facultative methane-utilizing bacterium, *Methylobacterium* sp. (CRL-26).

Table V shows the distribution of the methane- and propylene-oxidizing activity in various fractions derived from these organisms. About 85–90% of the total activity was detected in the P(40) (40,000 g particulate) fraction and 10% was detected in the P(80) fraction. The soluble fraction S(80) did not contain any activity. The specific activities for the methane and the propylene oxidation in fractions P(40) and P(80) did not vary greatly among the various organisms examined. Epoxidation of propylene and hydroxylation of methane were both dependent upon the presence of oxygen and NADH. Both reactions were linear during the first 15 minutes as measured by detection of product by gas chromatography.

The particulate fractions, P(40) and P(80), from various organisms also

TABLE V
RATE AND DISTRIBUTION OF METHANE- AND PROPYLENE-OXIDIZING ACTIVITIES IN THE CELL FRACTIONS OF METHYLOTROPHS

Microorganism	Propylene oxidation[a] (%) in fraction:		Methane oxidation[a] (%) in fraction:	
	P(40)	P(80)	P(40)	P(80)
Type I obligate methylotroph				
Methylomonas sp. CRL 17	2.2 (85)	2.0 (15)	2.9 (87)	2.7 (13)
Methylococcus capsulatus (Texas)	2.6 (89)	2.0 (11)	3.8 (90)	3.9 (10)
Type II obligate methylotroph				
Methylosinus sp. CRL 15	3.8 (87)	3.7 (13)	4.8 (88)	4.2 (12)
Methylosinus trichosporium OB3b	2.8 (82)	2.5 (18)	3.1 (83)	3.0 (13)
Facultative methylotroph				
Methylobacterium sp.	1.2 (85)	1.1 (15)	2.7 (82)	2.8 (18)

[a] The product of the reaction was estimated by gas chromatography after 5, 10, and 15 minutes of incubation of reaction mixtures at 30°C on a rotary shaker. The rate of oxidation is expressed as micromoles of product formed per hour per milligram of protein. Parentheses indicate percentage of the total enzyme activity. From Patel *et al.* (1979a).

TABLE VI
OXIDATION OF n-ALKENES AND n-ALKANES BY
P(40) PARTICULATE FRACTION OF *Methylosinus* SP. CRL 15

Substrate	Product	Rate of product formation[a] (μmol/hour/mg of protein)
Ethylene	Ethylene oxide	1.27
Propylene	Propylene oxide	4.1
1-Butene	Epoxy butane	2.18
Butadiene	Epoxy butene	0.63
Methane	Methanol	4.8
Ethane	Ethanol	3.2
Propane	n-Propanol / 2-Propanol	1.18 / 0.75
Butane	n-Butanol / 2-Butanol	0.72 / 0.45

[a] The product of the reaction was estimated by gas chromatography after 5, 10, and 15 minutes of incubation of reaction mixture at 30°C on rotary shaker. From Patel *et al.* (1979a).

catalyzed the epoxidation of other n-alkenes (ethylene, 1-butene, and 1,3-butadiene) to the corresponding 1,2-epoxides and the hydroxylation of n-alkanes (ethane, propane, and butane tested) to the corresponding alcohols. Both primary and secondary alcohols were detected as the products of oxidation. Table VI shows the rate of oxidation of various n-alkanes and n-alkenes by the P(40) particulate fraction of *Methylosinus* sp. CRL 15. The product of oxidation was identified by gas chromatography after incubating P(40) fraction with various substrates at 30°C for 10 minutes. Tonge *et al.* (1975, 1977) have reported the purification of a membrane-bound methane monooxygenase from the particulate fraction (precipitated between 10,000 and 150,000 g centrifugation) of *M. trichosporium* OB3b. Colby *et al.* (1977) and Stirling *et al.* (1979) have demonstrated a unique soluble methane monooxygenase from *Methylococcus capsulatus* (Bath strain), *Methylosinus trichosporium*, and *Methylomonas methanica* that catalyzes the oxidation of n-alkanes, n-alkenes, and other aromatic hydrocarbons. The strains from the three distinct groups of methane-utilizing bacteria that we have examined all catalyze the epoxidation of gaseous alkenes (C_2 to C_4) and the hydroxylation of gaseous alkanes (C_1 to C_4). In contrast to *M. capsulatus* (Bath), the methane monooxygenase enzyme systems in the methylotrophs we examined were located in the cell-free particulate fractions precipitated between 10,000 and 40,000 g.

II. Microbial Methyl Ketone Formation

A. Introduction

As early as 1932, Hopkins and Chibnall suspected that methyl ketones were involved in paraffin oxidation. *Aspergillus versicolor*, which grew at the expense of long-chain paraffins, was also able to grow at the expense of related methyl ketones. Methyl ketone formation is well known in mammals and in fungi. In these cases, the ketone is formed by decarboxylation of a β-keto acid and has, therefore, one less carbon atom than the precursor. In contrast, bacterial formation of methyl ketones from alkanes is a unique α-oxidation, with no change in the carbon skeleton.

The first bacterial ketone formation from gaseous alkanes was demonstrated by Leadbetter and Foster (1960) with methane-grown *Pseudomonas methanica*. This strain was able to oxidize but not assimilate propane and butane, in the presence of the growth substrate (methane). Products of this cooxidation of propane were n-propanol, propionic acid, and acetone; butane yielded n-butanol, butyric acid, and 2-butanone. Subsequently, Lukins and Foster (1963) reported that propane-grown *Mycobacterium smegmatis* produced much more methyl ketones and less neutral volatile substances calculated as n-propanol.

Leadbetter and Foster (1960) postulated free radical equilibrium between the C_1 and C_2 positions, followed by formation of 1-alkyl and 2-alkyl hydroperoxides to account for the cooxidative formation of methyl ketones and fatty acids of substrate chain length. Attack on multiple sites (Fig. 6) of free radicals is typical of chemical oxidation of alkanes. If the hydroperoxide is a transitory intermediate in ketone formation, it could also be postulated that the 2-alcohol need not be an intermediate because the ketone could be formed directly from the 2-alkyl hydroperoxide. Apparently, 2-alcohols were not searched for in the work reported from Foster's laboratory. Results obtained from subsequent studies suggested that secondary alcohols might be intermediates of n-alkane oxidation.

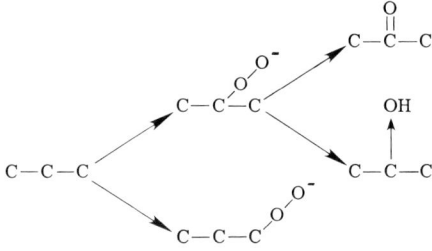

FIG. 6. Postulated mechanism of alkane oxidation. From Leadbetter and Foster (1960).

Phenazine methosulfate (PMS)-dependent methanol dehydrogenase has been reported from many methylotrophic bacteria (Anthony and Zatman, 1967; Patel et al., 1972, 1978; Patel and Felix, 1976; Yamanaka and Matsumoto, 1977; Bellion and Wu, 1978). This enzyme oxidizes primary alcohols from C_1 to C_{10} but does not oxidize secondary alcohols. Nicotinamide adenine dinucleotide (NAD)-dependent alcohol dehydrogenases have been reported from liver and from yeast (Branden et al., 1975). These alcohol dehydrogenases oxidize primary alcohols and acetaldehyde but have no activity on methanol. In addition, the alcohol dehydrogenases from yeast and liver also oxidize some secondary alcohols at a very low rate. An NAD(P)-dependent alcohol dehydrogenase was also reported in *Pseudomonas* (Tassin et al., 1973; Neihaus et al., 1978), *E. coli* (Hatanaka et al., 1971a), and *Leuconostoc* (Hatanaka et al., 1971b). However, these enzymes were active only toward long-chain primary alcohols or hydroxy fatty acids.

In our studies on the microbial oxidation of gaseous hydrocarbons, we found that resting-cell suspension of methane-grown methylotrophs oxidized n-propane and n-butane to their corresponding methyl ketones. In addition, we demonstrated for the first time the conversion of secondary alcohols to their corresponding methyl ketones by resting cell suspensions of C_1-utilizing microbes (Hou et al., 1979b; Patel et al., 1979b). A novel NAD-linked, secondary alcohol-specific alcohol dehydrogenase (SADH) was purified from both a bacterium (Hou et al., 1979c) and a yeast (Patel et al., 1980).

B. Assay Method

Cultures additional to those mentioned previously (Section II, A) were obtained from either the Northern Regional Research Laboratories (Peoria, Illinois) or the American Type Culture Collection (Rockville, Maryland). The organisms were maintained or grown on mineral-salt agar plates (or medium) with hydrocarbons as the sole carbon and energy source. When alcohols (at 0.3%) were used, the gaseous phase of the flask was air.

When whole cells were used for activity assay, the harvested cells from shake-flask cultures were washed twice with 0.05 M phosphate buffer, pH 7.0. One-half milliliter of this washed cell suspension (containing a given amount of dry cell mass) was placed in a 10-ml vial. Ten microliters of liquid substrate were added, and the vial was sealed with a rubber cap to minimize evaporation. The reaction mixture was incubated at 30°C on a water bath rotary shaker (New Brunswick Scientific Co., Edison, New Jersey) at 300 rpm. A 3-μl sample was removed with a syringe and was assayed with flame ionization gas chromatography using a stainless steel column (20 ft by ⅛ in.) packed with 10% Carbowax 20M on 80/100 Chromosorb W (Supelco, Inc.,

Bellefonte, Pennsylvania). The column temperature was maintained isothermally at 130°C, and the carrier gas flow was 35 ml of helium per minute. The methyl ketones were identified by retention time comparisons and cochromatography with authentic standards. The amount of methyl ketones accumulated was determined from the peak area using a standard curve that had been constructed with authentic samples. Duplicate measurements were performed for each assay.

For the cell-free system, the washed cells were disrupted with a Wave Energy Ultrasonic Oscillator, Model W201 (Wave Energy System, Inc., Newtown, Pennsylvania) and centrifuged at 20,000 g for 30 minutes. The clear supernatant was used for the enzyme assay. The enzyme activity was measured with a fluorescence spectrophotometer (Perkin-Elmer, Model MPF 44A) by following the formation of reduced NAD (EX 340 nm, Em 460 nm). The formation of reduced NAD was also followed with an absorption spectrophotometer at 340 nm. The assay system (3 ml) contained: potassium phosphate buffer, pH 7.0, 150 μmol; NAD, 1 μmol; a given amount of enzyme preparation; and secondary alcohol, 10 μmol. The reaction was started by the addition of substrate. One unit of enzyme activity represents the reduction of 1 nmol NAD per minute. Protein concentrations were determined by the method of Lowry *et al.* (1951).

C. Oxidation of 2-Butanol to 2-Butanone

Resting-cell suspensions of methane- and methanol-grown microbial cells oxidized secondary alcohols to their corresponding methyl ketones. After the incubation, the reaction mixture was centrifuged to remove the cells. The product methyl ketones were found to be accumulated extracellularly. Control experiments with heat-killed cells indicated that the methyl ketones were produced enzymatically. A comparison of the activity for converting 2-butanol to 2-butanone by C_1-utilizing microbes is listed in Table VII. Secondary alcohol dehydrogenase activity was found in all of the C_1 utilizers tested.

Secondary alcohol dehydrogenase activity was also found in cell suspensions of the methanol-grown or methylamine-grown microbes listed in Table VII. However, SADH is not a constitutive enzyme; the enzyme activity was not found in succinate-grown facultative C_1 utilizers.

Optimal conditions for the production of 2-butanone were compared among five distinct types of methane or methanol utilizers: *Methylosinus trichosporium* OB3b (type I, obligate), *Methylococcus capsulatus* CRL M1 (type II, obligate), *Methylobacterium organophilum* CRL 26 (facultative), *Hansenula polymorpha* ATCC 26012, and *Pseudomonas* sp. ATCC 21439 (obligate methanol utilizer).

The production of 2-butanone from 2-butanol reached a maximum after 14

TABLE VII
Conversion of 2-Butanol to 2-Butanone by Cell Suspensions of C_1-Utilizing Microbes[a]

Microbes	Conversion rate (μmol/hour/mg of protein)
Obligate methane utilizers (methane grown)	
Type II membrane structure	
Methylosinus trichosporium OB3b	4.8
Methylosinus sp. CRL 15	4.5
Methylosinus sporium 5	3.0
Methylosinus sp. CRL 16	2.5
Methylocystis parvus OBBP	1.1
Methylocystis sp. CRL 18	1.0
Type I membrane structure	
Methylomonas methanica S_1	0.4
Methylomonas albus BG8	3.5
Methylomonas sp. CRL 4	2.5
Methylomonas sp. CRL 8	3.0
Methylomonas sp. CRL 10	0.5
Methylomonas sp. CRL 17	2.0
Methylomonas sp. CRL 20	2.0
Methylomonas sp. CRL 21	2.8
Methylomonas sp. CRL 22	1.5
Methylomonas sp. CRL M6P	1.4
Methylobacter sp. CRL M6	1.4
Methylobacter sp. CRL 23	1.0
Methylobacter sp. CRL M1Y	3.0
Methylobacter sp. CRL 19	1.8
Methylobacter sp. CRL 5	2.0
Methylococcus capsulatus CRL M1	5.0
M. capsulatus Y	0.8
Methylococcus sp. CRL 25	0.9
Methylococcus sp. CRL 24	2.5
Facultative methane utilizers (methane grown)	
Methylobacterium organophilum CRL 26	2.5
Methylobacterium organophilum XX	2.5
Obligate methanol utilizers	
Pseudomonas sp. CRL 75 ATCC 21439	5.4
Methylomonas methylovora ATCC 21852	2.5
Facultative methanol utilizers (methanol grown)	
Pseudomonas sp. CRL 74 ATCC 21438	3.2
Pseudomonas Ms. ATCC 25262	3.5
Yeasts (methanol grown)	
Candida boidinii NRRL Y 2332	6.0
Candida utilis ATCC 26387	6.8
Hansenula polymorpha ATCC 26012	5.8
Hansenula anomala NRRL Y336	5.5
Pichia pastoris NRRL Y55	5.8
Pichia sp. CRL 72	4.0

[a] Strains with CRL numbers are newly isolated cultures in our lab. From Hou *et al.* (1979b).

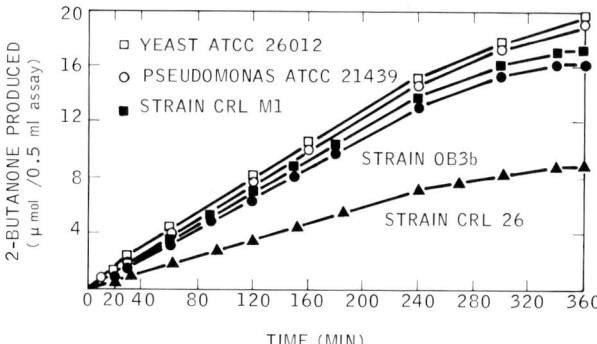

FIG. 7. Time course of 2-butanone production by resting-cell suspensions of methanol-grown cultures from five distinct types of C_1-utilizing microbes. Resting-cell suspensions (0.7 mg protein in 0.5 ml) were incubated with 10 μl of 2-butanol at 30°C for various times. A 3-μl sample was assayed at a given time with gas chromatography.

hours of incubation in batch experiments in all the microbes tested. The amount of 2-butanone had not declined after 30 hours of incubation. The rate of 2-butanone production was linear for the first 4 hours (Fig. 7). Therefore, the production of 2-butanone was measured within this interval whenever the effect of a variable was tested.

1. *pH*

The effect of pH on the production of 2-butanone was studied with tris(hydroxylmethyl) amino methane–HCl buffer (0.05 M) for pH values of 8.0–10.0, and 0.05 M potassium phosphate buffer for pH values from 5.0 to 8.0. A pH around 8.0 was found to be the optimum for 2-butanone formation in all the five distinct types of microbes tested (Fig. 8). Strains ATCC 21439 and CRL 26 showed high activity at both pH 8 and pH 9. In the case of strain OB3b, the 2-butanone formation was significantly lower at pH 9 than that at pH 8. Yeast cells appeared less affected by pH in the production of 2-butanone. Authentic samples of 2-butanol and 2-butanone, final concentrations 8 μmol/0.5 ml, was added to heat-killed cell suspensions of strain OB3b in 0.05 M buffer at pH 5.0, 7.0, and 10.0 to test for nonenzymatic oxidation and degradation of 2-butanol and 2-butanone, respectively. The concentration of 2-butanol and 2-butanone in these heat-killed cell suspensions were not decreased after 16 hours of incubation, indicating that nonenzymatic oxidation or hydrolysis of 2-butanol or 2-butanone did not occur under the assay conditions.

2. *Temperature*

The temperature optimum for the production of 2-butanone by cell suspensions was about 35°C except for the yeast culture, which had an optimum of about 40°C (Fig. 9).

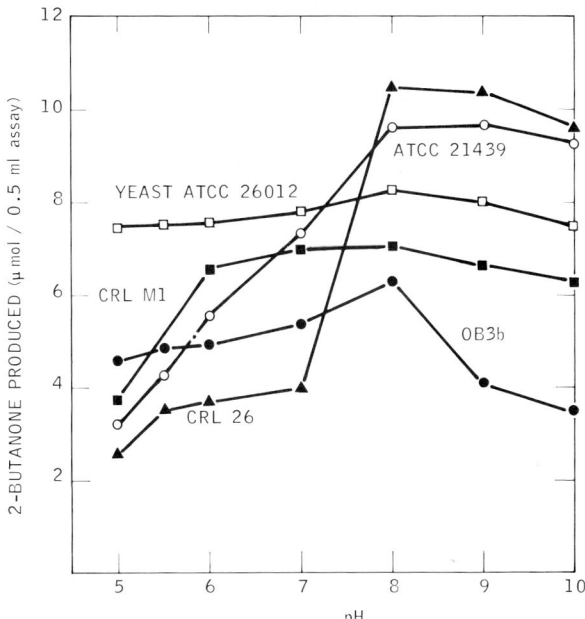

FIG. 8. Effect of pH on the production of 2-butanone by resting-cell suspensions of methanol-grown cultures from five distinct types of C_1-utilizing microbes. Product 2-butanone was assayed after 2 hours of incubation.

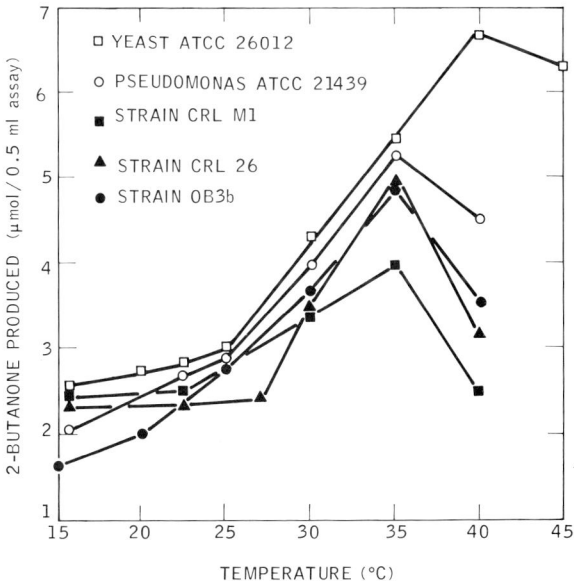

FIG. 9. Effect of temperature on the production of 2-butanone by resting-cell suspensions of methanol-grown cultures from five distinct types of C_1-utilizing microbes. Reaction conditions were the same as those described in the legend to Fig. 8.

3. Substrate Concentration

Various concentrations of 2-butanol were added to cell suspensions of yeast and of strain ATCC 21439. The production of 2-butanone was assayed after 35 minutes of incubation. The amount of 2-butanone produced was dependent on the amount of substrate initially added. A 2-butanol concentration of about 50 μmol supported maximum 2-butanone production (Fig. 10).

4. Product Inhibition and Further Oxidation

Examination of the time course of 2-butanone production revealed that the rate decreased after 4 hours of incubation, suggesting, among other possibilities, either product inhibition or further oxidation of 2-butanone. To test these possibilities, 8 μmol of 2-butanone was added to viable or heat-killed cell suspensions and incubated under the standard conditions. No decline was observed in 2-butanone concentration in all the heat-killed cell suspensions, but 2-butanone slowly disappeared in the presence of viable cells of all five strains. When 2-butanol (5 μl/0.5 ml reaction mixture) was added to viable cell suspensions along with the exogenously supplied 2-butanone, a net increase in 2-butanone production was detected (Fig. 11). The reaction rates were identical to those shown in Fig. 7 and were not affected by the presence of the exogenously supplied 2-butanone. These data indicate that there is no product inhibition in the production of 2-butanone. A small amount of further oxidation of 2-butanone by viable cell suspensions was observed. Therefore, the decrease in 2-butanone production rate after 4 hours of incubation may result from the depletion of other requirement(s), e.g., a cofactor(s).

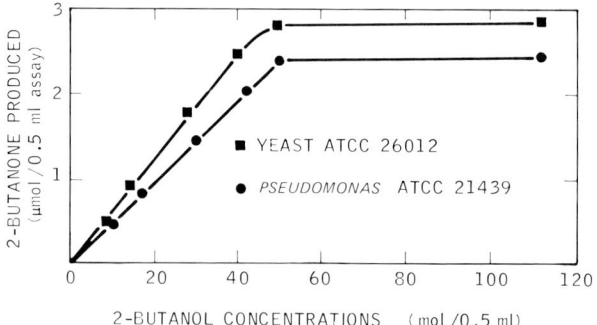

FIG. 10. Effect of 2-butanol concentration on the production of 2-butanone by resting-cell suspensions of strains ATCC 21439 and 26012 grown on methanol. Product 2-butanone was assayed after 35 minutes of incubation. Cell concentrations used were 0.92 mg per 0.5 ml.

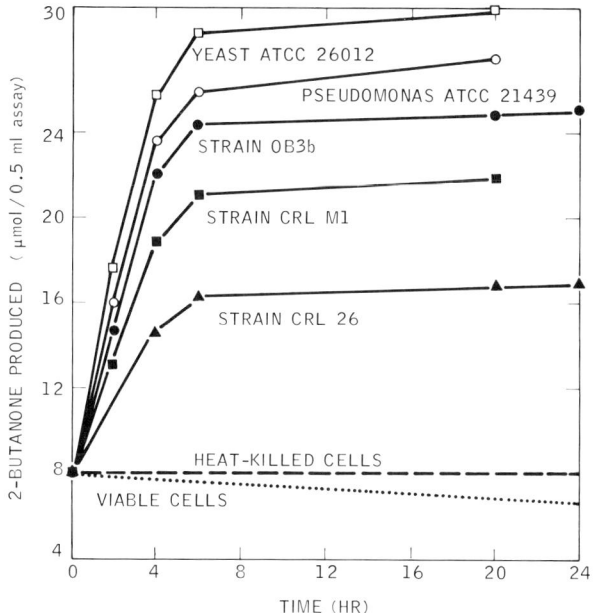

FIG. 11. Product inhibition and degradation of 2-butanone by resting-cell suspensions of methanol-grown cultures from five different C_1-utilizing microbes. Heat-killed cells were prepared by boiling the cell suspension for 5 minutes. 2-Butanone (8 μmol) and 2-butanol (10 μl) were added to the vials before the incubation.

5. Inhibition Studies

The production of 2-butanone from 2-butanol by cell suspensions of all five strains was inhibited by metal-chelating agents, such as 1,10-phenanthroline and α,α-dipyridyl. However, the activity was not inhibited by sodium cyanide or thiourea (Table VIII); this suggests metal involvement.

TABLE VIII
EFFECT OF METAL-CHELATING AGENTS AND OTHER INHIBITORS ON THE PRODUCTION OF 2-BUTANONE BY CELL SUSPENSIONS OF METHANOL-GROWN *Methylococcus capsulatus* CRL M1

Metal-chelating agents[a]	Inhibition (%)
Sodium cyanide	0
Sodium azide	10
Ethylenediaminetetraacetic acid	70
1,10-Phenanthroline	95
α,α-Dipyridyl	75
Thiourea	0

[a] Concentration of 1 mM. From Hou *et al.* (1979b).

TABLE IX
OXIDATION OF SECONDARY ALCOHOLS BY CELL SUSPENSIONS OF C_1 UTILIZERS GROWN ON METHANOL[a]

C_1 utilizers	Oxidation rate (μmol/hour per mg of protein)			
	2-Propanol to acetone	2-Butanol to 2-butanone	2-Pentanol to 2-pentanone	2-Hexanol to 2-hexanone
Methylosinus trichosporium OB3b	0.30	4.8	2.7	0.09
Methylococcus capsulatus CRL M1	2.0	5.0	0.24	0.08
Methylobacterium organophilum CRL 26	0.72	2.5	1.0	0.09
Hansenula polymorpha ATCC 26012	5.9	5.8	1.4	0.72
Pseudomonas sp. ATCC 21439	3.5	5.4	0.05	0.03

[a] Data from Hou *et al.* (1979b).

6. Substrate Specificity

The substrate specificity for the oxidation of secondary alcohols by the five strains of C_1 utilizers was studied. Among the secondary alcohols, 2-propanol and 2-butanol were oxidized at higher rates; 2-pentanol, 2-hexanol, and 2-heptanol were oxidized at a much slower rate (Table IX). The oxidation products of these secondary alcohols were the corresponding methyl ketones, as determined by gas chromatographic retention time comparisons with authentic standards.

D. PURIFICATION AND PROPERTIES OF A SECONDARY ALCOHOL-SPECIFIC ALCOHOL DEHYDROGENASE

Cell-free soluble extracts from sonically disrupted cells also oxidized 2-butanol to 2-butanone (Table X). However, the cell-free system required the addition of a cofactor, NAD, for its activity. Other cofactors tested [including NAD(P)H, NADP, phenazine methosulfate, GSH, FAD, potassium ferricyanide, and dichlorophenol indophenol] were not effective. The stoichiometry for the consumption of 2-butanol, the reduction of NAD, and the formation of 2-butanone was obtained (Table XI).

1. Purification of SADH

a. From Pseudomonas *sp.* Secondary alcohol dehydrogenase (SADH) from an obligate methanol utilizer, *Pseudomonas* sp. ATCC 21439, has been

TABLE X
CONVERSION OF 2-BUTANOL TO 2-BUTANONE BY
SOLUBLE CRUDE EXTRACTS OF METHYLOTROPHS[a]

Microbes	Growth substrate	Conversion (units/mg protein)
Obligate methylotrophs		
Type II membrane structure		
Methylosinus trichosporium OB3b	CH_4	4.5
Methylosinus trichosporium OB3b	CH_3OH	2.4
Type I membrane structure		
Methylococcus capsulatus CRL M1	CH_4	3.2
Methylococcus capsulatus CRL M1	CH_3OH	2.0
Facultative methylotrophs		
Methylobacterium organophilum CRL 26	CH_4	1.8
Methylobacterium organophilum CRL 26	CH_3OH	2.5
Others		
Pseudomonas sp. ATCC 21439	CH_3OH	25
Hansenula polymorpha ATCC 26012	CH_3OH	23
Candida utilis ATCC 26387	CH_3OH	78
Pichia sp. NRRL Y11328	CH_3OH	105
Torulopsis sp. strain A_1	CH_3OH	62
Kloeckera sp. strain A_2	CH_3OH	90

[a] Cells wre disrupted sonically as described in the text. The supernatant of 10,000 g centrifugation was used for the enzyme assay.

purified. The cells, suspended in 300 ml 0.05 M sodium phosphate buffer, pH 7.0, with 0.5 mM dithiothreitol (buffer A) were disrupted sonically (5 × 1 minute). The crude extract was separated by centrifugation. The crude extract was heat treated at 50°C in a water bath for 10 minutes. The resulting precipitate was removed by centrifugation. To the supernatant solution, 25

TABLE XI
STOICHIOMETRY OF THE PRODUCTION OF 2-BUTANONE FROM 2-BUTANOL BY CELL-FREE
EXTRACTS OF STRAIN ATCC 21439[a]

Experiment	2-Butanol consumed (nmol)	NAD consumed (nmol) XX	2-Butanone X produced (nmol)
1	260	270	250
2	530	540	520

[a] The reaction mixtures (1.0 mg of protein in 3 ml) were incubated at 30°C for 10 minutes (experiment 1) and for 20 minutes (experiment 2) in the presence of 1.0 μmol of NAD and 10 μmol of 2-butanol. X, Determined gas chromatographically; XX, determined fluorescence spectrophotometrically, and endogenous consumption of NAD was corrected. From Hou *et al.* (1979b).

ml of protamine sulfate solution (2% solution in 0.1 M Tris base) was added dropwise with continuous stirring. After it had stood for 30 minutes, the extract was centrifuged. The supernatant solution was fractionated with solid ammonium sulfate. Material precipitating between 30 and 60% saturation was collected and was dialyzed overnight against buffer A. The dialyzed material was applied to a DEAE–cellulose column (3 × 35 cm) that had been equilibrated with buffer A. The secondary alcohol dehydrogenase activity was eluted in the void volume. This DEAE–cellulose eluate was concentrated by ammonium sulfate fractionation. Material precipitating between 30 and 50% ammonium sulfate saturation was collected by centrifugation and dialyzed overnight against A. This fraction was further washed and filtered through an Amicon unit with XM 50 membrane. The concentrated fraction (6 ml) inside the Amicon unit was applied to an Affi-Gel Blue column (0.8 cm × 18 cm) that had been equilibrated with buffer A for affinity chromatography. The column was washed overnight with buffer A (0.18 ml/minute) and then was eluted with buffer A containing 5 mM NAD. Each 1-ml fraction was collected. Secondary alcohol dehydrogenase activity was located in tube No. 8-12. A summary of the purification steps is given in Table XII.

b. *From* Pichia *sp.*   Large-scale cultures of *Pichia* sp. NRRL Y 11328 were grown with aeration for 48 hours at 30°C in a 14-liter New Brunswick fermenter in a mineral-salt medium (Foster and Davis, 1966) containing methanol (0.4%, v/v) as the sole carbon source. The crude extract prepared from 200 gm cells (wet weight) in buffer A was treated with protamine

TABLE XII
PURIFICATION OF SECONDARY ALCOHOL DEHYDROGENASE FROM *Pseudomonas* SP. ATCC 21439[a]

Procedures	Volume (ml)	Protein (mg)	Specific activity (units/mg protein)	Total units	Yield (%)
Crude extract	250	2,698	25	67,450	100
Heat treatment	245	949	67.5	64,080	95
Protamine sulfate	260	526	103.8	54,640	81
$(NH_4)_2SO_4$ (30–60% sat.)	30	232	200	46,450	69
DEAE–cellulose column	150	42.2	875	37,160	55
Amicon filtration (XM-50)	6	22.0	1,500	33,050	49
Affi-Gel Blue column	5	0.34	65,600	22,300	33

[a] From Hou *et al.* (1979c).

sulfate as described in Section II,D,1,a. The supernatant solution thus obtained was fractionated with solid ammonium sulfate. Material precipitating between 50 and 70% saturation was collected by centrifugation and dissolved in buffer A. This preparation was dialyzed overnight against buffer A, and the dialyzed material was applied to a DEAE-cellulose column (5 × 40 cm) that had been equilibrated with buffer A. The sample was washed with 200 ml of buffer A and eluted with 2 liters of buffer A that contained NaCl in a linear concentration gradient running from 0 to 0.5M. Fractions of 15 ml were collected. Fractions containing SADH activity were pooled and were termed DEAE-cellulose eluate. The DEAE-cellulose eluate was concentrated by ammonium sulfate fractionation. Material precipitating between 50 and 70% of ammonium sulfate saturation was collected by centrifugation and dissolved in buffer A. This preparation was dialyzed overnight against buffer A, and 4-ml samples were passed through a Bio-Gel agarose A-1.5 column (2.5 × 100 cm) that had been equilibrated with buffer A. Fractions containing constant specific activity of enzyme were pooled and concentrated by Amicon ultrafiltration using an XM 50 filter. A summary of the purification steps is given in Table XIII.

2. *Properties of SADH*

The molecular weights of the purified SADH from *Pseudomonas* and *Pichia*, as estimated by a Bio-Gel agarose A-1.5 column, are 95,000 and 98,000, respectively. Acrylamide gel electrophoresis of the purified SADH fractions from both sources showed a single protein band (Fig. 12). The K_m values for 2-butanol and NAD are 0.25 and 0.11 mM, respectively, for the bacterial SADH. The pH optimum for SADH activity was around 8–9 for bacterial SADH and 8.0 for yeast SADH (0.05 M sodium phosphate buffer for pH 5–8; 0.05 M sodium pyrophosphate buffer for pH 8–11).

TABLE XIII
PURIFICATION OF SECONDARY ALCOHOL DEHYDROGENASE FROM *Pichia* SP.

Step	Volume (ml)	Protein (mg)	Units	Specific activity (units/mg protein)	Yield (%)
Crude extracts	875	21,875	2,391,375	109	100
Protamine sulfate treatment	890	21,360	2,370,960	111	99
Ammonium sulfate fraction (50–70% saturation)	117	3,090	1,820,010	589	76
DEAE-cellulose eluate	55	200	706,800	3,534	29
Bio-Gel chromatography	19	52	312,624	6,012	13

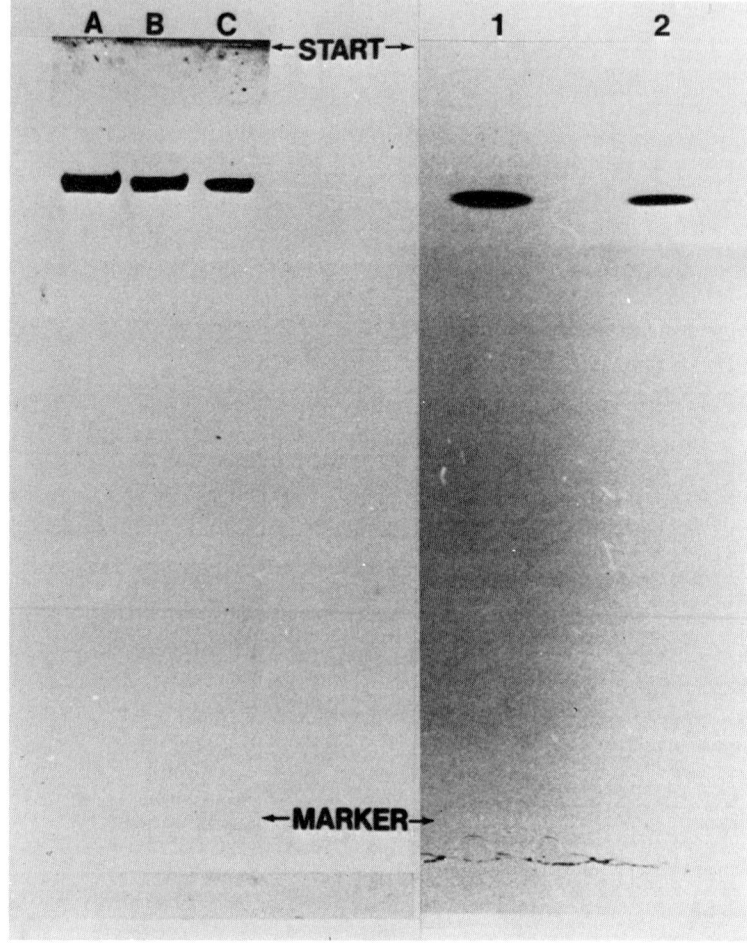

FIG. 12. Gel electrophoretic pattern of secondary alcohol dehydrogenases from *Pseudomonas* sp. ATCC 21439 and *Pichia* sp. NRRL Y-11328. Bacterial enzyme 40 μg (A), 20 μg (B), and 10 μg (C) and yeast enzyme 25 μg (2), and 50 μg (1) were applied to each slot. Protein migrated to the anode (bottom) at a constant 300 mV for 2H in 0.05 M Tris–glycine buffer (pH 9.0) [A,B,C from Hou *et al.* (1979c); 1,2 from Patel *et al.* (1979c)].

3. Substrate Specificity

Secondary alcohol dehydrogenase oxidizes secondary alcohols with the following relative percentage rates (*Pseudomonas* and *Pichia* SADH, respectively): 2-propanol (85, 78%), 2-butanol (100, 100%), 2-pentanol (5, 20%), 2-hexanol (2, 10%), acetaldehyde (4, 0%), propanol (2, 0%), cyclohexanol (4, 0%), butane 1,3-diol (2, 0%), and butane 2,3-diol (2.5, 0%). The following compounds tested were not oxidized by SADH: 2-heptanol to 2-decanol, formaldehyde, butanal to decanal, benzaldehyde, methanol to n-decanol, isobutanol, phenol, butane, 1,2-diol, and succinic acid. It seems that a hydrophobic carbon moiety adjacent to the secondary alcohol is required for the enzyme activity.

4. Inhibition Studies

Activity of SADH from both sources was inhibited by metal-chelating agents in the following order (percentage inhibition): 1,10-phenanthroline (95%), α,α-bipyridyl (70%), ethylenediaminetetraacetic acid (EDTA) (63%), and sodium azide (10%). This suggests possible metal involvement. However, the activity was not inhibited by sodium cyanide or thiourea. The enzyme activity was also inhibited by storing thio inhibitors, such as p-hydroxymercuribenzoate (100%) and 5,5'-dithiobis(2-nitrobenzoic acid), and was not inhibited by less potent thio inhibitors, such as iodoacetic acid or N-ethylmaleimide. Neither ethanol or n-propanol inhibited SADH activity, despite their structural similarity to 2-butanol with respect to potentially competing with 2-butanol for the alkyl binding site(s).

Secondary alcohol dehydrogenase also catalyzed the reductions of 2-butanone to 2-butanal and of acetone to 2-propanol in the presence of NADH.

5. Metal Content

Metal analyses carried out by X-ray fluorescence showed 2 mol of zinc per mole of SADH, or one zinc per subunit. This is different from primary alcohol dehydrogenase, which consists of two catalytically active zincs and two catalytically nonactive zincs.

III. Conclusion

Both the *Pseudomonas aeruginosa* system demonstrated by Van der Linden (1963) and the *Pseudomonas oleovorans* system demonstrated in our laboratory (May and Abbott, 1972, 1973; Abbott and Hou, 1973) epoxidized liquid 1-alkenes from C_6 to C_{12}, but not gaseous alkenes. We now describe the epoxidation of ethylene, propylene, 1-butene, and butadiene by cell suspensions of all three groups of methane-utilizing bacteria. The epoxidation of alkenes and the hydroxylation of methane were not found under anaerobic conditions or in methanol-grown cells, suggesting that the enzyme system was inducible. The product 1,2-epoxides accumulated extracellularly. The nonenzymatic degradation of propylene oxide in our standard assay system was not significant even after a prolonged incubation time. Further enzymatic metabolism of propylene oxide also was not observed.

Van der Linden (1963) reported that a large quantity of 1-octene was metabolized by *P. aeruginosa* via methyl group oxidation. In the epoxidation of propylene by cell suspensions of methane-utilizing bacteria, however, no formation of 3-hydroxy-1-propene was detected.

We have confirmed the report of Lukins and Foster (1963) that *n*-alkanes are cooxidized to their corresponding methyl ketones by methylotrophs in the presence of methane. We have also demonstrated that resting-cell suspensions of methane-grown cells oxidized *n*-alkanes to their methyl ketones in the absence of growth substrate. In addition, we demonstrated for the first time the conversion of secondary alcohols to their corresponding methyl ketones by resting cell suspensions of either alkane-grown or alcohol-grown cells. Succinate-grown cells do not have SADH activity, suggesting that either alkane or alcohol is required for inducing the enzyme. The reason that methylotrophs possess a secondary alcohol dehydrogenase is not well understood. However, this enzyme is of great advantage to the organism as its growth yield, when growing on gaseous alkanes as the sole source of carbon and energy, could be exclusively NAD(P)H dependent. Methane monooxygenase was reported to be a nonspecific system that oxidized *n*-alkanes to both their primary and secondary alcohols (Colby *et al.*, 1977). Therefore, it is not a complete surprise to see secondary alcohol dehydrogenase activity in the methylotrophic bacteria. One thing not clarified by these studies is the ability of the obligate methylotrophs to oxidize extensively substrates that they are incapable of utilizing for growth. The metabolism of the obligate methylotrophs is uniquely dependent on a one-carbon compound (formaldehyde) for the biosynthesis of certain essential cellular constituents. This compound can be obtained from methane and methanol but is unobtainable from the non-growth-supporting compounds.

The extent of resemblance between SADH and the well-characterized alcohol dehydrogenases from liver and yeast as to their catalytic sites and enzyme protein conformations requires further investigation.

REFERENCES

Abbott, B. J., and Hou, C. T. (1973). *Appl. Microbiol.* **26**, 86–91.
Anthony, C., and Zatman, L. J. (1967). *Biochem. J.* **104**, 953–959.
Bellion, E., and Wu, G. T. S. (1978). *J. Bacteriol.* **135**, 251–258.
Branden, C. I., Jornvall, H., Eklund, H., and Furugren, B. (1975). In "The Enzymes" (P. D. Boyer, ed.), 3rd ed., Vol. 11, pp. 103–190. Academic Press, New York.
Cardini, G., and Jurtshuk, P. (1970). *J. Biol. Chem.* **245**, 2789–2796.
Colby, J., and Dalton, H. (1978). *Biochem. J.* **171**, 461–468.
Colby, J., Stirling, D. I., and Dalton, H. (1977). *Biochem. J.* **165**, 395–402.
Coon, M. J., Strabel, H. W., Autor, A. P., Heidema, J., and Duppel, W. (1972). In "Biological

Hydroxylation Mechanisms" (G. S. Boyd and R. M. S. Smellie, eds.), p. 45. Academic Press, New York.
Ferenci, T. (1974). *FEBS Lett.* **41**, 94–98.
Foster, J. W., and Davis, R. H. (1966). *J. Bacteriol.* **91**, 1924–1931.
Hatanaka, A., Adachi, O., Chiyonobu, T., and Ameyama, M. (1971a). *Agric. Biol. Chem.* **35**, 1142–1143.
Hatanaka, A., Adachi, O., Chiyonobu, T., and Ameyama, M. (1971b). *Agric. Biol. Chem.* **35**, 1304–1306.
Higgins, I. J., and Quayle, J. R. (1970). *Biochem. J.* **118**, 201–208.
Hopkins, S. J., and Chibnall, A. C. (1932). *Biochem. J.* **26**, 133–142.
Hou, C. T., Patel, R. N., Laskin, A. I., and Barnabe, N. (1979a). *Appl. Environ. Microbiol.* **38**, 127–134.
Hou, C. T., Patel, R. N., Laskin, A. I., Barnabe, N., and Marczak, I. (1979b). *Appl. Environ. Microbiol.* **38**, 135–142.
Hou, C. T., Patel, R. N., Laskin, A. I., Barnabe, N., and Marczak, I. (1979c). *FEBS Lett.* **101**, 179–183.
Leadbetter, E. R., and Foster, J. W. (1959). *Nature (London)* **184**, 1428–1429.
Leadbetter, E. R., and Foster, J. W. (1960). *Arch. Mikrobiol.* **35**, 92–104.
Lowry, O. H., Rosebrough, N. J., Farr, A. I., and Randall, R. J. (1951). *J. Biol. Chem.* **193**, 255–275.
Lukins, H. B., and Foster, J. W. (1963). *J. Bacteriol.* **85**, 1074–1086.
May, S. W., and Abbott, B. J. (1972). *Biochem. Biophys. Res. Commun.* **48**, 1230–1234.
May, S. W., and Abbott, B. J. (1973). *J. Biol. Chem.* **248**, 1725–1730.
Neihaus, W. G., Jr., Frielle, T., and Kingsley, E. A., Jr. (1978). *J. Bacteriol.* **134**, 177–183.
Patel, R. N., and Felix, A. (1976). *J. Bacteriol.* **128**, 413–424.
Patel, R. N., Bose, H. R., Mandy, W. J., and Hoare, D. S. (1972). *J. Bacteriol.* **110**, 570–577.
Patel, R. N., Hou, C. T., and Felix, A. (1978). *J. Bacteriol.* **133**, 641–649.
Patel, R. N., Hou, C. T., Laskin, A. I., Felix, A., and Derelanko, P. (1979a). *J. Bacteriol.* **139**, 675–679.
Patel, R. N., Hou, C. T., Laskin, A. I., Derelanko, P., and Felix, A. (1979b). *Appl. Environ. Microbiol.* **38**, 219–223.
Patel, R. N., Hou, C. T., Laskin, A. I., Derelanko, P., and Felix, A. (1979c). *Eur. J. Biochem.* **101**, 401–406.
Patt, T. E., Cole, G. C., and Hanson, R. S. (1976). *Int. J. Syst. Bacteriol.* **26**, 226–229.
Ribbons, D. W. (1975). *J. Bacteriol.* **122**, 1351–1363.
Ribbons, D. W., and Michalover, J. L. (1970). *FEBS Lett.* **11**, 41–44.
Stirling, D. I., Colby, J., and Dalton, H. (1979). *Biochem. J.* **177**, 361–364.
Tassin, J. P., Celier, C., and Vandecasteele, J. P. (1973). *Biochim. Biophys. Acta* **315**, 220–232.
Tonge, G. M., Harrison, D. E. F., Knowles, C. J., and Higgins, I. J. (1975). *FEBS Lett.* **58**, 293–299.
Tonge, G. M., Harrison, D. E. F., and Higgins, I. J. (1977). *Biochem. J.* **161**, 333–344.
Van der Linden, A. C. (1963). *Biochim. Biophys. Acta* **77**, 157–159.
Whittenbury, R., Phillips, K. C., and Wilkinson, J. F. (1970). *J. Gen. Microbiol.* **61**, 205–218.
Yamanaka, K., and Matsumoto, K. (1977). *Agric. Biol. Chem.* **41**, 467–475.

Oxidation of Hydrocarbons by Methane Monooxygenases from a Variety of Microbes

HOWARD DALTON

Department of Biological Sciences, University of Warwick, Coventry, England

I.	Introduction	71
II.	Substrates of Methane Monooxygenases	73
	A. Alkanes and Substituted Methane Derivatives	74
	B. Alkenes	75
	C. Ethers	76
	D. Alicyclic, Aromatic, and Heterocyclic Hydrocarbons	77
	E. General Conclusions on Substrate Specificities	78
III.	Methane Monooxygenase	78
	A. *Methylosinus trichosporium*	78
	B. *Methylococcus capsulatus* Bath	80
	C. *Methylomonas methanica*	85
IV.	Whole Cell Oxidation of Propene	85
	References	86

I. Introduction

Methane monooxygenase is the enzyme responsible for the initial oxygenation of methane to methanol. It is a monooxygenase because only one atom of dioxygen is incorporated into methanol. The other atom ends up in water; viz.

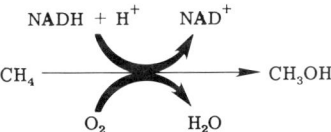

Organisms that grow on methane are capable of completely oxidizing methane to carbon dioxide. An intermediate in the oxidation pathway is formaldehyde which can either be assimilated into the cell to provide carbon for growth or further oxidized in a dissimilatory fashion to CO_2 and provide energy for the assimilatory functions. In fact, both the assimilatory and dissimilatory reactions occur simultaneously in the cell, i.e.,

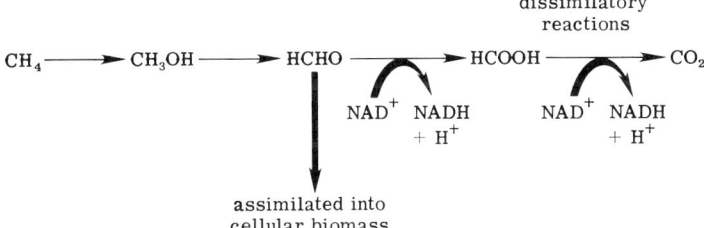

Carbon is assimilated either by the ribulose monophosphate pathway (RMP) in which formaldehyde is condensed with ribose 5-phosphate in a cyclic series of reactions leading to the synthesis of one triose phosphate molecule from 3 moles of formaldehyde, or by the serine pathway. In the latter case formaldehyde is converted to a tetrahydrofolate derivative which is then condensed with the acceptor molecule glycine to form serine. Carbon dioxide is also assimilated in the pathway via phosphoenol pyruvate carboxylase with the net result that two C_1 compounds (CO_2 and HCHO) are added to form a 2C compound, acetyl CoA. Glycine is regenerated by a cyclic series of reactions with variations occurring in the method of regeneration due to the presence or absence of isocitrate lyase.

Generally speaking the methanotrophs (methane-oxidizing bacteria) are divided into two types depending on whether they assimilate carbon via the RMP (type I) or serine pathway (type II), although there have been reports that both pathways may operate in the same organism (Trotsenko, 1976; Whittenbury *et al.*, 1976).

Although the pathways for carbon assimilation were fairly well understood and the oxygen in methanol had been shown to have originated from dioxygen (using $^{18}O_2$ fed to whole cells), it was not until 1970 that a cell-free system was successfully prepared that showed methane-oxidizing activity. Ribbons and Michalover (1970) demonstrated methane-stimulated respiration and methane-stimulated NADH oxidation by a membrane preparation of *Methylococcus capsulatus* (Texas).

Similar findings were also reported by Ferenci using *Pseudomonas methanica* (Ferenci, 1974; Ferenci *et al.*, 1975). Although the stoichiometries of methane oxidation suggested a monooxygenase reaction, no methanol accumulation was demonstrated, presumably due to the presence of methanol- and formaldehyde-oxidizing activities in the extracts, formate did accumulate, but only to the extent of 60% of the expected yield (Ribbons, 1975; Wadzinski and Ribbons, 1975). The first report of methanol accumulation from methane in a cell-free system appeared in 1975 when it was shown that particulate extracts of *Methylosinus trichosporium* OB3b accumulated methanol from methane in the presence of 150 mM phosphate (Tonge *et al.*, 1975). The stoichiometry of the oxidation, methane utilization:methanol formation:NADH utilization:O_2 consumption, was 1.0:0.9:1.6:1.3 and, as such, would be inconsistent with a monooxygenase-type reaction. However, the authors did suggest that such a stoichiometry would be consistent if methane caused a redirection of existing electron flow to oxygen into the methane oxygenase system (Higgins *et al.*, 1976). Similarly, Colby *et al.* (1975) prepared extracts from *Methylomonas methanica (Pseudomonas methanica)* that catalyzed the active disappearance of the methane analog bromomethane by the methane monooxygenase. The stoichiometry of

NADH and bromomethane disappearance also suggested a monooxygenase-catalyzed reaction in that organism.

In the following year Colby and Dalton (1976) obtained extracts from *Methylococcus capsulatus* Bath in which the methane-oxidizing activity resided in the soluble portion of the cell extract. The only effective electron donor for the system was NADH or NADPH. Methanol accumulation from methane was readily measured if 0.5 mM potassium cyanide was present in the assay mixture. The cyanide at this concentration had no effect upon the methane monooxygenase but completely inhibited any methanol oxidase present; unlike the system from *M. trichosporium* high phosphate concentrations did not inhibit methanol oxidase. Furthermore, the monooxygenase could be resolved into two components by DEAE–cellulose chromatography.

A complete purification scheme for the particulate enzyme from *M. trichosporium* OB3b was later reported by the Kent group (Tonge *et al.*, 1977) in which the solubilized protein was resolved into three components by DEAE–cellulose, Sephadex gel, and hydroxyapatite chromatography. The purified components were a soluble CO-binding cytochrome of molecular weight 13,000, a copper-containing protein of molecular weight 45,000, and a colorless protein of molecular weight 9,400. The purified system used only ascorbate or methanol in the presence of partially purified methanol dehydrogenase from the same organism as the electron donor by directly reducing the CO-binding cytochromes. NADPH would only serve as a donor for the crude or partially purified enzyme. It seemed to the authors that electrons were recycled *in vivo* from the further oxidation of methanol in a non-NADH-linked reaction. The inability of NADH to act as an electron donor in the purified monooxygenase system was said to be due to the lack of electron transport chain proteins.

Recently, work in our laboratory has revealed that the enzyme system from *M. capsulatus* Bath was capable of oxidizing a wide range of compounds other than methane, including gaseous and nongaseous hydrocarbons. The purpose of this article is to review what we now know about the substrate versatility of methane monooxygenases, the properties of the enzyme systems, and their potential use in commercial situations.

II. Substrates of Methane Monooxygenases

Using crude cell extracts from three different methane-oxidizing organisms we have shown that the monooxygenase system is capable of inserting an atom of oxygen into a wide range of hydrocarbon substrates (Colby *et al.*, 1977; Stirling *et al.*, 1979). Because crude extracts were used initially as the source of enzyme, it was necessary to have a number of controls to verify that the reactions were catalyzed by the monooxygenase. These included assays done with boiled extracts, in the absence of O_2 and NADH and in the presence of 0.2 ml ethyne, a specific and potent inhibitor of methane monooxygenase which does not inhibit any other oxidation sequence in the pathway of methane oxidation (Dalton and Whittenbury, 1976). Assays were also performed in the presence of KCN which permitted accumulation of alcohols from alkanes (see Section I).

A. ALKANES AND SUBSTITUTED METHANE DERIVATIVES

Alkanes up to pentane were oxidized at rates comparable to methane, yielding both primary and secondary alcohols. Longer chain alkanes were

TABLE I
OXIDATION OF ALKANES BY EXTRACTS OF *M. capsulatus*, *M. trichosporium*, AND *M. methanica*[a]

Substrate	Products	*M. capsulatus*		*M. trichosporium*		*M. methanica*	
		Amount (μmol)	Specific activity	Amount (μmol)	Specific activity	Amount (μmol)	Specific activity
Methane	Methanol	2.02	84	1.88	51	2.79	73
Ethane	Ethanol	1.64	68	0.16		1.68	87
	Ethanal			0.25	33	1.65	
	Acetate			0.83			
Propane	1-Propanol	0.65	69	0.04			
	2-Propanol	1.00		0.66	29	1.27	44
	Propanal			0.42		0.41	
Butane	1-Butanol	1.10	77				
	2-Butanol	0.92		0.60	33	0.54	17
	n-Butanal			0.42		0.13	
Pentane	1-Pentanol	0.49	73				
	2-Pentanol	1.26		1.12	35	0.38	17
	n-Pentanal			0.3		0.28	
Hexane	1-Hexanol	0.60	40	0.17			
	2-Hexanol	0.36		0.28	25		
	n-Hexanal			0.48			
Heptane	1-Heptanol	0.14	27				
	2-Heptanol	0.51		0.68	19		
	n-Heptanal			0.05			
Octane	1-Octanol	0.04	9				
	2-Octanol	0.39					

[a] The amount formed was measured after 10 minutes incubation using 3.8 mg extract protein. From Colby *et al.* (1977) and Stirling *et al.* (1979) with permission.

TABLE II
OXIDATION OF SUBSTITUTED METHANE DERIVATIVES BY EXTRACTS OF M. capsulatus BATH[a]

Substrate (μmol/reaction flask)	Specific activity (munits/mg protein)
Chloromethane (1)	84
Bromomethane (1)	66
Iodomethane (1–3)	0
Dichloromethane (1)	82
Trichloromethane	35
Tetrachloromethane (1–3)	0
Cyanomethane (1)	33
Nitromethane (2)	45
Methanethiol (2)	64
Methanol (5)	246
Trimethylamine (2–4)	0
CO (134)	61

[a] From Colby et al. (1977) with permission.

oxidized less readily, indicating that the size of the substrate was important in the process. Clearly the enzyme is not a terminal hydroxylase, but appeared to be specific for the 1- and 2-alkyl carbon atoms, although there was negligible formation of 3- and 4-alcohols with the longer chain alkanes (Table I). The importance of the size of the substrate with respect to its facility for oxidation was demonstrated clearly when substituted methane derivatives were investigated. Increasing either the size or the number of the substituent halogen decreased the rate at which they were oxidized. No products were identified from the oxidation of these derivatives although we have some preliminary evidence that carbon from bromomethane can be incorporated into cellular biosynthetic pathways (Stirling and Dalton, 1980). One interesting feature of the results presented in Table II is the oxidation of methanol by the monooxygenases from M. capsulatus and M. trichosporium. In the case of M. capsulatus the K_m for methanol was 0.95 mM whereas for methane it was 0.16 mM. Under normal assay conditions, and presumably in vivo, methanol would not accumulate sufficiently to be competitive with methane. Using purified methane monooxygenase from M. capsulatus, we have shown that the oxidation product of methanol is formaldehyde; in the crude system just described, however, formaldehyde does not accumulate.

B. ALKENES

Ethene, propene, and 1-butene were oxidized to the corresponding 1,2-epoxides by extracts from all three strains investigated (Table III). Internal

TABLE III
OXIDATION OF ALKENES BY EXTRACTS OF *M. capsulatus*, *M. trichosporium*, AND *M. methanica*[a]

Substrate	Products	M. capsulatus		M. trichosporium		M. methanica	
		Amount (μmol)	Specific activity	Amount (μmol)	Specific activity	Amount (μmol)	Specific activity
Ethene	Epoxyethane	3.54	148	2.20	59	6.02	158
Propene	1,2-Epoxypropane	2.1	83	2.03	53	3.10	81
1-Butene	1,2-Epoxybutane	1.19	49	1.69	44	3.83	100
trans-2-Butene	*trans*-2,3-Epoxybutane	0.77	141	0.24	41	0.82	68
	trans-But-2-en-1-ol	0.52				0.38	
	trans-But-2-en-1-al			1.30		1.39	
cis-2-Butene	*cis*-2,3-Epoxybutane	0.61	57	0.52	37	2.34	
	cis-But-2-en-1-ol	0.57					
	cis-But-2-en-1-al			0.63		0.28	
	Butanone	0.2		0.27		0.06	

[a] See Table I. From Colby *et al.* (1977) with permission.

alkenes, however, were oxidized both terminally at the methyl group and internally across the double bond. Both *cis*- and *trans*-2-butene yielded products that had retained their configuration, and thus eliminated the possibility that racemizable intermediates could be formed.

C. ETHERS

There have been many reports that dimethyl ether (DME) would serve as a growth substrate for methane oxidizers (Wilkinson, 1971; Davey, 1971; Hazeu, 1975; Ribbons, 1975; Patel *et al.*, 1976) and that dimethyl ether also was possibly an intermediate in the pathway of methane oxidation (Davey, 1971; Thomson, 1974; Wilkinson, 1975).

We have recently reinvestigated oxidation of dimethyl ether in whole cells and cell-free extracts, from which several points have emerged. Saturated solutions of dimethyl ether are heavily contaminated with methanol (in excess of 5 mM). Therefore the gas must be rigorously scrubbed before use (Stirling and Dalton, 1980; Meyers and Ribbons, 1978). In the reports by Davey (1971), Hazeu (1975), Ribbons (1975), and Patel *et al.* (1976), dimethyl ether oxidation was measured by oxygen consumption after saturated solutions of the substrate had been added to the cells—no indication was given as to the purity of the DME used. In our hands oxidation of DME by whole cells of *M. capsulatus* requires an electron donor such as methanol, formaldehyde, or formate. Dimethyl ether was oxidized stoichiometrically by purified preparations of the methane monooxygenase to methanol and formaldehyde but not in equal amounts (Stirling and Dalton, 1980).

After 5 minutes 0.38 μmoles of methanol and 1.75 μmoles of formaldehyde were produced from 1.05 μmoles of DME. The appearance of excess formaldehyde was probably due to the subsequent oxidation of the methanol formed by the monooxygenase which would normally oxidize DME to equimolar amounts of methanol and formaldehyde.

Theoretically, from whole cell and enzyme studies, it should be possible to grow the organism on DME in the presence of an exogenous electron donor. Attempts have yielded inconsistent results with sparse growth being observed only in flasks containing low concentrations of the donor sodium formate. The reasons for poor growth may be ascribed to the toxic nature of formate or the fact that the *in vivo* products of DME oxidation are not the same as the *in vitro* products.

D. ALICYCLIC, AROMATIC, AND HETEROCYCLIC HYDROCARBONS

The enzyme system appears to oxidize aromatic and cyclic alkanes by attacking the aromatic nucleus and alkyl side chains (Table IV). In the case of toluene there is simultaneous hydroxylation of the ring and the methyl side chain to produce benzyl alcohol and cresol. One particularly interesting observation is the oxidation of cyclopropane to cyclopropanol by *M. cap-*

TABLE IV
OXIDATION OF ALICYCLIC, AROMATIC, AND HETEROCYCLIC COMPOUNDS BY EXTRACTS OF *M. capsulatus* AND *M. trichosporium*

Substrate	Products	*M. capsulatus*		*M. trichosporium*	
		Amount (μmol)	Specific activity	Amount (μmol)	Specific activity
Cyclohexane	Cyclohexanol	3.0	62	0.74	20
Benzene	Phenol	3.0	62	1.16	31
Toluene	Benzyl alcohol	1.5	53	0.39	
	Cresol	1.0		0.12	
Styrene	Styrene epoxide	2.3	47	0.75	20
Pyridine	Pyridine-*N*-oxide	ND[a]	29	ND	19

[a] ND = Not determined; therefore pyridine oxidation was measured by following pyridine disappearance. From Colby *et al.* (1977) with permission.

sulatus extracts (B. W. Waters, B. T. Golding, and H. Dalton, unpublished observation). Chemically this is a reaction that, as far as the author is aware, has not been achieved under mild conditions and opens up the possibility of a mechanistic study using substituted cyclopropane derivatives. One reaction that we are currently interested in is the enzymic oxidation of methylcyclopropane. If oxidation by the monooxygenase proceeds via a radical attack (Hutchinson *et al.*, 1976), one would expect the methylcyclopropyl radical to be formed which would cleave spontaneously (rate constant 1.3×10^8 sec^{-1} at 25°C) to the 3-butenyl radical. If oxidation was by some other mechanism, perhaps a concerted oxygen insertion, then an alcohol would be the likely oxidation product. Such a reaction would not entirely exclude the radical attack mechanism. We must await the results of these experiments before any meaningful conclusions can be made.

E. General Conclusions on Substrate Specificities

The range of substrates oxidized and products observed from *M. capsulatus* and *M. trichosporium* extracts suggests that the monooxygenases from these two organisms essentially are similar. In certain instances our observations differ with other published observations. For example, Thomson (1974) claimed that whole organisms of *M. trichosporium* oxidized *n*-propane and *n*-butane to the corresponding 1-alcohol, aldehyde, and acid; in our hands considerable amounts of the corresponding 2-alcohol was formed. Tonge *et al.* (1977) reported that dimethyl ether was not oxidized at all by *M. trichosporium* extracts and Wilkinson (1975) observed that the products of diethyl ether oxidation in this organism were 2-ethoxy ethanol, 2-ethoxy ethanal, and 2-ethoxy acetic acid. Our observations suggest that oxidation of dimethyl ether yields methanol and methanal; oxidation of diethyl ether occurs at the subterminal carbon atom to yield ethanol and ethanal.

The monooxygenase from *M. methanica* differs from that of the other two organisms in that it does not oxidize *n*-alkanes with more than 6 carbon atoms, or alicyclic, heterocyclic, and aromatic compounds. However, the products formed from those oxidized substrates were almost identical to those observed using the extracts from *M. capsulatus* and *M. trichosporium*, suggesting that any differences between *M. methanica* and the other organisms are probably only minor.

III. Methane Monooxygenase

A. Methylosinus Trichosporium

Although the early work of Tonge *et al.* (1975, 1977) and Higgins *et al.* (1976) maintained that ascorbate was the electron donor for the enzyme from

M. trichosporium OB3b, work in our laboratory on this enzyme could not substantiate this claim (Stirling *et al.*, 1979), since the only effective electron donor was NADH. This observation, coupled with similar substrate and product specificity of the enzymes from *M. trichosporium* OB3b and *M. capsulatus* Bath, suggested that these enzyme systems may be quite similar. Further investigations in our laboratory (Stirling and Dalton, 1979) revealed that the crude enzyme systems from *M. trichosporium* OB3b and *M. capsulatus* Bath were indeed similar in the following respects: Both were soluble (i.e., not sedimented after centrifugation at 150,000 *g* for 90 minutes), only NADH or NADPH would serve as effective electron donors in the methane monooxygenase assay, high phosphate concentrations were not necessary for methanol accumulation from methane, and methanol was a very poor electron donor for *M. trichosporium* and was completely ineffective for *M. capsulatus*. Both extracts were rather unstable at 4°C but were completely stable on freezing to $-70°C$; 1 mM cyanide had no inhibitory effect on methane oxidation and the pattern of inhibitors in general on methane oxidation by both enzymes was very similar (Table V). To further substantiate our belief that the two systems are similar we also observed that purified components from *M. capsulatus* would restore activity to DEAE–cellulose-fractionated components from *M. trichosporium* (Table VI). For the most part these findings on the OB3b system differ with published reports by the Kent group (Tonge *et al.*, 1977; Higgins *et al.*, 1976) and further work is in progress to resolve this anomalous situation.

TABLE V

EFFECT OF VARIOUS POTENTIAL INHIBITORS ON METHANE OXIDATION BY CELL-FREE EXTRACTS OF *Methylococcus capsulatus* (BATH) AND *Methylosinus trichosporium* OB3b[a]

Inhibitor (0.1 mM)	Rate of methanol accumulation (% control)	
	M. capsulatus	*M. trichosporium*
None	100	100
Thiourea	90	103
Thioacetamide	93	84
Thiosemicarbazide	92	104
2,2-Bipyridyl	99	92
Neocuproine	100	94
8-Hydroxyquinoline	29	0
Imidazole	81	104
Carbon monoxide (15% in air)	95	81
Ethyne (3% in air)	0	0

[a] Assays were performed in the presence of 100 μM KCN. From Stirling and Dalton (1977, 1979) with permission.

TABLE VI

CROSS REACTIVITY OF PURIFIED COMPONENTS OF THE METHANE MONOOXYGENASE FROM *M. capsulatus* WITH FRACTIONS OF THE MONOOXYGENASE FROM *M. trichosporium*[a]

Test protein[b]	Specific activity (nmol epoxyethane produced/minute/ milligram protein)
Cell-free extract	83
F_1	0
$F_1 + MC_A$	0
$F_1 + MC_B$	0
$F_1 + MC_C$	23
$F_1 + (MC_A + MC_B)$	0
$F_1 + (MC_A + MC_B)$	21
$F_1 + (MC_B + MC_C)$	161
$(MC_B + MC_C)$	0
$F_1 + (MC_A + MC_C)$	160
F_2	0
$F_1 + F_2$	20

[a] From Stirling and Dalton (1979) with permission.

[b] F_1 and F_2 were produced by DEAE-cellulose chromatography of *Methylosinus trichosporium* OB3b. MC_A where present (0.5 mg), MC_B (0.5 mg), and MC_C (0.5 mg) are the purified components of the methane monooxygenase from *Methylococcus capsulatus* (Bath).

B. METHYLOCOCCUS CAPSULATUS BATH

The crude soluble extract from *M. capsulatus* Bath can be resolved into three fractions by DEAE-cellulose chromatography (Colby and Dalton, 1978). Protein A is not bound to the column and is purified by pH fractionation, molecular sieve, and further DEAE-cellulose chromatography. We have found that the protein appears to be slightly sensitive to oxygen when it is diluted and normally we purified the protein using anaerobic column techniques after pH fractionation. Protein B is eluted from the DEAE-cellulose column with 0.2 M NaCl and purified, as indicated in Fig. 1, in the presence of 50 μM phenylmethyl sulfonyl fluoride, a stabilizing agent. Protein C is easily purified in the presence of 5 mM sodium thioglycollate chromatography on 5′-AMP Sepharose 4B followed by Sephadex G100 column chromatography (Colby and Dalton, 1979).

Protein A has a molecular weight of about 220,000 and can be resolved into two subunits in the presence of SDS of molecular weights 47,000 and 68,000, respectively. It contains 2 gm atoms of iron and acid-labile sulfide per mole. Upon reduction with sodium dithionite (NADH is ineffective) an electron paramagnetic resonance (EPR) signal appears which is enhanced

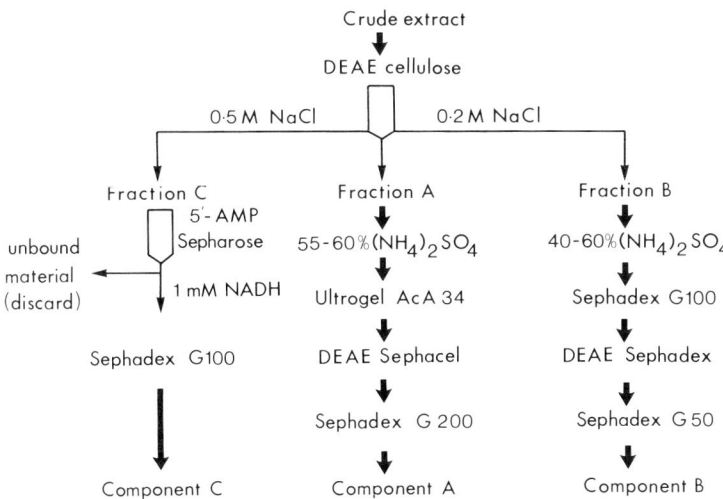

FIG. 1. Purification scheme for the methane monooxygenase components from *Methylococcus capsulatus* Bath.

4-fold in the presence of the substrate ethene (Fig. 2). Protein A has no catalytic activity alone but in the presence of proteins B and C will catalyze the oxygenation reactions observed in crude extracts. Protein C is an iron flavoprotein of molecular wieght 44,000. It contains 1 mole of FAD and 2 gm atoms each of nonheme iron and acid-labile sulfur. The protein is directly reduced by NADH under anaerobic conditions, and is fully reduced when 1.2 moles of NADH per mole of protein is added, suggesting that protein C accepts 2 electrons per mole of protein (Fig. 3). Upon admission of air to the cuvette the protein is rapidly reoxidized. Reoxidation of protein C also occurs anaerobically upon addition of stoichiometric amounts of protein A, the latter presumably acting as an electron acceptor from reduced protein C.

During the anaerobic titration of protein C with NADH a new spectral species appears in the 570–630 nm region, which presumably indicates the formation of the neutral flavin semiquinone of FAD. This is seen distinctly in the EPR spectrum of the protein when reduced under similar conditions, the semiquinone appearing as a free-radical signal at $G = 2.002$. This signal could be distinguished clearly from the reduced Fe-S* center (curve c on Fig. 4), which had G values of 2.047, 1.960, and 1.864 ($G_{av} = 1.957$) and could be observed upon complete reduction of the protein with sodium dithionite. Under these conditions the EPR-detectable semiquinone would be converted to the EPR nondetectable hydroquinone. Protein C will catalyze the reduction of potassium ferricyanide (specific activity 230 units/mg protein), 2,6-dichlorophenol-indophenol (50 units/mg protein), and horse heart cytochrome c (180 units/mg protein) when NADH is used as the electron donor. NADPH will also serve as electron donor for these reactions but with 10-fold lower V_{max} value for DCPIP reductase and a higher K_m value (50 μM for NADH and 15.5 mM for NADPH at pH 7) (Colby and Dalton, 1979).

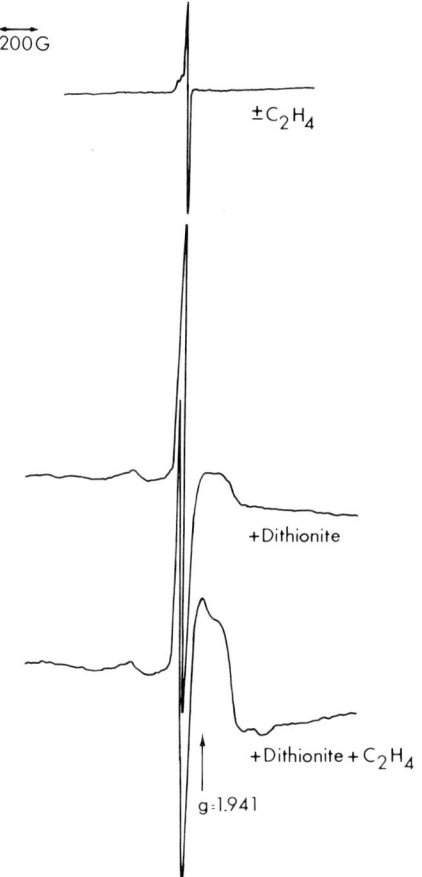

FIG. 2. EPR spectra of component A. The spectrum of purified component A (0.12 μmol/ 0.3 ml 20 mM Tris-HCl buffer pH 7.4) was recorded in the presence and absence of ethylene (a). A second sample of A was made anaerobic by adding a few crystals of sodium dithionite and the spectrum recorded (b). Two milliliters ethylene was then bubbled through sample (b) at room temperature for 2 minutes and the spectrum again recorded (c). The spectra were recorded at 18°K at a microwave frequency of 9.170 GH$_z$, a modulation amplitude of 10 G, and a microwave power of 10 mW. The receiver gain for spectrum (a) was 2×10^2 and for (b) and (c) was 3.2×10^3. The author wishes to thank D. J. Lowe and R. C. Bray of the University of Sussex, Brighton, England for help with providing these spectra.

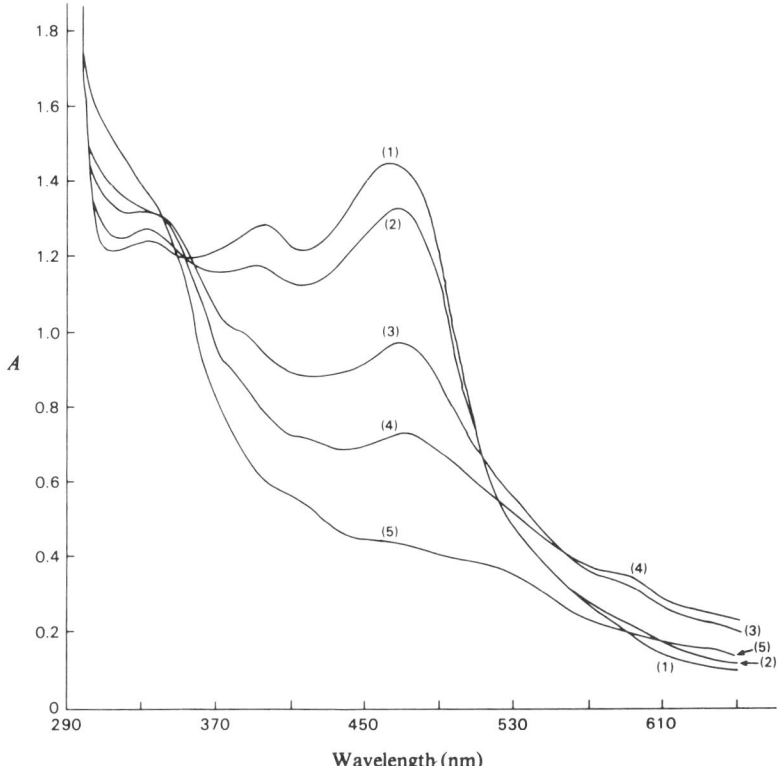

FIG. 3. A cuvette containing purified component C (1 ml, 5.5 mg of protein) was fitted with a Suba-seal stopper and made anaerobic by flushing with N_2 for 15 minutes. NADH (3.5 mM) was added through the stopper and the absorption spectrum measured after each addition. Spectra have not been corrected for the change in volume. (1) No addition; (2) +10 μl of NADH; (3) +30 μl of NADH; (4) +40 μl of NADH; (5) +60 μl of NADH. The spectrum returned to curve (1) within 2 minutes of readmitting air to the cuvette. Reproduced from Colby and Dalton (1979) with permission.

Protein C is the only component of the monooxygenase system for which a catalytic role other than in the methane oxidation system can be ascribed.

At present the role of protein B is uncertain. It is a colorless protein of molecular weight about 15,000. Use of purified preparations of A and C appears to be necessary for catalytic activity. An earlier report (Colby and Dalton, 1978) that B simply stimulated activity of A and C about 3-fold now is known to be due to a small amount of protein B in the protein A preparation.

These findings can be incorporated into a tentative scheme for the oxidation of methane by the enzyme system (Fig. 5).

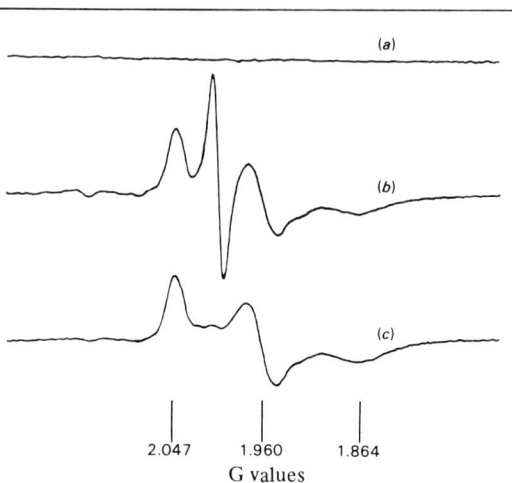

FIG. 4. Component C (0.07 μmol; 0.3 ml) in 50 mM Tris-HCl buffer, pH 7.0, was made anaerobic by repeated evacuation and flushing with argon (a). NADH (5 μmol) was added anaerobically to the sample and the spectrum recorded after incubation for 1 minute at room temperature (b). The fully reduced spectrum (c) was obtained by adding a few crystals of solid sodium dithionite to sample (b) and leaving it at room temperature for 1 minute. The spectra were recorded at 19°K at a microwave frequency of 9.232 GHz and modulation of 1.0 mT. The microwave powers were (a) 10 mW; (b) 1 mW; and (c) 1 mW. Reproduced from Colby and Dalton (1979) with permission.

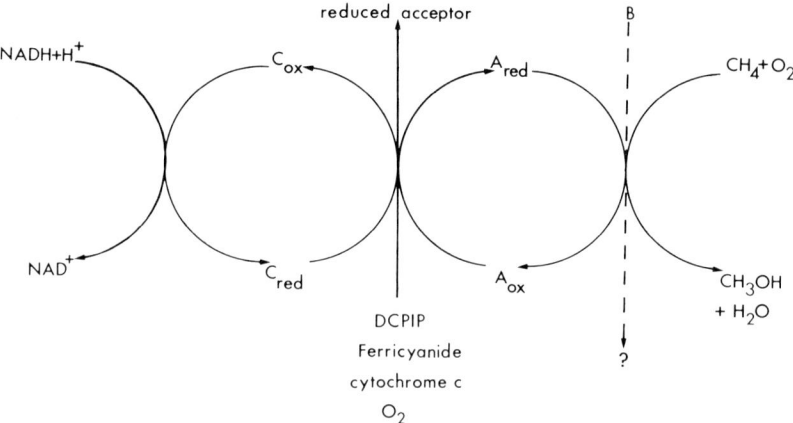

FIG. 5. Proposed mechanism for electron transfer within the methane monooxygenase complex from *Methylococcus capsulatus* Bath. $A_{ox/red}$ oxidized or reduced component A; $C_{ox/red}$ oxidized or reduced component C.

C. Methylomonas Methanica

The preparation of a crude extract from *M. methanica* capable of methane-stimulated NADH oxidation was first reported by Ferenci (1974). The enzyme was located mostly in a particulate fraction of the cell that could be sedimented from crude extracts at 38,000 g for 1 hour. It was first suggested (Ferenci, 1974) and then confirmed (Ferenci *et al.*, 1975) that the enzyme would also effect the oxidation of carbon monoxide to carbon dioxide. A similar finding had also been observed in whole cells using radiolabeled carbon monoxide as substrate (Hubley *et al.*, 1974). At about the same time Colby *et al.* (1975) reported that extracts of *M. methanica* would catalyze the O_2 and NADPH-dependent disappearance of bromomethane. They presented evidence that the methane monooxygenase was responsible for the catalysis by finding that the enzyme was particulate (75% of the activity being sedimented at 18,000 g for 30 minutes) and stable to freezing but unstable at 2°C.

Further attempts in our laboratory to purify the enzyme from the particulate preparation have met with failure because the enzyme cannot be released from the membranes by sonication, detergents, or phospholipases.

IV. Whole Cell Oxidation of Propene

Propene is oxidized by the methane monooxygenase to 1,2-epoxypropane. The same reaction also occurs in whole cells. Since the product of the reaction cannot be utilized by the cells it is excreted. All commercial efforts to mimic this reaction have met with economic failure. Existing commercial processes are rather inefficient since they give rise to a variety of unwanted side products. Therefore, it would be an interesting system to study with a view to possible commercial exploitation. Certainly no commercial system would ever be viable using cell extracts because of the instability of the enzyme and the exorbitant price of the electron donor.

We have studied the reaction using 12 different methane-oxidizing strains. Clearly some strains are much better at producing epoxypropane than others (Fig. 6). The electron donor for the reaction can be supplied externally as methanol, formaldehyde, or formate, all of which generate NADH for the monooxygenase when oxidized within the cell to CO_2. Most strains, however, can catalyze the epoxidation reaction in the absence of an electron donor for about 3 hours, presumably by utilizing endogenous reserves of poly-β-hydroxybutyrate. Once the reserve material has been utilized an external electron donor must be supplied for the reaction to continue.

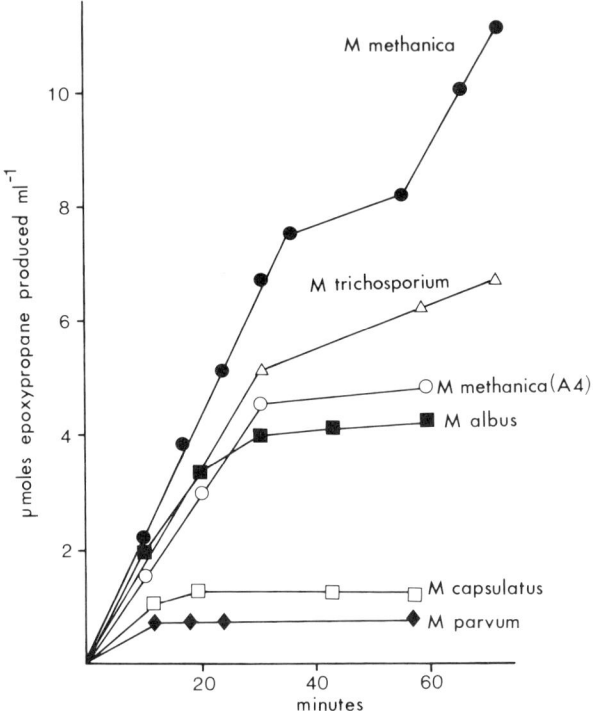

FIG. 6. Whole cell oxidation of propylene to propylene oxide by strains of methane-oxidizing bacteria.

REFERENCES

Colby, J., and Dalton, H. (1976). *Biochem. J.* **157**, 495–497.
Colby, J., and Dalton, H. (1978). *Biochem. J.* **171**, 461–468.
Colby, J., and Dalton, H. (1979). *Biochem. J.* **177**, 903–908.
Colby, J., Dalton, H., and Whittenbury, R. (1975). *Biochem. J.* **151**, 459–462.
Colby, J., Stirling, D. I., and Dalton, H. (1977). *Biochem. J.* **165**, 395–402.
Dalton, H., and Whittenbury, R. (1976). *Arch. Microbiol.* **109**, 147–151.
Davey, J. F. (1971). Ph.D. Thesis, University of Edinburgh, Edinburgh, Scotland.
Ferenci, T. (1974). *FEBS Lett.* **41**, 94–98.
Ferenci, T., Strom, T., and Quayle, J. R. (1975). *J. Gen. Microbiol.* **91**, 79–91.
Hazeu, W. (1975). *Ant. van. Leeuw.* **41**, 121–134.
Higgins, I. J., Knowles, C. J., and Tonge, G. M. (1976). *In* "Microbial Production and Utilization of Gases" (H. G. Schlegel, G. Gottschalk, and N. Pfennig, eds.), p. 389. Goltze, Göttingen.
Hubley, J. H., Mitton, J. R., and Wilkinson, J. F. (1974). *Arch. Microbiol.* **96**, 365–368.
Hutchinson, D. W., Whittenbury, R., and Dalton, H. (1976). *J. Theor. Biol.* **58**, 325–335.
Myers, A. J., and Ribbons, D. W. (1978). *Abstr. 78th Annu. Meet. Am. Soc. Microbiol.*, p. 184.

Patel, R. N., Hou, C. T., and Felix, A. (1976). *J. Bacteriol.* **126**, 1017–1019.
Ribbons, D. W. (1975). *J. Bacteriol.* **122**, 1351–1363.
Ribbons, D. W., and Michalover, J. L. (1970). *FEBS Lett.* **11**, 41–44.
Stirling, D. I., Colby, J., and Dalton, H. (1979). *Biochem. J.* **177**, 361–364.
Stirling, D. I., and Dalton, H. (1979). *Eur. J. Biochem.* **96**, 205–212.
Stirling, D. I., and Dalton, H. (1980). *J. Gen. Microbiol.* **116**, 277–283.
Thomson, A. W. (1974). Ph.D. Thesis, University of Edinburgh, Edinburgh, Scotland.
Tonge, G. M., Harrison, D. E. F., and Higgins, I. J. (1977). *Biochem. J.* **161**, 333–344.
Tonge, G. M., Knowles, C. J., Harrison, D. E. F., and Higgins, I. J. (1975). *FEBS Lett.* **44**, 106–110.
Trotsenko, Y. A. (1976). In "Microbial Production and Utilization of Gases" (H. G. Schlegel, G. Gottschalk, and N. Pfennig, eds.), p. 329. Goltze, Göttingen.
Wadzinski, A. M., and Ribbons, D. W. (1975). *J. Bacteriol.* **122**, 1364–1374.
Whittenbury, R., Colby, J., Dalton, H., and Reed, H. L. (1976). In "Microbial Production and Utilization of Gases" (H. G. Schlegel, G. Gottschalk, and N. Pfennig, eds.), p. 281. Goltze, Göttingen.
Wilkinson, J. F. (1971). *Symp. Soc. Gen. Microbiol.* **21**, 15–46.
Wilkinson, J. F. (1975). In "Microbial Growth on C_1-Compounds" (The Organizing Committee, G. Terui, eds.), p. 45. Society of Fermentation Technology, Japan.

Propane Utilization by Microorganisms

Jerome J. Perry

*Department of Microbiology,
North Carolina State University,
Raleigh, North Carolina*

I.	Introduction	89
II.	Isolation and Types	90
	A. Occurrence of Gaseous Alkane Utilizers	90
	B. Microbes That Utilize Methane and Other Gaseous Alkanes	91
	C. Propane-Utilizing Organisms	92
	D. Isolation of Gaseous Alkane-Utilizing Organisms	94
III.	Induction of the Propane-Oxidizing System	95
	A. Involvement of Molecular Oxygen	95
	B. Direct Incorporation of Oxygen-18	95
	C. Specificity of the Methane Monooxygenase	96
	D. Site of Initial Oxidative Attack on n-Alkanes	97
	E. Oxidation of Propylene	98
IV.	Metabolism of Propane and Related Compounds	99
	A. Propane	99
	B. Propylene	101
	C. Transition from Growth on Nonhydrocarbons to Growth on Hydrocarbons	102
	D. Lipids of Propane-Utilizing Organisms	103
	E. Metabolism of Other Three-Carbon Substrates by Propane Utilizers	104
V.	Growth Yield for Microorganisms Utilizing Hydrocarbon Substrates	106
	A. Yield with Propane and Related Compounds as Substrate	106
	B. Growth with Higher Alkanes	108
VI.	Products from Propane-Oxidizing Bacteria	108
	A. Production of Single-Cell Proteins	109
	B. Water-Soluble Vitamin Content of Propane-Grown Cells	109
	C. Production of Fatty Acids	109
VII.	Cooxidations Involving Propane	110
	A. Propane as Cosubstrate	110
	B. Propane as Substrate	110
VIII.	Toxicity of Propane and Related Compounds	111
IX.	Prospecting with Propane-Oxidizing Microbes	111
	References	112

I. Introduction

There are a number of advantages in employing gaseous alkanes as substrate for the growth of microorganisms in fermentation processes. Gaseous hydrocarbons can be obtained at a relatively low cost, are abundant in occurrence, and are available with a high degree of purity. They are free of objectionable coal tar products and any residual substrates not utilized in the fermentation cycle can be removed rather readily as they are mostly insoluble in water (McAulifee, 1966). As such they are a potential source of single cell protein and may also be utilized to produce an array of valuable metabolic intermediates, e.g., organic acids, amino acids, purines, pyrimidines, and B vitamins.

There are a number of disadvantages, however, in use of gaseous hydrocarbons as substrate for fermentation and these should be overcome if one is to establish an economical process. Generally, microorganisms that utilize gaseous hydrocarbons have a long doubling time and the total growth yield is low. The cell concentration attained under optimal conditions is generally such that although mutants may be selected to produce various metabolic intermediates, they may not yield economically feasible amounts. Stable mutants are not readily obtained and, for the most part, hydrocarbon-utilizing wild-type cultures yield little or no metabolic products.

The reduced hydrocarbon molecule requires plentiful amounts of oxygen if it is to serve as growth substrate and a combination of gaseous alkane and O_2 can result in an explosive mixture. This hazardous feature presents a definite challenge to the process engineers.

Propane is of widespread occurrence in petroleum deposits and represents 1–2% of natural gas. Propane would be of considerable advantage as a raw material in the fermentation industry as it is (1) relatively low in cost, (2) yields a weight increase on conversion proportional to the amount of oxygen incorporated, and (3) is relatively insoluble in water, so that unused gaseous substrate could readily be recycled in continuous fermentations. Microorganisms capable of utilizing propane are also of common occurrence (Fuhs, 1961). It is, however, the most water soluble of the gaseous alkanes (McAuliffe, 1966).

There have been a number of books and reviews written regarding hydrocarbons as substrates for microorganisms. A number of these are concerned with the metabolism of gaseous alkanes, including propane (see Foster, 1963; Fuhs, 1961; Einsele and Feichter, 1971; ZoBell, 1946; Watkinson, 1978; Shennan and Levi, 1974; Nyns and Wiaux, 1969; McKenna and Kallio, 1965; Davis, 1967; Bemmann and Troger, 1975; Klug and Markovetz, 1971; Beerstecher, 1954; Perry, 1979).

II. Isolation and Types

A. Occurrence of Gaseous Alkane Utilizers

The ability of microorganisms to grow with propane as source of carbon and energy was originally suggested by Tausz and Donath (1930), in a "methanbakterium" that utilized the n-alkanes from methane to n-hexane as substrate. This organism also utilized propylene. Since this report many organisms with the capacity have been described and the recalcitrance of gaseous alkanes relative to other low molecular weight hydrocarbons has been suggested as: C_{11}–C_{18} n-alkanes < C_{11}–C_{18} alkenes < gaseous alkanes < C_5–C_9 n-alkanes, etc. (Perry and Cerniglia, 1973). This would indicate that propane utilizers are of widespread occurrence in microbial populations, and of the organisms in our collection that utilize C_{10}–C_{20} n-alkanes about 50% will grow on propane (J. J. Perry, unpublished).

Others have reported that gaseous alkanes are not readily utilized by hydrocarbonoclastic microbes (Ratledge, 1978). Although this may be true in part, there is some misconception as to the relative number of hydrocarbon utilizers that grow on gaseous alkanes based on substrate specificity tests in which the gaseous alkanes often were not included. Many pseudomonads and other gram-negative bacteria are obligate methylotrophs or grow on a range of hydrocarbon substrates but appear to be less effective in growth on ethane, propane, and butane than the mycobacteria and nocardial types.

Fuhs (1961) published an extensive list of microorganisms that had been reported to that time as capable of growth on hydrocarbons. Although somewhat outdated, it could be considered as representative of the types of microorganisms involved in hydrocarbon biodegradation. Most of the microorganisms listed are gram positive and the number that can grow on gaseous alkanes (ethane, propane, and n-butane) is about equal to the total number that utilize the longer chain normal paraffinic hydrocarbons (C_{10} through C_{18}).

A source of error in assessing the ability of microbes to utilize propane is the length of time required for induction to metabolism of this substrate. Mueller (1969) suggested that methane is preferentially utilized by mixed cultures growing on natural gas based on the inability of such cultures to demonstrate growth after a 4-hour exposure to propane. Others have shown (Vestal and Perry, 1971; Blevins and Perry, 1971) that considerably longer exposure is required before growth is initiated.

B. Microbes That Utilize Methane and Other Gaseous Alkanes

There are several reports in the literature that both gram-positive and gram-negative methanotrophic bacteria can also utilize other gaseous alkanes as substrate: Nechaeva (1949) reported that *Mycobacterium flavum* v. *methanicum* and *Mycobacterium methanicum* grew with either methane or propane as substrate and the latter also utilized n-heptane; Slavina (1948) described a methanotrophic organism that would grow on ethane or propane: Tanaka *et al.* (1972) isolated a gram-positive organism described as *Brevibacterium ketoglutamicum* that grew with methane, ethane, propane, *n*-butane, isobutane, propylene, butylene, or mixtures of these as substrate; Hutton and ZoBell (1953) isolated several methanotrophic cultures and stated that "part of the cultures oxidized ethane, ethylene and propane as readily as methane" (p. 217). These workers (Hutton and ZoBell, 1949) also isolated methane-utilizing organisms form soil samples, and of 299 1-gm soil samples, 74% of the soils yielded methane oxidizers. They reported that either methane, ethane, or propane could support the growth of these organisms in the presence of O_2 and CO_2. Studies with soils taken near oil deposits led Smirnova and Taptykova (1967) to suggest that there are large numbers of organisms of the genera *Pseudomonas* and *Mycobacterium* that utilize gaseous alkanes.

Strawinski and Tortorich (1955) isolated 11 methane-oxidizing cultures from soils taken in an oil producing area. The organisms oxidized methane and "some of the higher hydrocarbon gases (ethane, propane and butane)" (p. 27) but the authors did not clarify further. None of 11 cultures isolated on methane from soil obtained in a nonproducing area could utilize the higher gaseous hydrocarbons.

C. Propane-Utilizing Organisms

Propane is utilized by *Mycobacterium rhodochrous* strains OFS (ATCC No. 29672) (Dunlap and Perry, 1968); *Mycobacterium convolutum* strain R-22 (ATCC No. 29671) and *M. rhodochrous* strain A-78 (ATCC No. 29670) (Blevins and Perry, 1972); *M. rhodochrous* strain OC2A (ATCC No. 29675) (Blevins and Perry, 1971); *M. convolutum* strain NPA-1 (ATCC No. 29674) (Cerniglia and Perry, 1975); *Mycobacterium vaccae* strain JOB5 (ATCC No. 29678) and *M. rhodochrous* strain 7E1C (ATCC No. 19067) (Perry, 1968); and *Mycobacterium album* strain 7E1B1W (ATCC No. 29676). Other than the above named species, over 50% of the mycobacterial and nocardial strains in our culture collection are able to utilize propane as substrate for growth (J. J. Perry, unpublished). These organisms have been isolated by enrichment with *n*-alkanes, alkylamines, carbohydrates, and other substrates. A predilection for utilization of hydrocarbons as sole source of carbon and energy by microbes designated *Nocardia* and *Mycobacterium* was suggested by Foster (1963; Lukins and Foster, 1963a,b) and Foster considered that these two species along with *Corynebacterium* and *Brevibacter* were the major utilizers of aliphatic hydrocarbon compounds in nature.

McLee *et al.* (1972) isolated 15 bacterial and four fungal cultures that utilized ethane, propane, or *n*-butane as source of carbon and energy. The bacteria were considered to be members of two genera: *Brevibacter* and *Arthrobacter*, and the fungi were identified as strains of *Penicillium nigri-*

cans, Alleschenia boydii, Graphium cuneiform, and an unknown species. *Mycobacterium paraffinicum* was isolated (Davis *et al.*, 1956) by enrichment with ethane and could utilize propane or *n*-butane as growth substrate. Lukins and Foster (1963a) found that four of nine strains of *Mycobacterium* sp. would grow on propane and a culture of *Mycobacterium smegmatis* would grow on propane or *n*-butane but not ethane, ethylene, or propylene, although nonproliferating cells, following growth on propane, would oxidize these substrates.

A *Mycobacterium* sp. similar to *Mycobacterium phlei* has been described (O'Brien and Brown, 1968) that will grow on propane and *n*-butane but not on ethane or methane. Members of the genus *Nocardia* have been examined for the ability to utilize gaseous alkanes (Smirnova and Taptykova, 1967). These workers examined 22 nocardial strains belonging to 15 different species and reported that two strains grew on methane (poorly), two on ethane (poorly), and nine abundantly on propane. Among the nine were two strains of *Nocardia asteroides*, two of *Nocardia brasiliensis*, and one each of *Nocardia caviae, Nocardia coelicaca, Nocardia rubropertincta, Nocardia convoluta,* and *Nocardia madurae*. An ethane-oxidizing organism, *Mycobacterium perrugosum* var. *ethanicum*, was isolated (Bokova, 1954) by enrichment culture, which could grow on the *n*-alkanes from C_2 through C_7 and also on paraffin. This organism would not grow on methane, benzene, or naphthalene. She also isolated a coral colored organism, *Mycobacterium rubrum* var. *propionicum*, with propane as substrate that would not grow on methane, ethane, benzene, cyclohexane, or naphthalene. A number of propane-utilizing organisms were isolated from underground water and rocks in the Promoslovaya, USSR, natural gas beds by Smirnova (1962). The organisms were described as a *Mycobacterium* sp., *M. rubrum, Mycobacterium lacticolum, Pseudomonas caudatus, Pseudomonas liquefaciens, Pseudobacterium subluteum,* and four unclassified pseudomonads. Kuznetsov and Telegina (1957) found that mycobacteria and pseudomonads were major propane oxidizers among several isolates. Fungi have been described that utilize propane as substrate: *Cunninghamella elegans* and *Penicillium zonatum* (Cerniglia and Perry, 1973); *Ceratocystis ulmi* (J. J. Perry, unpublished), and several yeasts and fungi that gave trace growth on propane (Lowery *et al.*, 1968). Davies *et al.* (1973) isolated 12 fungi from raw sewage by enrichment with ethane as substrate. All were Hyphomycetes and grew with ethane, propane, or *n*-butane as the carbon source but none utilized methane. The fungi were tentatively identified as *Graphium* sp. (4), *Phialophora jeanselmei* (3), and five were similar to *Acremonium*. A yeast has been isolated by enrichment with methane (Wolf and Hanson, 1979) that will grow on ethane and *n*-butane but unfortunately was not tested for the capacity to grow on propane.

D. Isolation of Gaseous Alkane-Utilizing Organisms

Microorganisms that utilize gaseous alkanes as sole source of carbon and energy can be isolated readily from soil or water samples by enriching with the gaseous alkane as a 50:50 air to gas mixture (Kester and Foster, 1963). Davis et al. (1956) employed an atmosphere of ethane–O_2–N_2 in a ratio of 40:20:40 for isolating of *M. paraffinicum*. Maximum growth with the propane-oxidizing JOB5 was attained with a propane–O_2–N_2 ratio of 50:40:10 (Blevins and Perry, 1971). Kuznetsov and Telegina (1957) isolated several propane-oxidizing organisms from soil taken around gas wells in an atmosphere of 1:3 propane in air. This same level was employed by Bokova (1954) in the isolation of propane-oxidizing organisms and she reported that 0.01% propane in air was sufficient to obtain measurable growth. The methane-utilizing yeast was isolated by enrichment with a methane–air–CO_2 ratio of 70:20:10, respectively (Wolf and Hanson, 1979). It is evident that a range of gaseous alkane in the atmosphere of a growth chamber ranging between 30 and 50% is sufficient in enriching for gaseous hydrocarbon utilizing organisms. The atmosphere should contain approximately 20% oxygen to attain maximum growth (Blevins and Perry, 1971), although these factors will vary depending on the volume of the growth chamber and the amount of liquid mineral salts available.

Enrichment on solid substrates is an effective technique for obtaining a greater variety of gaseous alkane-utilizing organisms. Small particles of fertile soil or dilutions of the soil should be spread over the surface of agar in petri dishes and the plates incubated in a desiccator with a 50:50 air–propane atmosphere.

There is some evidence that nitrate is a preferred form of nitrogen for growth of hydrocarbon-utilizing microorganisms (Smirnova, 1962), although some of the organisms tested gave superior growth with NH_4Cl as N source. Bushnell and Haas (1941) described a basal-salts medium and suggested that either NH_4NO_3 or $(NH_4)_2SO_4$ at 0.1% could serve as N source. A basal-salts medium (L salts) with a broad array of trace elements has been described (Leadbetter and Foster, 1958) and this formulation contains $NaNO_3$ at 0.2%. We have found this to be a more effective basal-salts medium if the $NaNO_3$ level is lowered to 0.1% and NH_4Cl also added at 0.1%. This combination of N source yielded somewhat better growth with a wide variety of hydrocarbon-utilizing organisms. Generally, for isolation and substrate specificity tests, gaseous and liquid hydrocarbons should be 99 mol % minimal purity.

The temperature range for growth of propane-utilizing microbes is within that for most free-living soil organisms, 20–35°C. None of several obligately thermophilic hydrocarbon-utilizing organisms in our collection (Merkel *et al.*, 1978a,b) would grow on paraffinic hydrocarbons that were fewer than 10 carbons in length.

A requirement for growth factors, e.g., B vitamins or amino acids, is not apparent in microorganisms that utilize aliphatic hydrocarbons. Extensive experimentation to find hydrocarbon utilizers that required them were unsuccessful (J. W. Foster, personal communication), although such microbes may well exist in nature.

III. Induction of the Propane-Oxidizing System

A. INVOLVEMENT OF MOLECULAR OXYGEN

The initial oxidative attack on the reduced gaseous alkane molecule involves molecular oxygen and is an obligately aerobic process (for a more complete discussion, see Perry, 1979). The oxygenases are inducible (Foster, 1962; Hansen and Kallio, 1957), although some cases have been reported of microorganisms that have a constituent oxygenase. Davis *et al.* (1956) considered that *M. paraffinicum* had a constitutive oxygenase because the cells would oxidize ethane after growth on ethanol. The ability of a significant number of organisms isolated on a substrate that requires an oxygenase for metabolic attack, e.g., *o*-phthalic acid, to metabolize propane without lag has been observed (Perry and Scheld, 1968). Others have shown that an array of gratuitous inducers can induce the alkane-oxidizing system. Nyns *et al.* (1969) found that the *n*-hexadecane-oxidizing systems in *Candida lipolytica* can be induced by cetyl alcohol, palmitaldehyde, or palmitic acid. The induction of the alkane-oxidizing system in *Pseudomonas aeruginosa* by such nonhydrocarbon substrates as 1,2-dimethoxymethane has been reported (Van Eyk and Bartels, 1968). The methyl moiety of propane is probably the inducer of the oxygenase as acetone and isopropanol have been shown to be effective inducers of the propane oxidation system (Perry, 1968; Lukins and Foster, 1963a).

B. DIRECT INCORPORATION OF OXYGEN-18

The direct incorporation of oxygen during propane oxidation was demonstrated by Leadbetter and Foster (1959b). Results presented in Table I show that oxygen-18 accounts for a substantial portion of the total oxygen in a cell during gaseous alkane utilization. It is interesting to note that there is a dilution of oxygen-18 uptake during growth on longer chain alkanes as a greater part of the cellular oxygen in these cells is derived from water.

TABLE I
INCORPORATION OF MOLECULAR OXYGEN BY MICROORGANISMS DURING GROWTH ON METHANE, ETHANE, AND PROPANE[a]

Organism	Carbon source	Atom percentage excess oxygen-18		Percentage of cell material oxygen derived from O_2
		O_2	Cell material	
Pseudomonas methanica	Methane (50% v/v)	6.2	0.32	5.1
	Methanol (0.5% w/v)	6.2	0.02	0.32
"Propane bacterium"[b]	Propane (50% v/v)	6.0	0.16	2.8
	Tridecane (0.67% w/v)	6.0	0.10	1.8
	Glucose (0.5% w/v)	6.0	0.05	0.83
"Ethane bacterium"[b]	Ethane (50% v/v)	6.2	0.44	7.1
	Propane (50% v/v)	6.2	0.22	3.5
	n-Butane (30% v/v)	6.2	0.10	1.8
	Glucose (0.5% w/v)	6.2	0	0

[a] Reproduced with permission from *Nature (London)*, **184**, 1428 (1959).
[b] "Propane bacterium" and "ethane bacterium" are unidentified species isolated from propane and ethane enrichment cultures, respectively.

C. Specificity of the Methane Monooxygenase

Colby *et al.* (1977) isolated the methane monooxygenase from *Methylomonas capsulatus* and it effectively oxidized the C_1–C_4 gaseous alkanes and both ethylene and propylene (Table II). Later Stirling and Dalton (1979) measured oxidation rates by intact cells of this organism and found that an exogenous source of reducing power (formaldehyde) enhanced the rate of oxidation of various alkanes and ascribed this to the limited ability of the organism to regenerate reducing power with the hydrocarbon substrate. The methane monooxygenase has been isolated from *Methylosinus trichosporium* and "catalyzed the oxygenation of the higher homologs ethane, propane and n-butane to the corresponding alcohols" (Tonge *et al.*, 1977, p. 340).

The ability of a methanotroph *Methanomonas methanica* (*Pseudomonas*

TABLE II
OXIDATION OF GASEOUS HYDROCARBONS BY THE METHANE MONOOXYGENASE FROM *Methylomonas capsulatus*[a]

Substrate (134 μmol per reaction vessel)	Products (μmole formed in 12 minutes)
Methane	Methanol (2.02)
Ethane	Ethanol (1.64)
Propane	Propane-1-ol (0.65)
	Propane-2-ol (1.00)
Butane	Butane-1-ol (1.10)
	Butane-2-ol (0.92)
Ethylene	Epoxyethane (3.54)
Propylene	1,2-Epoxypropane (2.10)

[a] Reproduced with permission from *Biochem. J.*, **165**, 395–402 (1977).

methanicum) growing on methane to oxidize ethane, propane, and *n*-butane to homologous alcohols was originally described by Leadbetter and Foster (1960) and for this conversion Foster (1962) proposed the term "cooxidation." The broad specificity of the alkane oxygenase in *n*-alkane-utilizing mycobacteria has been described (Perry, 1968).

D. SITE OF INITIAL OXIDATIVE ATTACK ON *n*-ALKANES

The initial oxidative attack on *n*-alkanes greater than two carbons in length can occur either via terminal or subterminal oxygen insertion. Data accumulated to date suggest that oxidation occurs mainly at either the terminal carbon or the adjacent methylene carbon.

The accumulation of the homologous methyl ketones by *M. smegmatis* during the oxidation of propane, *n*-butane, *n*-pentane, or *n*-hexane was described by Lukins and Foster (1963a). These workers found that oxidation of propane to acetone by nonproliferating cells in a medium enriched with D_2O did not yield acetone with detectable levels of deuterium incorporated into the molecule. This would suggest that propylene is not an intermediate in the oxidation of propane.

Studies by Vestal and Perry (1969) indicate that subterminal oxidation of propane to acetone is the major pathway for utilization of this substrate in *M. vaccae* JOB5. Further studies (Vestal and Perry, 1971) suggest, however, that some terminal oxidation of propane does occur in *M. vaccae*. Fatty acid profiles in the lipids of other microorganisms, following growth on propane, also indicate that some terminal attack occurs during propane oxidation (Dunlap and Perry, 1967, 1968), although subterminal oxidation through acetone is probably the major one. *Mycobacterium vaccae* can utilize n-butane as substrate but apparently the pathway of oxidative attack is through terminal oxidation to butyric acid (Phillips and Perry, 1974). Butyric acid and n-butane are metabolized via cleavage to acetate, whereas 2-butanone is split in this organism to CO_2 and propionate.

Allen *et al.* (1971) reported that *Cunninghamella blakesleeana* utilized n-alkanes through oxidation to the homologous carboxylic acid and that a *Penicillium* sp. oxidized n-tetradecane to the 2-, 3-, and 4-tetradecanols. A strain of *Arthrobacter* that grew poorly on hexadecane would oxidize hexadecane to a series of monohexadecanones during growth on yeast extract, and the amount of the individual ketone produced depended on the distance from the terminal carbon (Klein *et al.*, 1968). The greater the distance, the less ketone produced. Souw *et al.* (1977) found that both terminal and subterminal oxidation occurred in a *Candida* sp. during oxidation of n-pentadecane as they isolated as products: pentadecanol,1-pentadecanoic acid, pentadecanedioic acid, and 2-pentadecanone. There is an apparent lack of specificity in the molecular oxygenase for paraffinic hydrocarbons and both terminal and subterminal attack will occur with these paraffinic substrates and the amount of each that will occur depends on both the substrate and the microorganism involved.

E. Oxidation of Propylene

The microorganisms in our culture collection that can grow with propane as sole source of carbon and energy cannot utilize propylene (Cerniglia *et al.*, 1976). However, many of these organisms under nonproliferating cell conditions can oxidize propylene. When nonproliferating cells of *M. convolutum*, grown on propane, were exposed to propylene there was a significant accumulation of acrylic acid as determined by gas–liquid chromatography. Apparently the oxidation of propylene by propane-grown cells occurs at the methyl end of the molecule, resulting in the generation of a toxic product. The hydrocarbon-utilizing mycobacteria have been shown to preferentially attack longer chain 1-alkenes (1-dodecene to 1-octadecene) at the saturated end of the molecule (King and Perry, 1975a,b). Evidence has been presented (de Bont *et al.*, 1979) that the monooxygenase for ethylene can be separated from the ethane monooxygenase in *Mycobacterium* E20 and, indeed, the monooxygenase for alkenes may not be present in alkane-utilizing organisms. A mixed culture and an axenic culture, isolated by enrichment with propylene as source of carbon and energy, apparently metabolize this substrate by an initial attack at the double bond (Cerniglia *et al.*, 1976).

IV. Metabolism of Propane and Related Compounds

A. PROPANE

Lukins and Foster (1963a) reported on propane metabolism in *M. smegmatis* 422 and they suggest that methyl ketones are an intermediate in utilization of this substrate. They grew *M. smegmatis* on propane, harvested the cells, and resuspended them in a mineral-salts medium devoid of a nitrogen source and added propane to a level of 30% in the atmosphere. After incubation for 15 hours the replacement liquid was distilled and a crystalline derivative of 2,4-dinitrophenylhydrazine was obtained. This derivative was analyzed and proved to be the 2,4-dinitrophenylhydrazone of acetone. The supernatant yielded 193 μg of acetone per milliliter. A propane growth medium from the same organism was analyzed and contained 59 μg/ml acetone. They also reported that acetone was an excellent source of carbon and energy for *M. smegmatis* 422 and further showed that organisms isolated by enrichment with acetone as substrate were generally able to utilize propane (for a discussion of acetone metabolism, see Forney and Markovetz, 1971). Growth of *M. smegmatis* on acetone or propane yielded cells that could grow or oxidize either substrate without induction, whereas *n*-propanol-grown cells could not oxidize propane or acetone without an induction period. Similar results have been reported in other organisms (Blevins and Perry, 1972; Vestal and Perry, 1969).

Acetone may also be produced during ethane oxidation by cell suspensions of the obligate methanotroph *Methylosinus trichosporium* OB3B. The putative intermediates in this reaction are acetoacetic acid and β-hydroxybutyric acid (Thomson *et al.*, 1976).

Some workers have proposed that propylene is an intermediate in the oxidation of propane (Klausmeier *et al.*, 1958; Pabst and Brown, 1968; Van Der Linden and Thijsse, 1965). This idea has been based largely on the ability of microorganisms that grow on propane to utilize propylene or that induction of the propane oxygenation system also induced the system for propylene. As mentioned previously, acetone is produced by *M. smegmatis* 422 during propane oxidation (Lukins and Foster, 1963a) and they determined whether propylene was an intermediate in this organism. The oxidation of propane was carried out in a medium containing 36.4% D_2O, and the amount of deuterium incorporated into the acetone produced under these conditions was determined. The acetone generated during the oxidation of propane had a negligible amount of deuterium incorporated into the molecule and this indicates that propylene is not an intermediate formed during propane metabolism. Although many of the propane-utilizing organisms in our culture collection can oxidize propylene, none can grow with it as sole substrate.

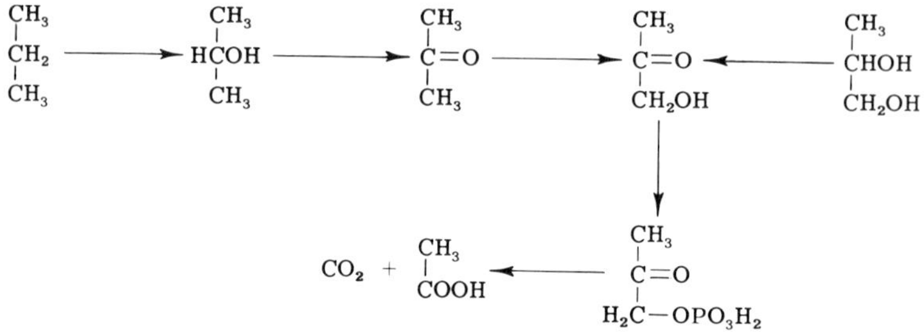

CH3
CH2
CH3
CH3
HCOH
CH3
CH3
C=O
CH3
CH3
C=O
CH2OH
CH3
CHOH
CH2OH
CO2 +
CH3
COOH
CH3
C=O
H2C—OPO3H2
FIG. 1. Probable pathways for propane utilization in *Mycobacterium vaccae* strain JOB5.

The probable pathway for the utilization of propane by *M. vaccae* is presented in Fig. 1. The presence of isocitrate lyase and malate synthase in propane-, isopropanol-, and acetone-grown cells and the absence of detectable levels of isocitrate lyase in cells following growth on propionate or glucose suggest a two-carbon intermediate in propane metabolism. The evidence from this study (Vestal and Perry, 1969) and that of Lukins and Foster (1963a) strongly implicates acetol (1-hydroxy-2-propanone) as the intermediate in propane metabolism. Earlier studies (Levine and Krampitz, 1952; Rudney, 1954) have suggested that acetol is also an intermediate in acetone oxidation. Propionic acid is metabolized in the propane-utilizing *M. vaccae* via the methyl malonate pathway (Fig. 2). Equivalent results (Blevins and Perry, 1972) were found for propane (acetol pathway) and propionate (methyl malonyl) metabolism in *M. album* (strains 7E4 and 7E1B1W), *M. rhodochrous* (strains OFS, A74, and 7E1C), and *M. convolutum* (strain R-22). Davies *et al.* (1974) investigated the intermediary metabolism of an *Acremonium* sp. following growth on propane. They exposed nonproliferating propane grown cells (Warburg respirometer) to putative intermediates of propane oxidation (1-propanol, propionaldehyde, and propionate) and reported that all were oxidized. They considered this evidence for monoterminal oxidation, but the study lacked the controls necessary for confirmation.

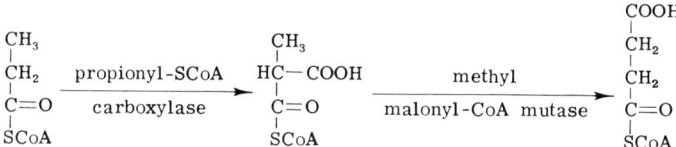

CH3
CH2
C=O
SCoA
propionyl-SCoA
carboxylase
CH3
HC—COOH
C=O
SCoA
methyl
malonyl-CoA mutase
COOH
CH2
CH2
C=O
SCoA
FIG. 2. Pathway for propionic acid utilization in *Mycobacterium vaccae* strain JOB5.

B. Propylene

An organism described as "methanbakterium" was isolated by Tausz and Donath in 1930 that would utilize the C_1-C_6 alkanes and also grew on propylene. Other workers have reported that gaseous alkane-utilizing organisms can grow on propylene (Tanaka et al., 1972; Klausmeier et al., 1958). Lukins and Foster (1963b) reported that a strain of M. smegmatis that grew with propane as substrate could not grow on propylene but nonproliferating cells could oxidize both ethylene and propylene following growth on propane.

The utilization of propylene as sole source of carbon and energy for bacteria has been studied in a mixed culture and in an axenic culture (Cerniglia et al., 1976). The mixed culture was composed of a dominant gram-negative rod-shaped organism (80-90%) along with a lesser number (15-20%) of a gram-positive yellow-pigmented rod. The mixed culture had significant levels of isocitrate lyase after growth on propylene and acetate but undetectable levels after growth on succinate or propionate. Carbon dioxide incorporation studies suggested that a carboxylation reaction occurred during propionate metabolism but not with propylene as substrate. These results indicate that propylene is cleaved to a C_2 + C_1, whereas propionate is metabolized through the methylmalonyl pathway. Neither of the organisms in the mixed culture would utilize propylene in the absence of the other, but the role of each in propylene utilization is unclear.

The axenic culture (strain PL-1) apparently utilizes propylene in a manner similar to that in the mixed culture (a C_2 + C_1 cleavage). The major difference is in the metabolic pathway for propionate utilization. The induction of isocitrate lyase (see Table IV) and relatively high level of even chain length fatty acids in the lipids of the propionate-grown cells is different from any of the other hydrocarbon-utilizing organisms investigated (Blevins and Perry, 1972; Cerniglia and Perry, 1975; Vestal and Perry, 1969, 1971). Carboxylation is apparently involved in the utilization of propionate in the mixed culture but not in strain PL-1. Complete elucidation of the pathway for propylene awaits further study.

As suggested previously, propylene is oxidized by nonproliferating cells of the propane-utilizing M. convolutum to acrylic acid and this is apparently toxic for the organism (Cerniglia et al., 1976). Other studies (Thijsse, 1964; Whitely and Ordal, 1957) have clearly shown that acrylate is an inhibitor of fatty acid oxidation. Attack at the methyl end of the molecule indicates that M. convolutum organism has an alkane monooxygenase and may lack an alkene monooxygenase similar to that reported by de Bont et al. (1979). There are reports that the monooxygenase present in cells following growth of the methylotroph Methylococcus capsulatus can oxidize propylene (Stirling and Dalton, 1979). A soluble methane monooxygenase has also been obtained from *M. capsulatus* that oxidizes propylene to 1,2-epoxypropane (134 μmol propylene yielded 12 μmol product in 12 minutes). Hou *et al.* (1979) reported that the ability to epoxidize propylene was widespread in methane-grown bacteria. The organisms employed in the latter study were both obligate and facultative methylotrophs.

C. Transition from Growth on Nonhydrocarbons to Growth on Hydrocarbons

Exposure of cells growing in glucose or acetate to propane as substrate resulted in a significant lag period before growth was initiated (Vestal and Perry, 1971). The sequence of events that occurs in *M. vaccae* strain JOB5 in conversion from a glucose-utilizing growth to growth on propane is depicted in Fig. 3. The inducible oxygenase for propane appears first, followed by induction of isocitrate lyase; the lipid/N ratio gradually increases about 50% after 5–7 hours of exposure to propane; and cell multiplication occurs after 12–15 hours (Vestal and Perry, 1971, and unpublished).

An increase in cellular lipids in *C. lipolytica* grown on *n*-hexadecane (17.1%) as compared with cells grown on glucose (6.2%) has been reported (Nyns *et al.*, 1968). Hug *et al.* (1974) found that during transient continuous culture experiments, a change of substrate from glucose to *n*-hexadecane resulted in an adaptation phase, during which the cellular lipid doubled in amount per cell. The organism studied was *Candida tropicalis* and they proposed that the "high cellular lipid concentration was necessary for hydrocarbon assimilation, and this is not just a reflection of the lipophilic nature of the substrate" (p. 983).

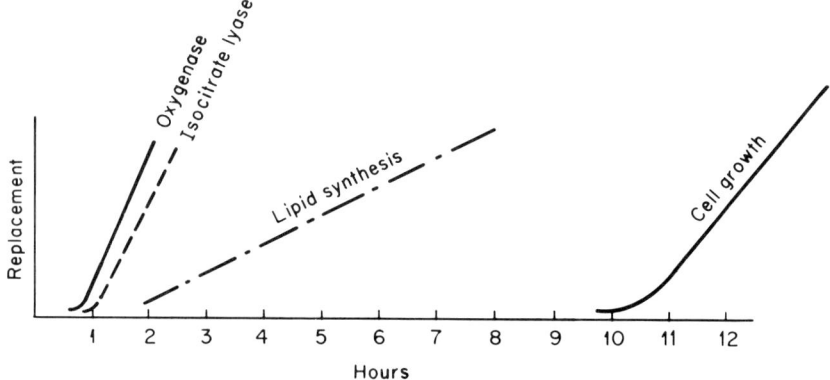

FIG. 3. Sequence of events involved in transition of *Mycobacterium vaccae* to growth with propane as substrate.

TABLE III
Lipid Composition of *Mycobacterium vaccae* Following Growth on Acetate and Propane[a]

	Growth substrate	
	Propane	Acetate
Phospholipid[b]	58.8	78.6
Cardiolipin[c]	46[c]	48
Phosphatidylethanolamine	30	34
Phosphatidylserine	17	14
Origin	3	2
Minor spots	4	2
Neutral lipid[b]	41.2	21.4
Isoprenologs (vitamin K)	3.0	1.3
Methyl ester	4.1	0.6
Triglyceride	15.3	3.3
Diglyceride	1.3	0.7
Fatty acid (free)	9.7	9.2
Unknown	6.4	6.3

[a] Reproduced by permission of the National Research Council of Canada from *Can. J. Microbiol.* 17, 445–449 (1971).
[b] Recorded as percentage of the total cell lipid.
[c] Recorded as percentage of the total phospholipid.

D. Lipids of Propane-Utilizing Organisms

As previously noted there is a gradual increase in the total lipid content of microbial cells during the transition from growth on acetate to propane utilization. *Mycobacterium vaccae* was grown on propane and on acetate and the lipid content was determined. The results are presented in Table III. The relative amount of phospholid in acetate-grown cells was significantly greater than in cells grown on propane, but the ratio of individual polar lipid in the fraction was not significantly different. The major difference was in the total amount of neutral lipid and the increase resulted from triglyceride. The possible role of neutral lipid in hydrocarbon assimilation is unknown.

A marked effect of substrate on the length of the fatty acid moieties in the cellular lipids of hydrocarbon-utilizing organisms has been reported. Growth on odd-chain length n-alkanes (C_{13}, C_{15}, C_{17}) results in a preponderance of odd-chain fatty acids in cells, whereas growth on even-chain length alkanes (C_{14}, C_{16}, C_{18}) results in mostly even-chain fatty acids in the cellular lipids (King and Perry, 1975a,b; Dunlap and Perry, 1967, 1968). These results suggest that n-alkanes of appropriate length can be incorporated directly into cellular fatty acids after monoterminal oxidation. Ascenzi and Vestal

(1979) have proved that direct incorporation does occur and that n-hexadecane or n-hexadecanoic acid inhibits fatty acid synthesis in *M. rhodochrous*. There is a decreased level of acetyl-CoA carboxylase in cells growing on n-hexadecane when compared with cells after growth on propane. It has also been noted that low molecular weight hydrocarbon and related substrates can influence the odd/even chain ratio of cellular fatty acids and this is dependent on the route by which the substrate is metabolized. Those metabolized via a two-carbon pathway, e.g., n-butane, yield even-chain fatty acids in cellular lipids and those substrates that are metabolized through a three-carbon pathway, e.g., n-propylamine, contain odd-chain fatty acids (see Tables IV, V, VI) (Phillips and Perry, 1974; Vestal and Perry, 1971). These results indicate that considerable knowledge of metabolic pathways can be obtained by determination of the presence or absence of isocitrate lyase and an analysis of the fatty acid composition of cells following growth on various hydrocarbon substrates and putative intermediates. The metabolic pathway for butadiene metabolism in *Nocardia* sp. 249 has been reported (Watkinson and Somerville, 1976) and in the course of this work fatty acid analysis and isocitrate lyase were determined. The results of these analyses were consistent with the two-carbon pathway demonstrated by more extensive studies.

E. Metabolism of Other Three-Carbon Substrates by Propane Utilizers

Microorganisms that utilize propane as sole source of carbon and energy, generally, grow very well on n-propylamine and other alkyl amines. It is of interest that n-propylamine, isopropylamine, and 1,3-propane diamine are

TABLE IV
Distribution of Odd/Even Fatty Acids in Cellular Lipids of the Propylene-Utilizing PL-1 Following Growth on Various Substrates[a]

Growth substrate	Chain length of fatty acids (% of total)		Isocitrate lyase[b]
	Even	Odd	
Acetate	98.3	1.7	1.3
Propylene	99.0	1.0	0.8
Propionate	83.6	15.6	1.0
Glucose	ND[c]	ND	0

[a] Isocitrate lyase levels were also determined.
[b] Units per milligram of protein in which one unit is the amount of enzyme necessary for the cleavage of 1 μmol isocitrate in 10 minutes.
[c] Not done.

TABLE V

CHAIN LENGTH OF THE FATTY ACIDS IN CELLULAR LIPIDS AND SPECIFIC ACTIVITY OF THE ISOCITRATE LYASE AFTER GROWTH OF *Mycobacterium vaccae* STRAIN JOB5 ON VARIOUS SUBSTRATES

Growth substrate	Chain length of fatty acids (% of total)		Isocitrate lyase[a]
	Even	Odd	
Acetate	99.8	0.2	0.8
Propionate	42.5	57.7	0
Isopropanol	99.8	0.2	1.6
Ethane	99.6	0.4	ND[b]
Propane	70.5	29.5	1.4
n-Butane	99.2	0.8	0.7
Butyrate	100	—	0.7
β-Hydroxybutyrate	100	—	0.5
2-Butanone	87.4	12.6	0
Glucose	ND	ND	0
n-Propanol	ND	ND	0

[a] See Table IV.
[b] Not done.

deaminated and the resultant oxidized intermediate is metabolized as might be anticipated. A study with an organism (*M. convolutum* NPA-1) isolated by enrichment on n-propylamine also grew with propane, 1,3-propane diamine, and isopropylamine as substrate. An analysis of the fatty acid profile in the organism and presence of isocitrate lyase is presented in Table VI. It is

TABLE VI

FATTY ACID COMPOSITION AND ISOCITRATE LYASE LEVEL IN *Mycobacterium convolutum* STRAIN NPA-1 FOLLOWING GROWTH ON VARIOUS SUBSTRATES

Growth substrate	Chain length of fatty acids (% of total)		Isocitrate lyase[a]
	Even	Odd	
Acetate	85.5	14.5	1.2
Propionate	4.2	95.8	0
Propane	ND[b]	ND	0.6
n-Propylamine	5.3	94.7	0
Isopropylamine	72.5	27.5	1.6
1,3-Propane diamine	89.6	9.2	1.3
Malonate	90.0	9.8	0.9
Isopropanol	77.1	22.9	ND

[a] See Table IV.
[b] Not done.

evident that propane, isopropylamine, 1,3-propane diamine, isopropanol, and malonate are metabolized via a $C_2 + C_1$ cleavage, whereas n-propylamine and propionate are not.

V. Growth Yield for Microorganisms Utilizing Hydrocarbon Substrates

There are a number of reports in the literature on the utilization of gaseous hydrocarbons as substrate and the cell yield obtained. Unfortunately the data are difficult to compare as there is considerable variation in substrate levels, oxygenation, and other growth parameters. There is also an inconsistency in the manner in which data are presented, e.g., grams per mole substrate vs. grams per mole O_2 consumed, substrate limiting vs. unlimited substrate.

A. Yield with Propane and Related Compounds as Substrate

Studies were conducted with *M. vaccae* JOB5 utilizing ethane, propane, or n-butane as substrate (Blevins and Perry, 1971) and the results are presented in Table VII. The yield from propane and ethane, on a gram per gram substrate basis, is quite high and in line with results for methane-utilizing organisms (Brown *et al.*, 1964; Van Dijken and Harder, 1975; Foster and Davis, 1966; Whittenbury *et al.*, 1970). The *M. vaccae* experiments were done with minimal levels of propane (5%) added and excess levels of O_2 (40%). The total cellular yield with this organism under optional conditions

TABLE VII
Cell Yield from *Mycobacterium vaccae* JOB5 Following Growth on Gaseous Alkanes and Related Substrates

Growth substrate	Cell yield (gm/gm substrate)
Ethane	1.22
Ethanol	0.58
Acetate	0.18
Propane	1.24
n-Propanol	0.70
Acetone	0.59
Propionate	0.45
n-Butane	0.72
n-Butanol	0.46
Butyrate	0.48

(40% O_2; 50% propane) was 0.65 gm/liter of medium (the experiment was done in closed flasks in a shake culture). Studies from other laboratories (Table VIII) indicate that considerably greater growth can be attained with a continuous flow of propane or butane into the system.

TABLE VIII
Cell Yield from Various Gaseous Alkane Substrates

Organism	Substrate and amount	Cell density (gm/liter)	(gm/gm substrate)	Reference
Arthrobacter simplex B-129	Propane 16% in air	0.3		Orgel et al. (1971)
Nocardia paraffinica Ky 4334	Propane 3 liters/minute	30	1.36	Sugimoto et al. (1972)
	n-Butane 3 liters/minute	22	0.95	Sugimoto et al. (1972)
Candida rigida Mo 113	n-Butane continuous vent gas, less than 0.10%	29		Imada et al. (1972)
Brevibacterium ketoglutamicum	n-Butane	13		Imada et al. (1972)
Mycobacterium cuneatum	Methane 33% in air	0.10		Iizuka et al. (1975)
Mycobacterium petroleophilum	Ethane 33% in air	0.19		Iizuka et al. (1975)
Mycobacterium cuneatum	6% Propane + 25% butane in air	0.30		Iizuka et al. (1975)
Mycobacterium petroleophilum	Propane 33% in air	0.25		Iizuka et al. (1975)
Nocardia asteroides 442	Propane 16% in air	0.3		Smirnova and Taptykova (1967)
Nocardia asteroides 443	Propane 16% in air	0.5		Smirnova and Taptykova (1967)
Nocardia rubropertincta	Propane 16% in air	0.6		Smirnova and Taptykova (1967)
Nocardia convolata	Propane 16% in air	0.9		Smirnova and Taptykova (1967)
Nocardia sp. 249	Butadiene 10-15% in air		1.5	Watkinson and Somerville (1976)
M45 mixed culture	Methane (fed continuously, gas limited)		0.62	Sheehan and Johnson (1971)
HR mixed culture	Methane 8% in air		0.62	Vary and Johnson (1967)
	Methanol (unknown)		0.33	Vary and Johnson (1967)
Nocardia sp.	Propane gas flow, 15% in air	12.8		Davis (1964)
	n-Butane gas flow, 15% in air	12.3		Davis (1964)

B. Growth with Higher Alkanes

A study by Gallo and Azoulay (1975) with *Candida tropicalis* as test organism and *n*-hexadecane as substrate yielded results similar to those with gaseous alkanes. They reported that 1.02 gm of cell mass was synthesized per gram alkane substrate utilized; 0.98 gm/gm palmitate and 0.31 gm/gm acetate. Blanch and Einsele (1973) obtained an equivalent yield from *n*-hexadecane in *C. tropicalis*. A mixture of C_{14}–C_{17} *n*-alkanes was employed as substrate in studies by Wagner *et al.* (1969) and they obtained results as follows: *Nocardia* NBC, 0.98; *Nocardia opaca*, 0.60; and *Mycobacterium phlei*, 1.050 gm/gm substrate.

VI. Products from Propane-Oxidizing Bacteria

The possible use of single-cell protein to alleviate the worldwide protein shortage has received considerable interest in recent years. That production of single-cell protein by hydrocarbon-utilizing microbes may be an economically feasible and practical solution to this problem has not been over-

TABLE IX
The Amino Acid Composition of Bacterial Cells Following Growth on Propane

	Mycobacterium petroleophilum[a]	*Arthrobacter*[b] *simplex* B129
Leucine	6.62	7.4
Isoleucine	3.74	3.5
Valine	5.17	5.5
Threonine	4.21	3.4
Methionine	1.38	0.5
Lysine	4.15	4.4
Phenylalanine	3.75	3.8
Tryptophan	—[c]	—
Arginine	5.58	—
Histidine	1.98	—
Cystine	0.32	—
Serine	3.87	2.4
Glutamic acid	11.00	8.7
Proline	3.92	2.9
Glycine	5.56	3.7
Alanine	8.34	7.3
Tyrosine	2.41	0.8
Aspartic acid	8.31	7.3

[a] Percentage by weight (Iizuka *et al.*, 1975).
[b] Percentage by weight (Orgel *et al.*, 1971).
[c] Data not given.

looked (Wilkinson, 1971). Natural gas is often a waste product in petroleum refining, and the conversion of this relatively low-cost substrate to utilizable food or chemicals would have considerable potential from both an economic and a humanitarian viewpoint.

A. Production of Single-Cell Proteins

Iizuka et al. (1975) reported that the protein content of *Mycobacterium cuneatum* cells grown on methane was 55% of the cell dry weight and that of *Mycobacterium petroleophilum* grown on propane was 56.3% by weight. The analysis of the amino acid content in the protein of the propane-grown organism is presented in Table IX. The amino acid composition of methane grown *M. cuneatum* was reported to be virtually the same. The composition of *Arthrobacter simplex* following growth on propane is also presented, and the total yield protein from this organism was 75–87% by weight assuming that all cellular nitrogen was protein.

B. Water-Soluble Vitamin Content of Propane-Grown Cells

The B vitamin content of *M. petroleophilum* following growth on propane was determined and the results are presented in Table X.

C. Production of Fatty Acids

The synthesis of a copolymer of β-hydroxybutyric and β-hydroxybutenoic (3-hydroxy-2-butenoic) acids by a *Nocardia* species during growth on propane or *n*-butane has been reported (Davis, 1964). The polymer represents about 13 and 4% of the dry weight in *n*-butane- and in propane-grown

TABLE X
A Profile of the B Vitamins Present in *Mycobacterium petroleophilum* Following Growth on Propane

	Dry wt. cells (mg/kg)
B_1 thiamin	17
B_2 riboflavin	48
$B_2 6$ pyridoxal	148
B_{12}	9.8
Nicotinic acid	178
Pantothenic acid	30

cells, respectively. The *Nocardia* species also can synthesize aliphatic waxes during growth on these substrates.

VII. Cooxidations Involving Propane

A. Propane as Cosubstrate

Leadbetter and Foster (1959a) were the first to demonstrate that an obligate methylotroph (*Pseudomonas methanica*), while growing on methane, would concomitantly oxidize other gaseous paraffinic hydrocarbons present in the growth flask. Foster (1962) termed this oxidative conversion cooxidation (for a discussion of hydrocarbon cooxidations, see Perry, 1979). Nonproliferating cellular suspensions of *P. methanica* did not oxidize propane or n-butane unless methane was present simultaneously to serve as growth substrate. During growth on methane, propane was oxidized to n-propanol, propionic acid, and acetone and from a culture containing n-butane they obtained n-butanol, n-butyric acid, and 2-butanone.

The use of microbes growing on gaseous alkanes to cooxidize nongrowth substrates to valuable industrial products may have potential. Foster stated in 1963, "in my opinion, co-oxidation provides an important new technique for industrial fermentations... the commercial appeal of the oxidative conversion of a large variety of non-growth hydrocarbons as co-substrates for organisms growing at the expense of cheap substrates, such as methane or natural gas, is obvious" (p. 251).

B. Propane as Substrate

The molecular oxygenase involved in oxidation of propane is an inducible one (Perry, 1979) and as a consequence the enzyme is present in cells only when a gratuitous or specific inducing substrate is present. A suggested role of cooxidation in the biodegradation of recalcitrant substrates has been reported (Perry, 1976a, 1977, 1979). The principle is illustrated in Fig. 4; the

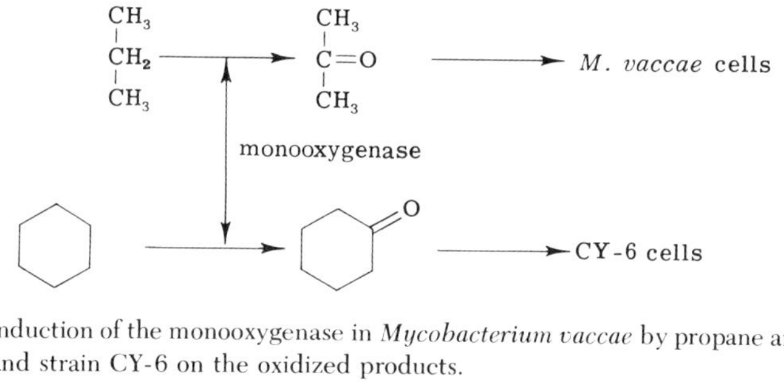

FIG. 4. Induction of the monooxygenase in *Mycobacterium vaccae* by propane and growth of *M. vaccae* and strain CY-6 on the oxidized products.

propane-utilizing *M. vaccae* cooxidizes cyclohexane to cyclohexanone, which is utilized as growth substrate by strain CY-6. Neither CY-6 nor *M. vaccae* can grow with cyclohexane as substrate and the latter cannot utilize the alkanone.

Mycobacterium vaccae can cooxidize many cycloparaffins to the homologous cycloalkanone during propane oxidation (Beam and Perry, 1973, 1974; Ooyama and Foster, 1965).

VIII. Toxicity of Propane and Related Compounds

When *P. methanica*, the obligate methylotroph, is grown on methane in the presence of ethane as cosubstrate, the concentration of ethane added to the atmosphere is critical. Ethane concentrations in the atmosphere of more than 7% along with methane at 30–45% and air at the remaining 30–50% completely inhibited growth of the organism (Leadbetter and Foster, 1960). Growth was obtained only when the ethane concentration was less than 5% of the atmosphere. Propane and n-butane were less inhibitory and concentrations up to 30% of the atmosphere were utilized in cooxidation experiments. Shushenacheva *et al.* (1975) reported that oxygen assimilation during methane oxidation by *Methylococcus ucrainicus* 21 is lower in the presence of propane, n-butane, isopropanol, and isobutanol.

Propane has been reported to effectively inhibit bacterial spore germination (Rode and Foster, 1965). These workers (Rode and Foster, 1966) also tested the C_1-C_5 n-alkane series for effects on a number of viable systems including bacteria, fungi, algae, plant seeds, chick embryos, and metamorphic stages of fruit fly development. Most were unaffected by methane; ethane generally had a slight effect and the toxicity increased with chain length from C_3 to C_5. The harmful effects of gas seeping into the soil above gas leaks on the vegetation and microflora of the area has been noted (Perry, 1976b). Generally, the problem is caused by oxygen depletion in the soil by microbial oxidation of the hydrocarbons. There is some indication that small amounts of propylene are present and inhibitory.

Propylene is readily converted to acrylic acid by propane-utilizing bacteria (Cerniglia *et al.*, 1976) and acrylic acid is an effective inhibitor of fatty acid synthesis (Thijsse, 1964; Whiteley and Ordal, 1957).

IX. Prospecting with Propane-Oxidizing Microbes

Prospecting for subterranean oil deposits by determining the relative number of organisms in the surface soil or water that are capable of growth on selected hydrocarbons has received considerable attention. Davis and Updegraff (1954) published a review outlining the published studies to that time (see also Davis, 1967; Brisbane and Ladd, 1965). The major objectives in prospecting by means of the microbiologic flora are in (1) selecting a substrate that is attacked by an array of hydrocarbon-utilizing organisms, but (2) not of immediate biogenic origin, and (3) should be of a nature that seepage to the earth's surface will occur. Propane would fulfill these requirements rather well. Kuznetsov and Telegina (1957) isolated a considerable number of propane-oxidizing bacteria and concluded that these organisms were markedly different from methane utilizers. They suggest that propane is more reliable in prospecting than methane as it is not generated biologically to a considerable extent. However, the production of small amounts of propane, propylene, butane, and butene from soil, roots, and grass has been demonstrated (Smith and Ellis, 1963). These workers state "to ascribe the presence of saturated hydrocarbons in the soil only to a petroleum source, is an error" (p. 1903). There are similar reports of propane production from thermal and biologic sources (Juranek, 1958; Veber and Turkel'taub, 1958; Davis and Squires, 1954; Horvat *et al.*, 1964).

Bokova *et al.* (1947) demonstrated that propane-oxidizing organisms accumulate in soil where hydrocarbon gases pass and the number doubled after bubbling through 2–3 liters of a gas–air mixture containing 13 ppm propane. The propane-oxidizing organisms isolated by these workers utilized more complex hydrocarbons but did not assimilate methane. Propane-oxidizing organisms were also isolated from soil obtained from oil fields in southern Slovakia (Dostalek, 1954). The cultures were similar to propane oxidizers isolated from other oil field soil and utilized gasoline vapors or aromatic hydrocarbons but not methane. Field water taken from above the Izbaskent gas oil deposit contained methane, ethane, propane, and butane down to a depth of 2680 m, indicating a migration of the oil. Soil above the pool contained methane- and propane-utilizing microorganisms in greater numbers than were present in water from areas outside the poor (Telegina *et al.*, 1963). Davis *et al.* (1959) and Dostalek and Spurny (1962) concluded that ethane- or propane-utilizing organisms are a better indicator of oil deposits as methanotrophs are more ubiquitous and evenly distributed in soil.

References

Allen, J. E., Forney, F. W., and Markovetz, A. J. (1971). *Lipids* **6**, 448–452.
Ascenzi, J. M., and Vestal, J. R. (1979). *J. Bacteriol.* **137**, 384–390.
Beam, H. W., and Perry, J. J. (1973). *Arch. Mikrobiol.* **91**, 87–90.
Beam, H. W., and Perry, J. J. (1974). *J. Gen. Microbiol.* **82**, 163–169.
Beerstecher, E. (1954). "Petroleum Microbiology." Am. Elsevier, New York.
Bemmann, W., and Troger, R. (1975). *Zentralbl. Bakteriol., Parasitenkd., Infektionskr. Hyg., Abt. 2* **129**, 742–752.

Blanch, H. W., and Einsele, A. (1973). *Biotechnol. Bioeng.* **15**, 861-877.
Blevins, W. T., and Perry, J. J. (1971). *Z. Allg. Mikrobiol.* **11**, 181-190.
Blevins, W. T., and Perry, J. J. (1972). *J. Bacteriol.* **112**, 513-518.
Bokova, E. N. (1954). *Mikrobiologiya* **23**, 15-21.
Bokova, J. N., Kuznetsova, V. A., and Kuznetsov, S. I. (1947). *Dokl. Akad. Nauk SSSR* **56**, 755-757.
Brisbane, P. G., and Ladd, J. N. (1965). *Annu. Rev. Microbiol.* **19**, 351-364.
Brown, L. R., Strawinski, R. J., and McClesky, C. S. (1964). *Can. J. Microbiol.* **10**, 791-799.
Bushnell, L. D., and Haas, H. F. (1941). *J. Bacteriol.* **41**, 653-673.
Cerniglia, C. E., and Perry, J. J. (1973). *Z. Allg. Mikrobiol.* **13**, 299-306.
Cerniglia, C. E., and Perry, J. J. (1975). *J. Bacteriol.* **124**, 285-289.
Cerniglia, C. E., Blevins, W. T., and Perry, J. J. (1976). *Appl. Environ. Microbiol.* **32**, 764-768.
Colby, J., Stirling, D. I., and Dalton, H. (1977). *Biochem. J.* **165**, 395-402.
Davies, J. S., Wellman, A. M., and Zajic, J. E. (1973). *Can. J. Microbiol.* **19**, 81-85.
Davies, J. S., Zajic, J. E., and Wellman, A. M. (1974). *Dev. Ind. Microbiol.* **15**, 256-262.
Davis, J. B. (1964). *Appl. Microbiol.* **12**, 301-304.
Davis, J. B. (1967). "Petroleum Microbiology." Am. Elsevier, New York.
Davis, J. B., and Squires, R. M. (1954). *Science* **119**, 381-382.
Davis, J. B., and Updegraff, D. M. (1954). *Bacteriol. Rev.* **18**, 215-238.
Davis, J. B., Chase, H. H., and Raymond, R. L. (1956). *Appl. Microbiol.* **4**, 310-315.
Davis, J. B., Raymond, R. L., and Stanley, J. P. (1959). *Appl. Microbiol.* **7**, 156-165.
de Bont, J. A. M., Attwood, M. M., Primrose, S. B., and Harder, W. (1979). *FEMS Microbiol. Lett.* **6**, 183-188.
Dostalek, M. (1954). *Cesk. Biol.* **3**, 173-181.
Dostalek, M., and Spurny, M. (1962). *Folia Microbiol. (Prague)* **7**, 141-149.
Dunlap, K. R., and Perry, J. J. (1967). *J. Bacteriol.* **94**, 1919-1923.
Dunlap, K. R., and Perry, J. J. (1968). *J. Bacteriol.* **96**, 318-321.
Einsele, A., and Feichter, A. (1971). *Adv. Biochem. Eng.* **1**, 169-194.
Forney, F. W., and Markovetz, A. J. (1971). *J. Lipid Res.* **12**, 383-395.
Foster, J. W. (1962). *In* "Oxygenases" (O. Hayaishi, ed.), pp. 241-271. Academic Press, New York.
Foster, J. W. (1963). *Antonie van Leeuwenhoek* **28**, 241-274.
Foster, J. W., and Davis, R. H. (1966). *J. Bacteriol.* **91**, 1924-1931.
Fuhs, G. W. (1961). *Arch. Mikrobiol.* **39**, 374-422.
Gallo, M., and Azoulay, E. (1975). *Biotechnol. Bioeng.* **17**, 1705-1715.
Hansen, R. W., and Kallio, R. E. (1957). *Science* **125**, 1198-1199.
Horvat, R. J., Lane, W. G., Ng, H., and Shepherd, A. D. (1964). *Nature (London)* **203**, 523-524.
Hou, C. T., Patel, R., Laskin, A. I., and Barnabe, N. (1979). *Appl. Environ. Microbiol.* **38**, 127-134.
Hug, H., Blanch, H. W., and Fiechter, A. (1974). *Biotechnol. Bioeng.* **16**, 965-985.
Hutton, W. E., and ZoBell, C. E. (1949). *J. Bacteriol.* **58**, 463-473.
Hutton, W. E., and ZoBell, C. E. (1953). *J. Bacteriol.* **65**, 216-219.
Iizuka, H., Seto, N., and Sakayanagi, S. (1975). U.S. Patent 3,888,736.
Imada, O., Hoshiai, K., and Tanaka, M. (1972). U.S. Patent 3,635,796.
Juranek, J. (1958). *Czech. Inst. Pet. Res., Trans.* **9**, 57-79.
Kester, A. S., and Foster, J. W. (1963). *J. Bacteriol.* **85**, 859-869.
King, D. H., and Perry, J. J. (1975a). *Can. J. Microbiol.* **21**, 85-89.
King, D. H., and Perry, J. J. (1975b). *Can. J. Microbiol.* **21**, 510-512.

Klausmeier, R. E., Brown, L. R., Benes, E. N., and Strawinski, R. J. (1958). *Bacteriol. Proc.* **58**, 123.
Klein, D. A., Davis, J. A., and Casida, L. E. (1968). *Antonie van Leeuwenhoek* **34**, 495–503.
Klug, M. J., and Markovetz, A. J. (1971). *Adv. Microb. Physiol.* **5**, 1–43.
Kuznetsov, S. I., and Telegina, Z. P. (1957). *Mikrobiologiya* **26**, 513–518.
Leadbetter, E. R., and Foster, J. W. (1958). *Arch. Mikrobiol.* **30**, 91–118.
Leadbetter, E. R., and Foster, J. W. (1959a). *Arch. Biochem. Biophys.* **82**, 491–492.
Leadbetter, E. R., and Foster, J. W. (1959b). *Nature (London)* **184**, 1428.
Leadbetter, E. R., and Foster, J. W. (1960). *Arch. Mikrobiol.* **35**, 92–104.
Levine, S., and Krampitz, L. O. (1952). *J. Bacteriol.* **64**, 645–650.
Lowery, C. E., Foster, J. W., and Jurtshuk, P. (1968). *Arch. Mikrobiol.* **60**, 246–254.
Lukins, H. B., and Foster, J. W. (1963a). *J. Bacteriol.* **85**, 1074–1087.
Lukins, H. B., and Foster, J. W. (1963b). *Z. Allg. Microbiol.* **3**, 251–264.
McAuliffe, C. (1966). *J. Phys. Chem.* **70**, 1267–1275.
McKenna, E. J., and Kallio, R. E. (1965). *Annu. Rev. Microbiol.* **19**, 183–208.
McLee, A. G., Kormendy, A. C., and Wayman, M. (1972). *Can. J. Microbiol.* **18**, 1191–1195.
Merkel, G. J., Stapleton, S. S., and Perry, J. J. (1978a). *J. Gen. Microbiol.* **109**, 141–148.
Merkel, G. J., Underwood, W. H., and Perry, J. J. (1978b). *FEMS Microbiol. Lett.* **3**, 81–83.
Mueller, J. C. (1969). *Can. J. Microbiol.* **15**, 1114–1116.
Nechaeva, N. B. (1949). *Mikrobiologiya* **18**, 310–317.
Nyns, E. J., and Wiaux, A. L. (1969). *Agricultura (Louvain)* **17**, 3–56.
Nyns, E. J., Chiang, N., and Wiaux, A. L. (1968). *Antonie van Leeuwenhoek* **34**, 197–204.
Nyns, E. J., Auquiere, J. P., and Wiaux, A. L. (1969). *Z. Allg. Mikrobiol.* **9**, 373–380.
O'Brien, W. E., and Brown, L. R. (1968). *Dev. Ind. Microbiol.* **9**, 389–393.
Ooyama, J., and Foster, J. W. (1965). *Antonie van Leeuwenhoek* **31**, 45–65.
Orgel, G., Pietrusza, E. W., and Joris, G. G. (1971). U.S. Patent 3,622,465.
Pabst, G. S., and Brown, L. R. (1968). *Dev. Ind. Microbiol.* **9**, 394–400.
Perry, J. J. (1968). *Antonie van Leeuwenhoek* **34**, 27–36.
Perry, J. J. (1976a). In "The Role of Microorganisms in the Recovery of Oil," Workshop, 1975, National Science Foundation, Research Applied to National Needs, pp. 45–50.
Perry, J. J. (1976b). In "Vapor-Phase Organic Pollutants," National Research Council, National Academy of Sciences, pp. 259–262.
Perry, J. J. (1977). *Crit. Rev. Microbiol.* **5**, 387–412.
Perry, J. J. (1979). *Microbiol. Rev.* **43**, 59–72.
Perry, J. J., and Cerniglia, C. E. (1973). *Microb. Degradation Oil Pollut.*, Workshop, 1972 Louisiana State Univ. Publ. No. LSU-SG-73-01. pp. 89–94.
Perry, J. J., and Scheld, H. W. (1968). *Can. J. Microbiol.* **14**, 403–407.
Phillips, W. E., and Perry, J. J. (1974). *J. Bacteriol.* **120**, 987–989.
Ratledge, C. (1978). In "Developments in Biodegradation of Hydrocarbons-1" (R. J. Watkinson, ed.), pp. 1–46. Appl. Sci. Publ. Ltd., London.
Rode, L. J., and Foster, J. W. (1965). *Proc. Natl. Acad. Sci. U.S.A.* **53**, 31–38.
Rode, L. J., and Foster, J. W. (1966). *Z. Allg. Mikrobiol.* **6**, 353–360.
Rudney, H. (1954). *J. Biol. Chem.* **210**, 361–371.
Sheehan, B. T., and Johnson, M. J. (1971). *Appl. Microbiol.* **21**, 511–515.
Shennan, J. L., and Levi, J. D. (1974). *Prog. Ind. Microbiol.* **13**, 1–57.
Shushenacheva, E. V., Nesterov, A. I., and Netrusov, A. I. (1975). *Mikrobiologiya* **44**, 536–540.
Slavina, G. P. (1948). *Mikrobiologiya* **17**, 76–81.
Smirnova, Z. S. (1962). *Mikrobiologiya* **31**, 794–800.
Smirnova, Z. S., and Taptykova, S. D. (1967). *Mikrobiologiya* **36**, 311–314.

Smith, G. H., and Ellis, M. M. (1963). *Bull. Am. Assoc. Pet. Geol.* **47**, 1897–1903.
Souw, P., Luftmann, H., and Rehm, H. J. (1977). *Eur. J. Appl. Microbiol.* **3**, 289–301.
Stirling, D. I., and Dalton, H. (1979). *FEMS Microbiol. Lett.* **5**, 315–318.
Strawinski, R. J., and Tortorich, J. A. (1955). *Bacteriol. Proc.* p. 27.
Sugimoto, M., Yokoo, S., and Imada, O. (1972). *Ferment. Technol. Today, Proc. Int. Ferment. Symp., 4th, 1972* pp. 503–507.
Tanaka, K., Ohshima, K., Kimura, K., and Yamamoto, M. (1972). U.S. Patent 3,639,210.
Tausz, J., and Donath, P. (1930). *Hoppe-Seyler's Z. Physiol. Chem.* **190**, 141–168.
Telegina, Z. P., Subbota, M. I., and Nikitina, E. A. (1963). *Mikrobiologiya* **32**, 26–33.
Thijsse, G. J. E. (1964). *Biochim. Biophys. Acta* **84**, 195–197.
Thomson, A. W., O'Neill, J. G., and Wilkinson, J. F. (1976). *Arch. Microbiol.* **109**, 243–246.
Tonge, G. M., Harrison, D. E. F., and Higgins, I. J. (1977). *Biochem. J.* **161**, 333–344.
Van Der Linden, A. C., and Thijsse, G. J. E. (1965). *Adv. Enzymol.* **27**, 469–546.
Van Dijken, J. P., and Harder, W. (1975). *Biotechnol. Bioeng.* **17**, 15–30.
Van Eyk, J., and Bartels, T. J. (1968). *J. Bacteriol.* **96**, 706–712.
Vary, P. S., and Johnson, M. J. (1967). *Appl. Microbiol.* **15**, 1473–1478.
Veber, V. V., and Turkel'taub, N. M. (1958). *Geol. Nefti* **2**, 39–44.
Vestal, J. R., and Perry, J. J. (1969). *J. Bacteriol.* **99**, 216–221.
Vestal, J. R., and Perry, J. J. (1971). *Can. J. Microbiol.* **17**, 445–449.
Wagner, F., Kleemann, T., and Zahn, W. (1969). *Biotechnol. Bioeng.* **11**, 393–408.
Watkinson, R. J., ed. (1978). "Developments in the Biodegradation of Hydrocarbons-1." Appl. Sci. Publ. Ltd., London.
Watkinson, R. J., and Somerville, H. J. (1976). *In* "Proceedings of the International Biodegradation Symposium" (J. M. Sharpley and A. M. Kaplan, eds.), pp. 35–42. Appl. Sci. Publ. Ltd., London.
Whiteley, H. R., and Ordal, E. J. (1957). *J. Bacteriol.* **74**, 331–336.
Whittenbury, R., Phillips, K. C., and Wilkinson, J. F. (1970). *J. Gen. Microbiol.* **61**, 205–218.
Wilkinson, J. F. (1971). *In* "Microbes and Biological Productivity," pp. 15–46. Cambridge Univ. Press, London and New York.
Wolf, H. J., and Hanson, R. S. (1979). *J. Gen. Microbiol.* **114**, 187–194.
ZoBell, C. E. (1946). *Bacteriol. Rev.* **10**, 1–49.

Production of Intracellular and Extracellular Protein from n-Butane by Pseudomonas butanovora sp. nov.

JOJI TAKAHASHI

Institute of Applied Biochemistry,
University of Tsukuba,
Ibaraki, Japan

I.	Introduction	117
II.	Properties of *Pseudomonas butanovora* sp. nov.	118
	A. Taxonomic Properties and Identification	118
	B. Growth on Various Substrates	120
	C. Composition of Cells	121
III.	Cellular Growth and Accumulation of Extracellular Protein	121
	A. Culture Conditions	121
	B. Accumulation of Higher Concentrations of Cells	122
	C. Accumulation of Extracellular Protein	123
IV.	Properties of Intracellular and Extracellular Protein	124
V.	Concluding Remarks	126
	References	126

I. Introduction

Gaseous normal alkanes ranging from methane to *n*-butane have been known to be more advantageous than liquid alkanes for the production of biomass in several points. Gaseous alkanes are less expensive than liquid ones and occur abundantly in the world. They are pure and clean not only from the scientific but also from the psychological point of view.

Among the gaseous alkanes, methane is the cheapest and of the most abundant occurrence. Many notable results have been reported on the production of biomass from methane in pilot plant tests as well as in basic studies (Hamer *et al.*, 1967; Wolnak *et al.*, 1967; Whittenburg *et al.*, 1970; Sheehan and Johnson, 1971; Barnes *et al.*, 1976). Especially, very promising results have been obtained by Harrison *et al.* (1976), using a mixed culture of a *Methylomonas* sp. with a methanol utilizer. However, the production rates of biomass from gaseous substrates are limited by the transfer rates of those gases into culture fluids, and a fermenter having much higher rate of mass transfer than so far has been used is required to industrialize the process of biomass production from methane, because the transfer rate of methane is relatively low.

Gaseous normal alkanes, such as ethane, propane, and *n*-butane, are more advantageous than methane in terms of transfer rate. The transfer rates of those alkanes into water are 1.5–2 times as high as that of methane under the same conditions (Takahashi, 1970; Matsumura *et al.*, 1973), that is, the load on a fermenter for supplying the same mass of gaseous substrates is reduced to one-half or one-third when gaseous alkanes other than methane are used as the growth substrate. In addition to this advantage, the yields of biomass theoretically expected on those alkanes are about 1.4 times as high as that expected on methane, provided that the yield of biomass is proportional to the amount of ATP obtained by the complete oxidation of each substrate (Van Dijken and Harder, 1975). Therefore, those gaseous alkanes are considered to be promising carbon sources for the production of biomass, if a few difficulties reported by previous workers can be overcome. These difficulties are in isolating a strain having much higher rate of growth and in achieving the accumulation of much higher concentration of cells (McLee et al., 1972; Sugimoto et al., 1972; Kawakami et al., 1977).

From the above points of view, screening work has been carried out, searching the strains capable of growing on ethane, propane, and n-butane as sole carbon source, and several strains having higher rates of growth than ever reported have been newly isolated. The most promising of these, which is designated as *Pseudomonas butanovora* sp. nov., also has been found able to accumulate a large amount of extracellular protein in proportion to its cellular growth. The cultural conditions for the accumulation of extracellular protein as well as for higher concentration of cells at the expense of gaseous alkanes have been investigated.

In this contribution properties of the new isolate and cultural conditions for producing large amounts of intracellular and extracellular protein are described. The differences in the properties between extracellular and intracellular protein are also discussed.

II. Properties of *Pseudomonas butanovora* sp. nov.

A. TAXONOMIC PROPERTIES AND IDENTIFICATION

The strain employed in this study was newly isolated from activated sludge, sampled from an oil refining plant, by an enrichment culture technique using n-butane as the sole carbon source (Takahashi et al., 1980). It is a gram-negative, monotrichous rod and 0.6–0.8×1.1–2.4 μm in size. Catalase and oxidase were positive, whereas urease was negative. Nitrate reduction and denitrification were also positive. GC content in DNA calculated from the thermal denaturation temperature (Marmur and Doty, 1962) was 67.3%. These characteristics indicate that the new isolate belongs to the denitrifying group of genus *Pseudomonas*. However, none of the denitrifying species of *Pseudomonas* appearing in *Bergey's Manual of Determinative Bacteriology* (Buchanan and Gibbons, 1974) is identical with the new isolate, as shown in Table I. Accordingly, the new isolate is recognized as a new species and designated *Pseudomonas butanovora*.

TABLE I
CHARACTERISTICS OF DENITRIFYING SPECIES OF *Pseudomonas* AND NEW ISOLATE

Characteristics	New isolate	P. aeruginosa	P. fluorescens	P. chlororaphis	P. aureofaciens	P. stutzeri	P. mendocina	P. pseudomallei	P. caryophylli	P. mallei	P. solanacearum
Accumulation of PHB[a]	+	−	−	−	−	−	−	+	+	+	+
Production of diffusible pigments	−	+	+	+	+	−	−	−	−	−	D
Growth at 41°C	+	+	−	−	−	+	+	+	+	+	−
Arginine dihydrolase	−	+	+	+	+	−	+	+	+	+	−
Hydrolysis of gelatin	−	+	−	+	+	−	−	+	−	+	−
Hydrolysis of starch	−	−	−	−	−	+	−	+	−	D[b]	−
Carbon sources for growth											
D-Glucose	−	+	+	+	+	+	+	+	+	+	+
D-Xylose	−	+	+	+	ND[c]	−	−	−	+	+	−
D,L-Arginine	−	+	+	+	+	−	+	+	+	+	−
1,2-Ethanediol	−	+	−	ND	ND	+	+	−	−	ND	−
2,3-Butanediol	+	+	−	ND	ND	+	−	−	+	−	−
Glycolate	−	+	−	−	−	+	+	+	+	−	−
Geraniol	−	+	−	−	−	−	+	−	−	−	D
L-Valine	+	+	−	+	+	+	+	+	−	D	+
GC Content in DNA (%)	67.3	67.3	59.4–61.3	63.5	63.6	60.7–66.3	62.8–64.3	69.5	65.3	69.0	66.5–68.0

[a] Poly-β-hydroxybutyrate.
[b] Reaction differs.
[c] No description.

B. Growth on Various Substrates

Pseudomonas butanovora is able to utilize a variety of organic compounds for the growth, but not sugars. The results of assimilation tests of various substrates are summarized in Table II. These tests were carried out by using a shaking culture technique in which shaker flasks containing 20 ml of culture medium with 50 mg of each substrate were employed (Takahashi et al., 1980). The cells were harvested at the stationary phase of growth, and the yield factor, $Y_{x/s}$, on each substrate was calculated by dividing the amount of dried cells harvested not by the amount of substrate consumed but by the amount of substrate supplied. Accordingly, the values of yield factors shown in Table II are somewhat smaller than those expected from the common calculation based on the amounts of consumption.

Among the organic compounds tested, normal alkanes ranging from C_2 to C_9; primary alcohols and carboxylic acids of C_2, C_3, and C_4, and polyvalent alcohols of C_3 and C_4 were utilized for growth, whereas normal alkanes of C_{10} and more, C_1 compounds, normal alkenes, and sugars were not utilized. The yields of cells on gaseous alkanes, such as ethane, propane, and n-butane,

TABLE II
UTILIZATION OF VARIOUS SUBSTRATES BY *Pseudomonas butanovora* FOR GROWTH

Substrate	Yield factor[a]	Substrate	Yield factor
Methane	—[b]	Ethylene	—
Ethane	0.80	Propylene	—
Propane	0.82	1-Butene	—
n-Butane	0.83	2-Butene	—
n-Pentane	0.87	1,3-Butadiene	—
n-Hexane	0.73		
n-Heptane	0.49	1,2-Ethanediol	—
n-Octane	0.37	1,2-Propanediol	0.46
n-Nonane	0.04	2,3-Butanediol	0.43
n-Alkanes, C_{10}–$C_{1'}$	—	Glycerol	0.10
		D-Sorbitol	—
Methanol	—	Erythritol	—
Ethanol	0.57		
n-Propanol	0.95	L-Arabinose	—
n-Butanol	0.85	D-Xylose	—
		D-Glucose	—
Sodium acetate	0.16	D-Fructose	—
Sodium propionate	0.75		
Sodium butyrate	0.80		
Sodium lactate	0.43		

[a] Grams cell/gm substrate.
[b] Not utilized.

TABLE III
COMPOSITION OF DRY CELLS OF Pseudomonas butanovora

Component	Content (%)
Crude protein	77.69
(pure protein)	72.81
Carbohydrates	3.06
Crude fat	2.77
Crude fiber	0
Ash	11.07
Moisture	8.93

are rather high and the values of $Y_{x/s}$ attain to 0.8 and more. It is of interest that this new isolate can utilize normal alkanes of C_5–C_9 quite well; these have been thought hardly assimilable by microorganisms (Sharpley, 1966).

C. COMPOSITION OF CELLS

Table III shows the composition of cells of *P. butanovora* grown on *n*-butane as the sole source of carbon and energy and then dried. It is notable that the content of protein in the cells is very high and that of carbohydrates is extremely low. The content of pure protein, as high as 73%, is one of the highest values ever obtained in cells grown on gaseous alkanes, although slightly higher values have been reported in cells produced from methanol (Kono *et al.*, 1973). The elemental composition of the dried cells was C, 47.33%; H, 6.85%; and N, 13.37%, and the content of nonprotein nitrogen calculated from the difference between total and protein nitrogen was 1.72%. This corresponds to the cellular nucleic acid content of 11.05%, assuming all nonprotein nitrogen is in nucleic acids.

III. Cellular Growth and Accumulation of Extracellular Protein

A. CULTURE CONDITIONS

A culture medium of choice for the cellular growth was composed of $(NH_4)_2HPO_4$, 0.45%; $Na_2HPO_4 \cdot 12 H_2O$, 0.25%; KH_2PO_4, 0.2%; $MgSO_4 \cdot 7 H_2O$, 0.05%; $FeSO_4 \cdot 7 H_2O$, 30 ppm; $CaCl_2 \cdot 2 H_2O$, 60 ppm; $MnCl_2 \cdot 2 H_2O$, 0.06 ppm; $CuSO_4 \cdot 5 H_2O$, 0.015 ppm; and yeast extract, 50 ppm. No growth factor was required, although yeast extract was rather stimulative to the growth. Ammonium phosphate was the best nitrogen source among nitrates and ammonium salts tested.

The optimum pH for the growth was 7.5, and the optimum temperature was 33–35°C.

The n-butane–air mixture employed contained n-butane 6% in air so as to make the rate of transfer of n-butane limiting to the rate of growth. This value was slightly lower than the critical one, 7.0%, where the growth-limiting factor was shifted from being the rate of n-butane transfer to being that of oxygen transfer (Ichikawa et al., 1980).

B. Accumulation of Higher Concentrations of Cells

Figure 1 shows two growth curves on n-butane in a 2.6-liter laboratory fermenter containing 1.3 liter of culture medium. The culture system was mechanically agitated with an impeller of 1000 rpm, and the n-butane–air mixture described above was supplied at a rate of 0.5 ppm. The pH of the culture system was kept at 7.5 by using an automatic pH controller.

When the pH was controlled by feeding a 2.5 N aqueous ammonia, as shown by open circles, cellular growth attained a peak at the concentration of 15 gm/liter or so, and no further growth was observed. In contrast, when the aqueous ammonia to be fed was supplemented with the same kinds of mineral salts as in the culture medium, the cells continued to grow to the concentration of nearly 100 gm/liter, as shown by the closed circles, although the rate of growth decreased gradually and the data plots after 72 hours were omitted in Fig. 1. The ammonia–mineral-salts solution employed here contained $Na_2HPO_4 \cdot 12\ H_2O$, 27 gm; KH_2PO_4, 54.8 gm; $MgSO_4 \cdot 7\ H_2O$, 17.8 gm, $FeSO_4 \cdot 7\ H_2O$, 0.5 gm; $CaCl_2 \cdot 2\ H_2O$, 1.3 gm; $MnCl_2 \cdot 4\ H_2O$, 1 mg;

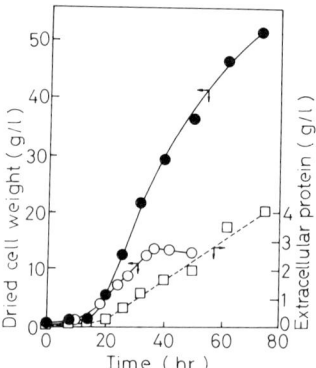

FIG. 1. Growth of *Pseudomonas butanovora* on n-butane. (●) Cellular growth, fed with aqueous ammonia containing mineral salts; (○) cellular growth, fed with aqueous ammonia free from mineral salts; (□) extracellular protein.

$CuSO_4 \cdot 5\ H_2O$, 0.1 mg; and yeast extract, 25 mg, in 1 liter of the 2.5 N aqueous ammonia; the ratio of each inorganic nutrient to nitrogen in this solution was nearly equal to that in the dried cells. It is clear from the above results that the accumulation of cells as high as 100 gm can be achieved by supplementing inorganic nutrients together with the nitrogen source in the course of the cultivation, and one of the difficulties reported by previous workers therefore has been overcome.

The maximum specific growth rate calculated from the growth curves is 0.25 per hour, is one of the highest rates ever obtained in pure cultures growing on gaseous hydrocarbons, although much higher rates have been reported in mixed cultures growing on methane (Sheehan and Johnson, 1971; Harrison et al., 1976).

The linear part of the growth curve, where the rate of growth is limited by the rate of transfer of n-butane, gives the maximum cellular productivity. It is as high as 1.2 gm/liter/hour.

C. Accumulation of Extracellular Protein

In the course of cellular growth, described above, the culture system was found to become more viscous than expected from the amount of cells accumulating; the increase in viscosity was confirmed as coming not from the accumulation of polysaccharides, as had been reported by several authors (Sugimoto et al., 1972; Tezuka et al., 1977), but from the extracellular accumulation of protein. The amount of accumulation of extracellular protein, as shown by open squares in Fig. 1, increased in proportion to the cellular growth and attained to 4 gm/liter, when assayed by Lowry's method (Lowry and Rosbrough, 1951), at the seventy-second hour of the cultivation. The extracellular protein thus accumulated was considered not to come from the autolysis of cells but to be secreted instead from the cells as one of the metabolic products, because very small amounts of nucleic acids were detected in the culture filtrate. It is noteworthy that a large amount of extracellular protein is accumulated by a gram-negative bacterium such as *Pseudomonas* sp. (Glenn, 1976) in the expense of a gaseous alkane, although Ryabushko (1979) has reported the formation of extracellular protein by representatives of *Pseudomonas* sp. during the growth on media containing ethanol.

Among the culture conditions tested for the accumulation of extracellular protein, the incubation temperature was found to have a considerable influence, whereas the pH of the culture system and the kinds of nitrogen sources did not.

Figure 2 shows the effects of temperature on the accumulation of extracellular protein and the rate of cellular growth. In this figure, $Y_{p/x}$ is the ratio of

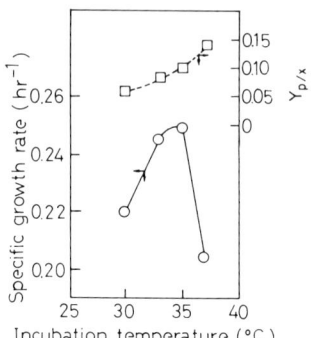

FIG. 2. Effect of temperature on accumulation of extracellular protein. $Y_{p/x}$, ratio of extracellular protein to cell mass.

the amount of extracellular protein to the cell mass. The value of $Y_{p/x}$ increased with the rise in the incubation temperature and attained to 0.14 at 37°C. However, the specific growth rate attained a peak as high as 0.25 per hour at 35°C and decreased to 0.21 at 37°C. The amount of accumulation of cell mass also decreased when the incubation temperature exceeded 35°C, and the cell mass accumulated at 37°C was one-third of that accumulated at 35°C. Therefore, the optimum temperature is 35°C for the accumulation of extracellular protein as far as the amount of protein is concerned.

IV. Properties of Intracellular and Extracellular Protein

In Table IV, the amino acid composition of the intracellular protein of *P. butanovora* is shown to be very similar to that of a mixed culture of methane utilizers resembling pseudomonads (Sheehan and Johnson, 1971). In the same table, the amino acid composition of the extracellular protein of *P. butanovora* is also compared with that of the intracellular protein. The extracellular protein is rich in acidic amino acids, such as glutamic and aspartic acids, and the contents of these amino acids are about 1.5 times as much as those in the intracellular protein. In contrast, the intracellular protein is rich in arginine, an alkaline amino acid, and the content is over 2 times as much as that in the extracellular protein.

Figure 3 shows the electrophoretic patterns of intracellular and extracellular protein in sodium dodecyl sulfate (SDS)–polyacrylamide gel. At least 16 peaks are observed in the pattern of intracellular protein, and these peaks correspond to protein fractions having molecular weights ranging from 10,000 to 100,000. On the other hand, only two peaks are observed in the pattern of extracellular protein. This indicates that the extracellular protein

TABLE IV
Amino Acid Composition of Intracellular and Extracellular Protein of *Pseudomonas butanovora*

Amino acid	Intracellular protein (%)	Extracellular protein (%)	Intracellular protein of M45 (%)[a]
Lysine	5.7	4.7	4.6
Histidine	2.2	1.3	1.7
Arginine	7.9	3.4	5.3
Aspartic acid	10.1	15.4	9.7
Threonine	5.0	5.1	5.2
Serine	4.1	4.3	4.2
Glutamic acid	12.1	18.4	14.7
Proline	4.0	3.8	4.1
Glycine	6.1	6.4	6.2
Alanine	8.9	7.3	7.8
Cystine	0.3	0.7	0.4
Valine	7.0	8.1	6.6
Methionine	2.9	1.7	3.2
Isoleucine	4.8	3.8	5.3
Leucine	9.0	6.4	8.4
Tyrosine	3.6	3.0	4.1
Phenylalanine	4.7	4.3	5.6
Tryptophan	1.5	1.8	2.9

[a] A mixed culture of methane utilizers reported by Sheehan and Johnson (1971).

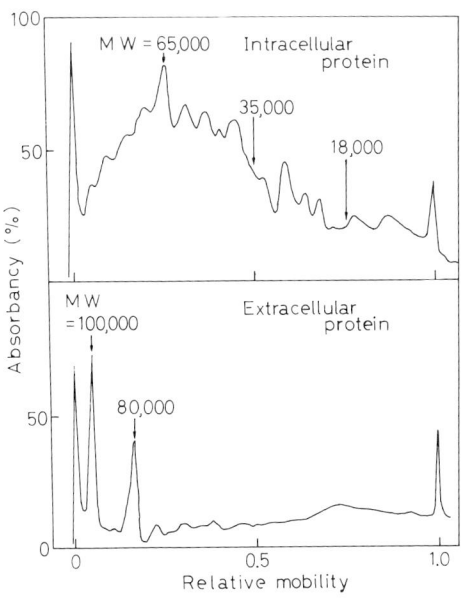

FIG. 3. Patterns of SDS-polyacrylamide gel electrophoresis.

is composed of two protein fractions having molecular weights of 80,000 and 100,000, respectively.

These differences in the amino acid composition and the electrophoretic pattern indicate that the extracellular protein is different from the intracellular one and does not come from the autolysis of cells; instead, it is secreted from the cells as a metabolic product.

With regard to the enzymatic activities of the extracellular protein, no activity was detected as far as the enzymes of practical interest, such as protease, amylase, and RNase, were concerned. Elucidation of other physiological activities and physicochemical properties of the extracellular protein is still pending.

V. Concluding Remarks

The new isolate designated as *Pseudomonas butanovora* sp. nov. is remarkable for its rate of growth on n-butane and ability to produce extracellular protein as well as intracellular protein. The total yield of intracellular and extracellular protein on n-butane is 66% by weight under the optimum conditions.

It may be concluded that this new isolate is one of the most promising strains for the production of protein from such gaseous alkanes as ethane, propane, and n-butane. Further investigations on increasing the amount of accumulation of extracellular protein are expected not only to be of practical interest but also to be informative about the physiological significance of the accumulation of extracellular protein at the expense of gaseous alkanes.

Acknowledgments

The author is deeply indebted to the members of the Central Research Laboratories, Ajinomoto Co., Ltd. for the identification of the new isolate and the determination of amino acid composition.

References

Barnes, L. J., Drozd, J. W., Harrison, D. E. F., and Hamer, G. (1976). *Proc. Symp. Microb. Prod. Util. Gases, 1975* p. 301.
Buchanan, R. E., and Gibbons, N. E., eds. (1974). "Bergey's Manual of Determinative Bacteriology," 8th ed. Williams & Wilkins, Baltimore, Maryland.
Glenn, A. R. (1976). *Annu. Rev. Microbiol.* **30**, 41.
Hamer, G., Heden, C. G., and Carenberg, C. O. (1967). *Biotechnol. Bioeng.* **9**, 499.
Harrison, D. E. F., Drozd, J. W., and Khosrovi, B. (1976). *Proc. Int. Ferment. Symp., 5th, 1976* p. 395.

Ichikawa, Y., Sato, S., and Takahashi, J. (1980). *J. Ferment. Technol.* (in press).
Kawakami, S., Shoji, H., Nonaka, N., Nakayama, M., and Hatano, T. (1977). *Bull. Jpn. Pet. Inst.* **19**, 187.
Kono, K., Oki, T., Nomura, H., and Ozaki, A. (1973). *J. Gen. Appl. Microbiol.* **19**, 11.
Lowry, O. H., and Rosbrough, N. J. (1951). *J. Biol. Chem.* **193**, 265.
McLee, A. G., Kormendy, A. C., and Wyman, M. (1972). *Can. J. Microbiol.* **18**, 1191.
Marmur, J., and Doty, P. (1962). *J. Mol. Biol.* **5**, 109.
Matsumura, M., Haraya, K., Yoshitome, K., and Kobayashi, J. (1973). *Abstr. Annu. Meet., Soc. Ferment. Technol.*, Japan, p. 71.
Ryabushko, T. A. (1979). *Prikl. Biokhim. Mikrobiol.* **15**, 157.
Sharpley, J. M. (1966). "Elementary Petroleum Microbiology." Gulf Publishing, Houston, Texas.
Sheehan, B. T., and Johnson, M. J. (1971). *Appl. Microbiol.* **21**, 511.
Sugimoto, M., Yokoo, S., and Imada, O. (1972). *Ferment. Technol. Today, Proc. Int. Ferment. Symp., 4th, 1972* p. 503.
Takahashi, J. (1970). *Pet. & Microorg.* **4**, 24.
Takahashi, J., Ichikawa, Y., Sagae, H., Komura, I., Kano, H., and Yamada, K. (1980). *Agric. Biol. Chem.* (in press).
Tezuka, C., Takeda, K., and Hachiya, Y. (1977). *Hakko Kogaku Kaishi* **55**, 349.
Van Dijken, J. P., and Harder, W. (1975). *Biotechnol. Bioeng.* **12**, 15.
Whittenburg, R., Phillips, K. C., and Wilkinson, J. F. (1970). *J. Gen. Microbiol.* **61**, 205.
Wolnak, B., Andreen, B. H., Chisholm, J. A., and Saadeh, M. (1967). *Biotechnol. Bioeng.* **9**, 57.

Effects of Microwave Irradiation on Microorganisms

JOHN R. CHIPLEY

*U.S. Tobacco Company,
Nashville, Tennessee*

I.	Introduction	129
II.	Effects of Microwave Irradiation on Microorganisms	130
III.	Effects of Microwave Irradiation on Other Biologic Systems	137
IV.	Thermal versus Nonthermal Effects of Microwave Irradiation	138
	References	143

I. Introduction

One of the more modern approaches to food cooking and preparation techniques has been the introduction of microwave irradiation. In the past few years, microwave ovens have flooded the consumer market, and microwave cooking is becoming more popular.

The U.S. Food and Drug Administration, responsible for verifying oven safety and radiation processes, currently certifies only 915 and 2450 MHz as usable power frequencies for commercial and home ovens. Use of these frequencies is based upon a 1956 triservice committee report of the U.S. Army, Air Force, and Navy, which proposed exposure limits of 10 mW/cm^2 from investigations of the limits of hazardous exposure to humans.

Microwaves are a form of electromagnetic radiation, as are ultraviolet and visible light and radio waves. They differ in frequency, wavelength, and energy content. Microwave ovens generate waves at either the 915- or the 2450-MHz frequency, with a wavelength of 12 cm, and contain energy sufficient to cause heating without breaking chemical bonds (Schiffmann, 1979).

Microwaves are a form of energy, not a form of heat. When they interact with such materials as water or oils in food products, heat is generated. Production of heat occurs primarily when microwaves either cause ions to accelerate and collide with other molecules or cause dipoles (such molecules as water, which have a net polarity) to try to rotate and line up with the rapidly alternating (915 or 2450 million times per second) electrical field. Polyatomic molecules may or may not act as dipoles, depending on their structures. This "coupling" of energy into materials to be heated leads to very efficient and rapid heating. Foods can be heated at one-quarter their normal heating time or less (Goldblith, 1966; Goldblith and Decareau, 1973; Schiffmann, 1979).

At the same time, development of microwave heating has not been without problems. Uneven heat distribution caused by standing wave patterns in ovens has been described (Craven and Lillard, 1974; Ringle and David, 1975). Other problems have included lack of browning and crispness in foods cooked by microwaves, selection of proper packaging, and development of accurate heating instructions to insure acceptable products. Paramount among these problems is the possibility of consumer abuse and the potential for health hazards from foods improperly cooked in microwave ovens.

The present article is not intended to be a comprehensive review of all research concerned with microwave irradiation. Instead, it is an attempt to familiarize the reader with the more pertinent information regarding the effects of microwave irradiation on microorganisms. Reference will also be made to other biologic systems and to the current controversy surrounding the mode of action of microwave irradiation. Proceedings of conferences concerning the biologic effects and health implications of microwave irradiation and the biologic effects of nonionizing irradiation have been published (Cleary, 1970; Tyler, 1975). A symposium on the practical applications of microwave energy has also been published (Fung and Cunningham, 1980). The reader is advised to consult these references for a more generalized review of literature involving microwave irradiation of biologic systems and foods.

II. Effects of Microwave Irradiation on Microorganisms

Microwave irradiation of microorganisms was studied infrequently prior to 1950, and most of the work was done in the radiofrequency range. Ingram and Page (1953) tested *Saccharomyces cerevisiae*, *Escherichia coli*, tobacco mosaic virus (TMV), and bacteriophage T_4 in high-frequency voltage fields. When samples were irradiated at 10 MHz or at 20 MHz for up to 12 minutes, no significant effects on viability of the two microorganisms and no effects in either plant lesions or plaque-forming ability were observed.

Cultures of *E. coli* have been used in microwave pasteurization experiments (Brown and Morrison, 1954). Both constant and pulsed irradiation of cultures resulted in a 30% loss of viability. Cultures diluted in sterile distilled water (1:100) and 5% NaCl lost between 55 and 75% viability. Unfortunately, the authors used three different powers (55 V/cm, 464 V/cm, and 548 V/cm) to irradiate the cultures, so no correlation was possible; moreover, the authors claimed that they had probably introduced other variables because they were electrical engineers and not bacteriologists.

Studies were later specifically involved with characterizing effects of microwaves on microorganisms (Goldblith and Wang, 1967). Cultures of *E. coli* and *Bacillus subtilis* were used to compare the effects of conventional thermal reactions to those of microwave energy at 2450 MHz. Samples were placed in an oven for designated times, diluted with phosphate buffer, and then serially diluted for plate counts. The results with *E. coli* agreed with a calculated hypothesis that high death rate would be seen once lethal temperatures were reached. Results of tests for viability of *B. subtilis* spores also showed identical death curves compared with those obtained by conventional heat-kill methods. The authors concluded that death was solely from heat produced by the microwave irradiation and that no destruction was possible below those lethal temperatures.

Delaney *et al.* (1968), in an attempt to determine the effects of microwaves other than those of heating, irradiated spores of *Bacillus stearothermophilus* and *Aspergillus niger* on paper strips and discs without any water. These organisms were also irradiated in five different media (custard, 0.1 M sucrose, reconstituted dry milk, physiological saline, and distilled water). The results indicated a high viability for microorganisms irradiated on the dry strips. No adverse effects on *B. stearothermophilus* and only slight decreases in *A. niger* occurred as a result of 2450-MHz irradiation. However, significant decreases in viability with large amounts of cellular destruction were seen in the five media tested. The authors concluded that survival of the two organisms in heated samples in the 2450-MHz treatment did not differ significantly from survival in conventionally heated samples.

Lechowich *et al.* (1969) investigated the effects of heat and microwave energy at 2450 MHz on viability of *Streptococcus faecalis* and *S. cerevisiae* and on respiratory rate changes in *S. cerevisiae*. Modifications of the Hotpoint Microwave Oven used in this study were made to control temperature. Suspensions of 10^8 and 10^9 cells in 6-ml samples were irradiated and plated. Continuous application of microwaves to microbial suspensions appeared to produce no lethal effects other than those produced by heat. Respiration rates of microwave-exposed *S. cerevisiae* were directly related to decreases in viable count produced by increased microwave exposure times. The authors concluded that all effects were solely caused by thermal inactivation generated during the irradiation periods, and that respiratory damage was related to viability loss.

Roberts (1972) also observed the effects of microwave irradiation on vegetative cells and ascospores of *S. cerevisiae*. Control of temperature during microwave irradiation resulted in 8 minutes for vegetative cell death and 12 minutes for ascospore death at 33°C, whereas only 4 minutes were required for death of both by conventional methods. With no control of temperature, the thermal death point was reached in less than 30 seconds for both forms of the yeast when suspended in 5 ml of medium.

Correlli *et al.* (1977) studied the effects of 2.6–4.0 GHz microwave irradiation on *E. coli* B, by using measurement of the colony-forming ability (CFA) of the cells and the alterations in the molecular structure by comparing the infrared spectrum of irradiated and nonirradiated cell cultures. At absorbed power levels of 20 mW in 1 ml of cellular suspension for 10-12 hour exposures, no effects were observed on either the molecular structure or the CFA of the cells.

The conclusion of all these researchers was that cell death was solely the result of heat produced by the microwave irradiation and that little or no destruction was possible below these lethal temperatures.

Some reports, in contrast, have indicated that microwave irradiation had no effects, either thermal or nonthermal, on microorganisms. Hamrick and Butler (1973) tested four strains of *E. coli* and one of *Pseudomonas aeruginosa* at 2450 MHz for 12 hours at an intensity of 60 mW/cm^2. Suspensions of 50 ml were irradiated and then plated. Growth curves showed no differences between the control and irradiated cells in any of the different species. Absorbed doses determined at 41, 259, and 450 mW/cm^2 gave similar results.

Carroll and Lopez (1969) studied the effects of irradiating *S. cerevisiae*, *E. coli*, and *B. subtilis* with radiofrequency waves (60 MHz). No effects on the viability of any of the cultures were observed when suspended in 0.06 M citrate-disodium phosphate buffer at various pH ranges. Similar tests in different liquid foods (reconstituted and fresh orange juice, tomato juice, and milk) failed to show any selective killing effect except in various dilutions of ethanol (2-10%), where a synergistic killing effect was observed at 48.8°C with *S. cerevisiae*.

Soil samples containing *E. coli*, spores of *B. subtilis*, various actinomycetes, and fungi were irradiated at 2450 MHz, with up to 40,000 J/cm^2. Growth curves and viability studies indicated little or no effects of the irradiation upon the populations until microbial cells were removed from their natural environment, and then these effects were destructive. A "heat-shock" effect on spores was observed (Vela *et al.*, 1976).

Examination of particular components of microbial cells exposed to microwave irradiation at different energy levels has been reported. Absorption spectra for DNA and RNA, as well as protein, RNA, and DNA synthesis in *E. coli* B, *Enterobacter aerogenes*, and *E. coli* B/r treated with microwave irradiation were determined (Webb and Booth, 1969). Results of studies using ^{14}C-labeled nutrients indicated that decreased metabolic uptake of ^{14}C labels at 66, 71, and 73 GHz occurred, whereas increased uptake occurred at 68 GHz. Each peak or depression for rate of metabolic change was identified with one or more absorption maxima of the different compounds. Results of another study indicated that increased RNA synthesis between 71 and 129 GHz but decreased protein and DNA synthesis and lowered cell growth occurred following irradiation of cells, when measured at 5-minute intervals postirradiation (Webb, 1975). Temperature was not considered a factor.

Microwave frequencies inhibited or stimulated protein, DNA, and RNA synthesis and cell growth, and both the frequencies absorbed and those which formed set series, each separated by a constant value, were controlled by the nutrition supplied in glucose, succinate, or chemically defined NH_4^+-salts media.

Webb and Dodds (1968) stated that *E. coli* exposed to microwave irradiation (136 GHz) at a constant temperature of 25°C did not divide, but no lethality occurred (as measured by plate counts) after 4 hours of incubation. These investigators concluded that the irradiation injured the cells, slowing down their rate of division and specifically inhibiting some undetermined metabolic processes in the early stages of growth of the cell.

Another study examined the effects of microwave irradiation to determine whether any genetic mutation could be induced in *E. coli* mutants deficient in excision and recombinant repair mechanisms. Berteaud *et al.* (1975) exposed cells to 70–75 GHz of irradiation for 15 minutes. No new mutants resulting from irradiation or lesions were seen in DNA accessible to the known repair mechanisms. Cultures exposed for 30 minutes at 10 mW/cm^2 also showed little effect of the irradiation exposure.

Wellman (1978) studied the effects of microwave heating on growth of *Staphylococcus aureus* and subsequent stability of nuclease and enterotoxin B. No viable cells could be detected after 18 minutes of heating. Nuclease, either crude or purified, was unaffected by the microwave heating process. Enterotoxin B levels were reduced with purified preparations being inactivated more rapidly than crude preparations. However, after 15 minutes of heating, 0.1 µg of toxin per gram of food product could still be detected.

Because water is dipolar in nature, its presence or absence may determine to a large degree the effectiveness of microwave irradiation upon microorganisms. For example, suspensions of *Clostridium sporogenes* spores were more susceptible to microwave heating than to conventional heating (Grecz *et al.*, 1964); however, dried spores were not affected. Similar results were obtained with spores of *A. niger*, *B. stearothermophilus* (Delaney *et al.*, 1968), and *Bacillus cereus* (Dreyfuss, 1978). In addition, lyophilized cultures of *E. coli*, *S. aureus*, and *Salmonella enteritidis* were unaffected by microwave treatment (Dreyfuss, 1978).

Page and Martin (1978) found that 5-, 2-, and 0-log unit reductions of viable cells occurred when air-dried films of *E. coli*, *S. cerevisiae*, and *B. subtilis* spores, respectively, were exposed to 2450-MHz irradiation for 10 minutes. Suspensions of cells or spores in phosphate buffer treated under similar conditions showed an 8-log unit cycle reduction of *S. cerevisiae* within 30 seconds, *E. coli* within 45 seconds, and *B. subtilis* spores by 10 minutes of exposure.

In a recent study (Dreyfuss *et al.*, 1980), cultures of *E. coli*, *B. cereus*, *S.*

aureus, and *S. enteritidis* were exposed to microwave irradiation in a microwave oven and examined for survival. Identical experiments were performed in a conventional heating system. Although generally similar results were seen in comparison of microwave and conventional heating systems when approaching the thermal death points, a significant enhancement in cell numbers ($P < 0.01$) was found after 20 seconds of exposure to microwave irradiation when cultures were grown in nutrient broth (Fig. 1). This enhancement was sustained beyond 2 hours postirradiation but was lost when cultures were suspended in phosphate buffer instead of nutrient broth.

Cultures of these four microorganisms were also exposed to microwave irradiation in the presence of reducing, oxidizing, and radiosensitizing agents. Cells were irradiated in media containing dithiothreitol, thioglycolate, cysteine, methionine, glutathione, potassium permanganate, and *p*-nitroacetophenone. The pH of culture media was also adjusted to 6.0 and 8.0 in order to determine the effects of acidic and alkaline environments upon cells exposed to microwave irradiation. The presence of reducing, oxidizing, and radiosensitizing compounds resulted in increases or decreases in viability of irradiated cultures. When cultures were irradiated in nutrient broth at an acidic pH, a significant reduction ($p < 0.01$) of growth was observed as measured by decreases in cell population (Fig. 2).

We have also recently compared the efficiencies of microwave irradiation and conventional water bath treatment for heat activation of *Bacillus* spores (Chipley *et al.*, 1980). Strains of *B. subtilis, B. licheniformis, B. cereus, B. brevis,* and *B. stearothermophilus* were used in these experiments. Spore suspensions from each of the above strains were either exposed to microwave irradiation for 30 seconds (the time required for the internal temperature of the suspensions to reach 60°C) or conventionally heated in a water bath for 60 minutes at 60°C. These spore suspensions, as well as untreated controls, were then immediately plated on nutrient agar, incubated, and colony-forming units were determined.

Ten representative colonies of each strain from each treatment (irradiated, conventionally heated, and control) were picked from plates and inoculated into 10 tubes, each containing 10 ml of nutrient broth, and incubated. Several biochemical media were subsequently inoculated with cultures from these isolates to determine if any mutations had occurred as a result of microwave irradiation or conventional heat treatment. The results of these studies indicated that microwave irradiation was as effective and, in some cases, more effective than conventional heat treatment for activation of *Bacillus* spores. No changes in biochemical activity were observed for either of these treatments. Thus, microwave irradiation could be employed as an effective method for heat activation of spores of this genus.

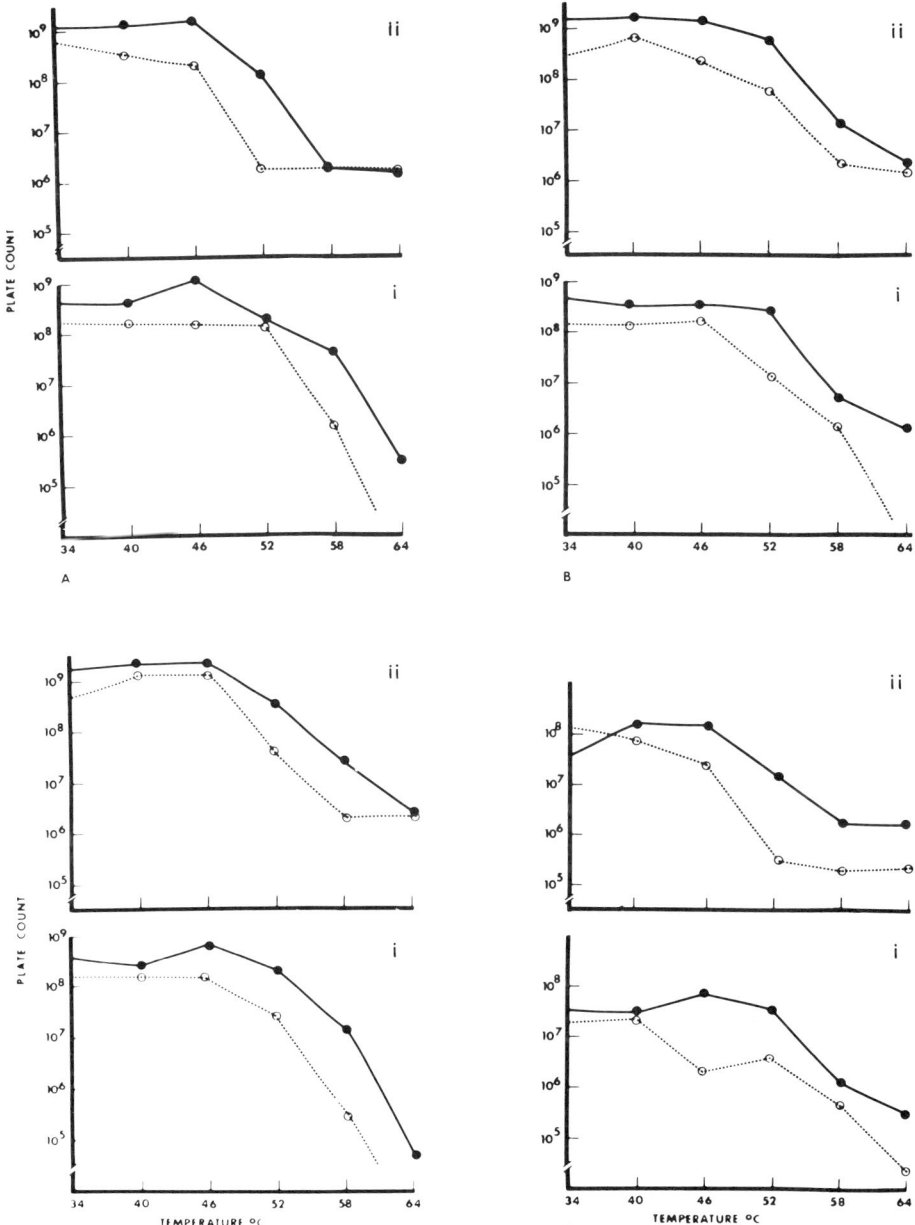

FIG. 1. Effects of microwave and conventional heating upon survival of microorganisms. (——) Microwave, (· · · ·) water bath; (i) immediately after irradiation, (ii) 2 hours postirradiation. (A) *Staphylococcus aureus*; (B) *Escherichia coli*; (C) *Salmonella enteritidis*; (D) *Bacillus cereus*. From Dreyfuss et al. (1980).

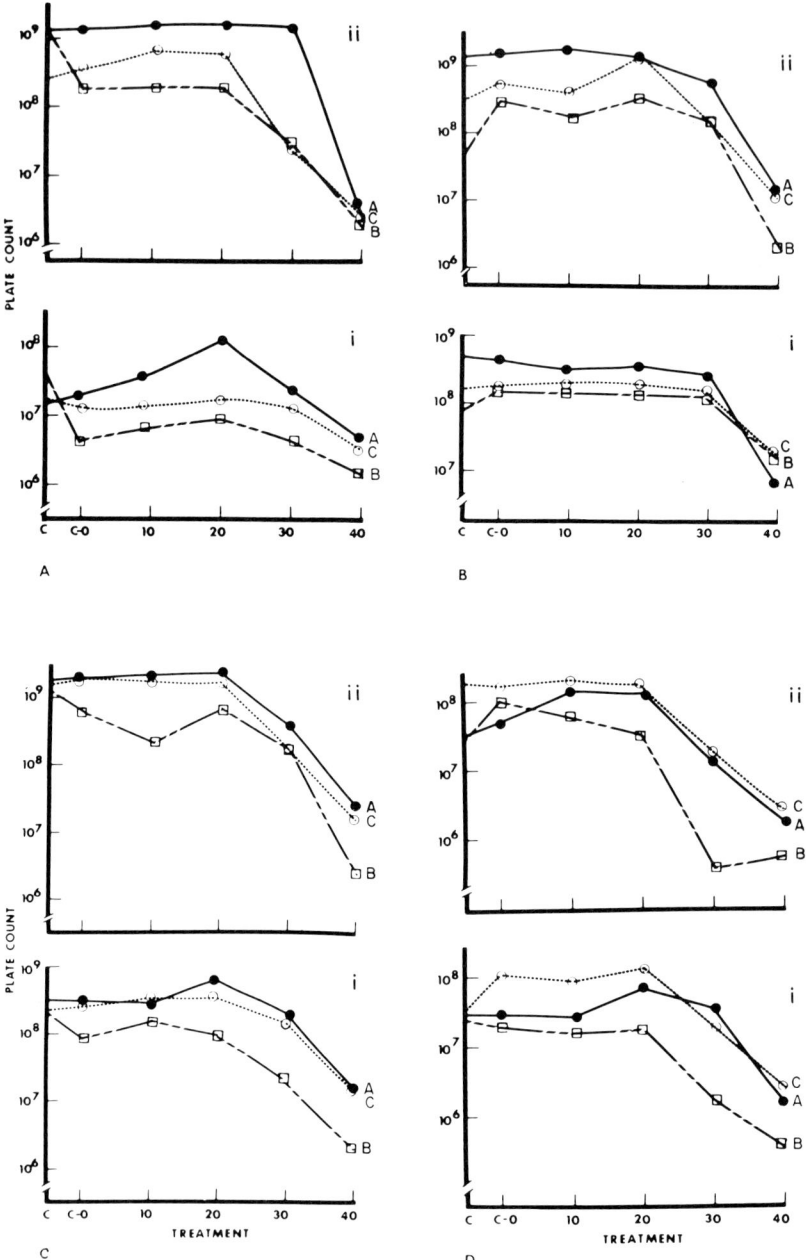

FIG. 2. Influence of pH on survival of microorganisms exposed to microwave irradiation. (A) *Staphylococcus aureus*, (B) *Escherichia coli*, (C) *Salmonella enteritidis*, (D) *Bacillus cereus*. (i) 1 hour postirradiation, (ii) 2 hours postirradiation; curve A (●), nutrient broth adjusted to pH 7.0; curve B (□), nutrient broth adjusted to pH 6.0; curve C (○), nutrient broth adjusted to pH 8.0; Treatment legend: C, cultures with no pH adjustment or irradiation; C-O, cultures with pH adjustment but no irradiation; 10-40, irradiation time in seconds. From Dreyfuss *et al.* (1980).

III. Effects of Microwave Irradiation on Other Biologic Systems

Other biologic systems have been tested for the effects of microwave irradiation. Horseradish peroxidase, one of the more heat-resistant enzymes known and an indicator of the end point of pasteurization, was irradiated between 5 and 40 minutes (2450 MHz, 62.5–375 W/cm^3). Although power densities greater than 60 W/cm^3 for more than 20 minutes inactivated 96.2% of the enzyme present, the authors concluded that this power range was too high for practical use (Henderson et al., 1975).

Studies involving microwave irradiation of other enzymes have yielded similar results (Ward et al., 1975). Yeagers et al. (1975) irradiated lysozyme and trypsin at 10^{-6} M concentration in phosphate buffer at levels of from 50 to 300 W. The temperature was allowed to rise to specific values between 30 and 95°C. Once the desired temperature was reached, the microwave power level was reduced to maintain that temperature for 1.5 hours. After this time, the enzymes were removed from the oven, allowed to cool to 25°C, and then assayed. The authors found essentially identical enzymatic activities when compared to conventionally heated solutions of the enzymes.

Tuengler et al. (1979), using an irradiation intensity near 10 mW/cm^2 and a frequency of 40–115 GHz, studied the effects of microwave irradiation on the activity of alcohol dehydrogenase and the binding of oxygen by hemoglobin. No effects of irradiation were observed in either system.

Belkhode et al. (1974a,b) studied the effect of 2.8 GHz microwave irradiation in vitro on human serum glucose 6-phosphate dehydrogenase, lactate dehydrogenase, and acid and alkaline phosphatases. Enzymes were irradiated at 37, 46.7, and 49.7°C for 4.5 or 18.5 minutes. There was a reduction in enzymatic activity at 49.7°C for both time exposures, but no effects at the lower temperatures. The authors concluded that no athermal effects were seen.

Researchers examined in vivo granulocyte system reactions in rabbits to microwave irradiation by challenging the system with acute staphylococcal infections (Szmigielski et al., 1975). Rabbits were irradiated at 3 mW/cm^2 for 6 hours daily for up to 3 months. Blood samples for the various tests were taken from irradiated and nonirradiated rabbits, which served as controls. Mature granulocyte reserves and bone marrow reserve pools were lowered, and their ability to combat the infection was decreased.

Low electrical current densities and frequencies generated by microwave irradiation (1.6–12 GHz) were used to examine differences in the Na–K ATPase from guinea pig brain, oxidative phosphorylation in rat liver mitochondria, and frog skin ion transport (Straub and Carver, 1975). Altered potential differences were observed, but few differences in electrical current across frog skin were seen. Increases in frequencies from 10 Hz to 100 Hz produced a peak of the electrical current at 45 Hz. No effects were seen on inorganic phosphate release by Na–K ATPase, whereas energy potentials were changed, at low, medium, or high current densities between 20 Hz and 20 KHz. Using a Clarke oxygen electrode, assays for adenosine diphosphate (ADP) and oxygen were made with irradiated and nonirradiated mitochondria. No changes were seen when ADP/O ratios were compared, suggesting no effects of irradiation.

Walker *et al.* (1974) studied the effects of 2450 MHz of microwave irradiation on bacteriophage T_4 r_{II} of *E. coli*. It was proposed that the irradiation would disrupt tail fibers of T_4 r_{II} by either an electrical–chemical change or some other mechanism such that plaque formation would be altered. However, no changes in plaque-forming ability or infection were observed.

In addition to the above studies, microwave irradiation has also been employed to study kinetics of the photoinduced electron paramagnetic resonance signal in *Rhodospirillum rubrum* (Corker and Sharpe, 1974), to determine the electron spin relaxation of iron–sulfur proteins (Rupp *et al.*, 1978), to quantitate metals in metalloenzymes (Kawaguchi and Vallee, 1975), to study the growth of cells in tissue culture (Peters *et al.*, 1979), to stimulate macrophage and T lymphocyte systems (Szmigielski et al., 1978), to decontaminate media and cultures (Latimer and Matsen, 1977), to sterilize (Foster, 1968; Boucher, 1972), and to disinfect seeds (Seaman and Wallen, 1966; Hankin and Sands, 1977).

IV. Thermal versus Nonthermal Effects of Microwave Irradiation

A great deal of controversy surrounds the effects of microwave irradiation. The two major considerations are based upon whether the effects are thermal or nonthermal (Tyler, 1975). All authors agree that heat does play a part in any effects seen, no matter how well minimized. General physiological effects are attributed to heat (Cleary, 1977; Milroy and Michaelson, 1971).

Initially, research involving microwave irradiation was conducted to determine the relationship of end-point temperature to viability of microorganisms following irradiation. This led to the proposal that effects of microwave irradiation upon microorganisms were thermal in nature as initially reported by Burton (1950), Jacobs *et al.*, (1950), Ingram and Page (1953), Brown and Morrison (1954), and Tomberg (1961). Other studies have been cited in Section II of this review. In addition, Lystsov *et al.* (1965) studied the effects of radiowaves on microbial vegetative cells, spores, and transforming DNA. These authors stated that they found no specific nonthermal lethal or mutagenic effects of irradiation upon these systems.

The effects of microwave irradiation upon virus replication in mammalian cells have been reported (Luczak et al., 1976; Szmigielski et al., 1977). Inhibition of replication was reportedly caused by thermal effects.

Vela and Wu (1979) exposed various bacteria, actinomycetes, fungi, and bacteriophages to microwaves of 2450 ± 20 MHz in the presence and in the absence of water. Microorganisms were inactivated only when in the presence of water, and dry or lyophilized organisms were not affected even by extended exposures. These authors emphatically stated that their data proved that microorganisms were killed by "thermal effect" only and that, most likely, there was no "nonthermal effect"; cell constituents other than water did not absorb sufficient energy to kill microbial cells. However, these authors did concede that "if there is a nonthermal effect, it is not a bactericidal one, although the possibility exists that water may be necessary to potentiate the nonthermal effect" (p. 553).

Nevertheless, much effort has been devoted to studies that have attempted to demonstrate the existence of nonthermal effects of microwave irradiation by containing end-point temperatures below thermal death points of the microorganisms under investigation. Early reports by Beckwith and Olsen (1931), Fabian and Graham (1933), Yen and Lui (1934), Fleming (1944), Nyrop (1946), Carpenter (1958), Susskind and Vogelhut (1959), and Van Ummerson (1961) initiated this hypothesis for both procaryotic and eucaryotic organisms.

Reports have also shown or suggested that there are nonthermal microwave effects in terms of the energy required to produce various types of molecular transformation and alterations. These include the breaking of hydrogen bonds and secondary linkages, alteration of protein hydration (Kalant, 1959), release of bound water (Ballario et al., 1975), electron tunneling (Cope, 1976), and realignment of polar molecules (Crews and Goertz, 1973).

Nonthermal effects of 2450-MHz microwave energy were postulated by Olsen (1965), who reported that conidia of *Penicillium* sp. and of *A. niger* were inactivated by microwave irradiation. This author stated that "the excellent results with microwaves are probably not due to conventional thermal kill" (p. 53). However, temperature control below 65°C was not achieved in these experiments, and no conventionally heated controls were employed.

Culkin and Fung (1975) demonstrated that *E. coli* and *Salmonella typhimurium* could not survive in soups cooked at 915 MHz by microwave irradiation. They found that microbial destruction occurred at lower temperatures and shorter time periods when compared to conventional heating methods. They also found varied effects based upon the intensity of the field strength at the top and bottom of the soup containers. They postulated that factors other than thermal effects might be involved in the effects of microwave irradiation on microorganisms.

The interdependence of heat and microwave effects have been demonstrated in inactivation studies of spores of B. subtilis (Wayland et al., 1977). Differences in inactivation rates shown by microwave irradiation versus conventional heating of the spores indicated a nonthermal effect, and the authors concluded that heat and electromagnetic effects were strongly interdependent. Nonthermal effects of microwave irradiation have also been observed in yeasts (Grundler and Keilmann, 1978).

Amannur (1979) studied the effects of microwave heating (2450 MHz) and conventional heating upon several microbial genera and concluded that conventional heating was more effective in reducing numbers of these organisms. Although the theory of nonthermal effects of microwave heating was questioned, the author conceded that at relatively lower temperatures, microwave heating was more lethal to microorganisms.

Moore et al. (1979) observed that when virulent cells of Agrobacterium tumefaciens strain B6 were exposed to low-level microwave irradiation at a frequency of 10,000 MHz and an intensity of 0.58 mW/cm^2 for 30–120 minutes, a 30–60% decrease occurred in their ability to produce tumors on potato and turnip disks. This microwave exposure did not affect the viability of these bacteria or their ability to attach to a tumor-binding site; neither did it induce thermal shock. Loss of virulence was reversible within 12 hours. The authors concluded that the exact nature of this effect was unknown, as is the case with most nonthermal microwave effects. However, the possibility that one or more metabolic processes affecting virulence were temporarily altered by microwave exposure could not be excluded.

Dreyfuss and Chipley (1980) attempted to characterize some of the effects of sublethal microwave irradiation in cells of S. aureus. Cultures were exposed to microwave irradiation. The activity of glucose 6-phosphate dehyd- 40 seconds. The effects of a conventional heat treatment were also compared by placing flasks containing cultures in a boiling water bath for the amount of time required to reach equivalent temperatures found in cultures exposed to microwave irradiation. Control, microwave-treated, and conventionally heat-treated cultures were centrifuged; pellets were resuspended in distilled water; and the resulting suspensions were passed through a French Pressure Cell. Cell lysates and walls were then isolated and assayed for enzymatic activity. Thermonuclease production was also determined at various levels of exposure of cells to microwave irradiation. When compared to control cells, activities of malate and α-ketoglutarate dehydrogenases, cytochrome oxidase, and cytoplasmic adenosine triphosphatase (ATPase) were higher in microwave-treated cells (Table I). Membrane ATPase, alkaline phosphatase, and lactate dehydrogenase activities were unaffected when cells were exposed to microwave irradiation. The activity of glucose 6-phosphate dehydrogenase was decreased by exposure of cells to microwave irradiation. In

TABLE I

EFFECTS OF MICROWAVE IRRADIATION AND CONVENTIONAL HEATING UPON ENZYMATIC ACTIVITY OF *Staphylococcus aureus*[a]

Enzyme	Specific activity[b]			Ratios	
	Control[c]	Irradiated[d]	Heat treated[e]	Irradiated/control	Heat treated/control
Glucose 6-phosphate dehydrogenase	6.16	4.57	8.51	0.74	1.38
Malate dehydrogenase	3.60	9.82	6.64	2.72	1.84
α-Ketoglutarate dehydrogenase	29.34	51.82	7.10	1.76	0.24
Lactate dehydrogenase	2.89	2.81	1.50	0.97	0.52
Alkaline phosphatase	16.27	16.31	14.61	1.00	0.90
Cytochrome oxidase	0.45	0.55	—[f]	1.23	—
ATPase[g]					
Cell lysate	0.81	1.26	1.35	1.55	1.66
Cell walls	0.28	0.29	—	1.06	—

[a] From Dreyfuss and Chipley (1980).
[b] Expressed as milliunits per milligram total protein.
[c] Cells incubated at 34°C.
[d] Cells irradiated for 20 seconds; internal temperature of media, 46°C.
[e] Cells conventionally heat treated for 39 seconds; internal temperature of media, 46°C.
[f] Not assayed.
[g] Expressed as moles of inorganic phosphate liberated per gram total protein.

conventionally heated cells, activities of glucose 6-phosphate and malate dehydrogenases and cytoplasmic ATPase increased; activities of α-ketoglutarate and lactate dehydrogenases decreased; and alkaline phosphatase activity remained unaffected. Increased levels of thermonuclease activity were observed when cells were exposed to microwave irradiation for 10 or 20 seconds (Table II). Data indicated that microwave irradiation affected S. aureus in a manner that could not be explained solely by thermal effects.

Theories involving nonthermal effects of microwave irradiation have been reported. Barnes and Hu (1977) presented a mathematical model for changes by which nonthermal effects of microwave irradiation could result in ion shifts across membranes and reorientation of long-chain molecules. Straub and Carver (1975) stated that increased active ion transport could be responsible for increases in potential differences and electrical current across membranes. They based this proposal on data that indicated an irreversible damage to membrane structures, with a decrease in ion transport and an increase in passive permeability to ions.

Another model was described by Cope (1976), who proposed that superconductivity, the passage of electron current without the generation of heat, may play a biologic role in microwave irradiation. Various observations have been made of changes in rates of metabolism of growing tissues in magnetic fields, electron tunneling in living systems that respond to very low magnetic fields, and nonlinear Arrhenius plots as reviewed by Cope (1976). Ahmed *et al.* (1975) reported that the enzyme lysozyme might be made superconductive from observations of dielectric changes in varied magnetic fields which were dependent upon temperature.

TABLE II
EFFECTS OF MICROWAVE IRRADIATION ON THERMONUCLEASE ACTIVITY OF S. *aureus*[a]

	Diameter of zones (mm)				
Time (hours)	Control	10 seconds[b]	20 seconds	30 seconds	40 seconds
0	0	0	0	0	0
0.5	3.25	3.50	3.00	0	0
1	4.75	4.40	4.25	0	0
1.5	5.50	6.25	5.75	2.00	0
2	6.20	6.50	6.00	2.00	0
2.5	6.75	6.75	6.25	3.00	0
3	7.50	7.25	7.25	3.75	0

[a] From Dreyfuss and Chipley (1980).
[b] Length of time of microwave irradiation.

In summary, microbiologic studies involving microwave irradiation have resulted in the following observations: (1) failure to distinguish microwave effects from thermal effects, (2) demonstration of microwave effects regardless of temperature, and (3) failure to observe any effects, either thermal or nonthermal, from microwave irradiation. Only with the passage of time and intensification of research involving microwave irradiation may the controversy surrounding the utilization of this technique be resolved.

ACKNOWLEDGMENTS

The author expresses sincere gratitude to M. S. Dreyfuss, whose M.S. thesis formed a significant part of this review; to the Ohio State University, for partial funding of research; and to Drs. D. Y. C. Fung and F. E. Cunningham, for access to their preprinted manuscript.

REFERENCES

Ahmed, N. A. G., Calderwood, J. H., Frolich, H., and Smith, C. W. (1975). *Phys. Lett. A* **53**, 129.
Amannur, I. (1979). M.A. Thesis, Mankato State University, Mankato, Minnesota.
Ballario, C., Bonincontro, A., and Cametti, C. (1975). *J. Colloid Interface Sci.* **51**, 191.
Barnes, F. S., and Hu, C. L. J. (1977). *IEEE Trans. Microwave Theory Tech.* **25**, 742.
Beckwith, T. D., and Olsen, A. R. (1931). *Proc. Soc. Exp. Biol. Med.* **29**, 362.
Belkhode, M. L., Johnson, D. L., and Muc, A. M. (1974a). *Health Phys.* **26**, 45.
Belkhode, M. L., Muc, A. M., and Johnson, D. L. (1974b). *J. Microwave Power* **9**, 23.
Berteaud, A. J., Dardalhon, M., Rebeyrotte, N., and Averback, D. (1975). *C. R. Hebd. Seances Acad. Sci.* **281**, 843.
Boucher, R. M. (1972). *Am. J. Hosp. Pharm.* **29**, 661.
Brown, G. H., and Morrison, W. C. (1954). *Food Technol.* **8**, 361.
Burton, H. (1950). *Nature (London)* **166**, 434.
Carpenter, R. L. (1958). *In* "Proceedings of the Second Annual Tri-Service Conference on Biological Effects of Microwave Energy," pp. 146–168. US Department of Commerce, Springfield, Virginia.
Carroll, D. E., and Lopez, A. (1969). *J. Food Sci.* **34**, 320.
Chipley, J. R., Rohlfs, L. A., and Ford, C. L. (1980). *Microbios* (submitted).
Cleary, S. F., ed. (1970). "Proceedings of the Biological Effects and Health Implications of Microwave Radiation, Richmond, Virginia, 1969," Bur. Radiol. Health Rep. 70-2. US Department of Health, Education, and Welfare, Washington, D.C.
Cleary, S. F. (1977). *Crit. Rev. Environ. Control* **7**, 121.
Cope, F. W. (1976). *J. Microwave Power* **11**, 267.
Corker, G. A., and Sharpe, S. A. (1974). *Photochem. Photobiol.* **19**, 443.
Correlli, J. C., Gutmann, R. J., Kahzi, S., and Levy, J. (1977). *J. Microwave Power* **12**, 141.
Craven, S., and Lillard, H. (1974). *J. Food Sci.* **39**, 211.
Crews, G. W., and Goertz, G. E. (1973). *Poult. Sci.* **52**, 1496.
Culkin, K. A., and Fung, D. Y. C. (1975). *J. Milk and Food Technol.* **38**, 8.
Delaney, E. K., Van Zonte, H. J., and Hartman, P. A. (1968). *Microwave Energy Appl. Newsl.* **1**, 11.

Dreyfuss, M. S. (1978). M.S. Thesis, Ohio State University, Columbus.
Dreyfuss, M. S., and Chipley, J. R. (1980). *Appl. Environ. Microbiol.* **39**, 13.
Dreyfuss, M. S., Chipley, J. R., and Kolodziej, B. J. (1980). *Appl. Environ. Microbiol.* (submitted for publication).
Fabian, F. W., and Graham, H. T. (1933). *J. Infect. Dis.* **53**, 76.
Fleming, H. (1944). *Electr. Eng. (Am. Inst. Electr. Eng.)* **63**, 18.
Foster, H. L. (1968). *Lab. Anim. Sci.* **18**, 356.
Fung, D. Y. C., and Cunningham, F. E. (1980). *J. Food Prot.* (in press).
Goldblith, S. A. (1966). *Adv. Food Res.* **15**, 277–301.
Goldblith, S. A., and Decareau, R. V. (1973). "An Annotated Bibliography on Microwaves: Their Properties, Production and Applications to Food Processing." MIT Press, Cambridge, Massachusetts.
Goldblith, S. A., and Wang, D. I. C. (1967). *Appl. Microbiol.* **15**, 1371.
Grecz, N., Walker, A. A., and Anellis, A. (1964). *Bacteriol. Proc.* p. 145.
Grundler, W., and Keilmann, F. (1978). *Z. Naturforsch. Teil C* **33**, 15.
Hamrick, P. E., and Butler, B. T. (1973). *J. Microwave Power* **8**, 227.
Hankin, L., and Sands, D. C. (1977). *Phytopathology* **67**, 794.
Henderson, H. M., Hergenroeder, K., and Stuchly, S. S. (1975). *J. Microwave Power* **10**, 27.
Ingram, M., and Page, L. J. (1953). *Proc. Soc. Appl. Bacteriol.* **16**, 69.
Jacobs, S. E., Thornley, M. J., and Maurice, P. (1950). *Proc. Soc. Appl. Bacteriol.* **2**, 161.
Kalant, H. (1959). *Can. Med. Assoc. J.* **81**, 575.
Kawaguchi, H., and Vallee, B. L. (1975). *Anal. Chem.* **47**, 1029.
Latimer, J. M., and Matsen, J. M. (1977). *J. Clin. Microbiol.* **6**, 340.
Lechowich, R. V., Beuchat, L. R., Fox, K. I., and Webster, F. H. (1969). *Appl. Microbiol.* **17**, 106.
Luczak, M., Szmigielski, S., Janiak, M., Kobus, M., and de Clerq, E. (1976). *J. Microwave Power* **11**, 173.
Lystsov, V. N., Frank-Kamenetskii, D. A., and Shchedrina, M. V. (1965). *Biofizika* **10**, 114.
Milroy, W. C., and Michaelson, S. M. (1971). *Health Phys.* **20**, 567.
Moore, H. A., Raymond, R., Fox, M., and Galsky, A. G. (1979). *Appl. Environ. Microbiol.* **37**, 127.
Nyrop, J. E. (1946). *Nature (London)* **157**, 51.
Olsen, C. M. (1965). *Food Eng.* **37**, 51.
Page, W. J., and Martin, W. G. (1978). *Can. J. Microbiol.* **24**, 1431.
Peters, W. J., Jackson, R. W., and Iwano, K. (1979). *J. Surg. Res.* **27**, 8.
Ringle, E. C., and David, B. D. (1975). *Food Technol.* **29**, 46.
Roberts, P. C. B. (1972). *J. Sci. Food Agric.* **23**, 544.
Rupp, H., Rao, K. K., Hall, D. O., and Cammack, R. (1978). *Biochim. Biophys. Acta* **537**, 255.
Schiffmann, R. F. (1979). *Food Proc. Dev.* **13**, 38.
Seaman, W. L., and Wallen, V. R. (1966). *Can. J. Plant Sci.* **47**, 39.
Straub, K. D., and Carver, P. (1975). *Ann. N.Y. Acad. Sci.* **247**, 292.
Susskind, C., and Vogelhut, P. O. (1959). *In* "Proceedings of the Third Annual Tri-Service Conference on Biological Effects of Microwave Radiation Equipment," pp. 46–53. US Department of Commerce, Springfield, Virginia.
Szmigielski, S., Jeljaszewicz, J., and Wiranowska, M. (1975). *Ann. N.Y. Acad. Sci.* **247**, 305.
Szmigielski, S., Luczak, M., Janiak, M., Kobus, M., Laskowska, B., de Clerq, E., and de Somer, P. (1977). *Arch. Virol.* **53**, 71.
Szmigielski, S., Janiak, M., Hryniewicz, W., Jeljaszewicz, J., and Pulverer, G. (1978). *Z. Krebsforsch.* **91**, 35.

Tomberg, V. T. (1961). In "Biological Effects of Microwave Radiation" (M. F. Peyton, ed.), Vol. 1, pp. 221–228. Plenum, New York.
Tuengler, P., Keilmann, F., and Genzel, L. (1979). Z. Naturforsch. Teil C **34**, 60.
Tyler, P. E. (1975). Ann. N.Y. Acad. Sci. **247**, 6.
Van Ummerson, C. (1961). In "Biological Effects of Microwave Radiation" (M. F. Peyton, ed.), Vol. 1, pp. 201–219. Plenum, New York.
Vela, G. R., and Wu, J. F. (1979). Appl. Environ. Microbiol. **37**, 550.
Vela, G. R., Wu, J. F., and Smith, D. (1976). Soil Sci. **121**, 44.
Walker, C. M. B., McWhirter, K. G., and Voss, W. A. G. (1974). J. Microwave Power **9**, 221.
Ward, T. R., Allis, J. W., and Elder, J. A. (1975). J. Microwave Power **10**, 315.
Wayland, J. R., Brannen, J. P., and Morris, M. E. (1977). Radiat. Res. **71**, 251.
Webb, S. J. (1975). Ann. N.Y. Acad. Sci. **247**, 327.
Webb, S. J., and Booth, A. D. (1969). Nature (London) **222**, 1199.
Webb, S. J., and Dodds, D. D. (1968). Nature (London) **218**, 374.
Wellman, J. E. (1978). M.S. Thesis, Ohio State University, Columbus.
Yeagers, E. K., Langley, J. B., Sheppard, A. P., and Huddleston, G. K. (1975). Ann. N.Y. Acad. Sci. **247**, 301.
Yen, A. C., and Lui, S. (1934). Proc. Soc. Exp. Biol. Med. **31**, 1250.

Ethanol Production by Fermentation: An Alternative Liquid Fuel

N. Kosaric, D. C. M. Ng, I. Russell,[1] and G. S. Stewart[1]

*Chemical and Biochemical Engineering,
Faculty of Engineering Science,
The University of Western Ontario,
London, Ontario, Canada*

I.	Introduction	148
II.	Alcohol as a Fuel	151
III.	Overview of Ethanol Production Processes	152
IV.	Ethanol from Sugars	154
	A. Brazil's Alcohol Program	154
	B. Continuous Fermentation	157
	C. Yeasts Tolerant to High Sugar Concentration	161
	D. The "ABE" Process	162
	E. Fermentation of Pentoses	164
	F. Other Developments	164
V.	Ethanol from Starch	165
	A. Ethanol from Corn	165
	B. Integration with Processing Corn for Feed	168
	C. Ethanol from Cassava Roots	171
	D. Ethanol from Potatoes	174
	E. Ethanol from Novel Starch Sources	175
VI.	Ethanol from Cellulosic Materials	176
	A. Ethanol from Wood via Acid Hydrolysis	177
	B. Acid Hydrolysis of Cellulosic Wastes	180
	C. Enzymatic Hydrolysis of Cellulose	182
	D. *Trichoderma* Cellulase	182
	E. Pretreatment of Cellulose	186
	F. Enzymatic Hydrolysis Processes	189
	G. Fermentation of Hydrolyzate to Ethanol	194
	H. Enzymatic Hydrolysis of Cellulose Waste from the Food Processing Industry	195
	I. One-Step Process for Production of Ethanol from Cellulosic Wastes	196
	J. Comparison of Acid and Enzymatic Hydrolysis	197
	K. Future Developments	197
	L. Cellulose vs. Corn	198
VII.	Ethanol from Other Wastes	199
	A. Cheese Whey	199
	B. Coffee Waste	201
	C. Potato Peel Waste	202
	D. Pineapple Wastes	202
	E. Spent Sulfite Liquor	202

---

[1]Brewing Research and Development Department, The Labatt Brewing Company, London, Ontario, Canada.

VIII. Economic Analysis of Ethanol Production Processes 204
    A. Ethanol from Chemical Synthesis 204
    B. Ethanol from Sugar and Cassava Root 206
    C. Ethanol from Corn 207
    D. Ethanol from Wood 212
    E. Ethanol from Cellulosic Wastes 213
    F. By-product Credits 215
    G. Comparison of Various Processes 216
    H. Potential of Ethanol as a Fuel 216
IX. Energy Considerations 218
X. Summary and Conclusions 220
References 224

I. Introduction

Fossil fuels have three outstanding virtues. They are a natural storage system from which energy can be recovered at will, almost anywhere at any time. Their energy can be released at high temperature, which means that they possess a high capacity for producing work or that they have a high thermodynamic availability. Their energy is stored at high density, characteristically between about 23,300 and 55,900 kJ/kg (10,000 and 24,000 Btu/lb) (Kemp and Szego, 1976). During the past 100 years, the tremendous increase in energy availability from fossil fuels has brought about a period of extremely rapid industrial growth with great additions to our technology and improvement of our living standard.

Fossil fuels have two insurmountable drawbacks, however, which limit their production and usage. One is that the supply of many of their more desirable embodiments is approaching exhaustion, and the second is the environmental problems associated with their continued and increasing use. Therefore, fossil fuels are rapidly becoming a precious commodity, which is reflected in their increasing price. Within the past few years, the price of gasoline at the pump has tripled and coal costs have increased over 50% (Cherimisinoff and Morresi, 1976). Natural gas prices have doubled and the price of petroleum products in general has been inflated.

A long-term practical solution to this problem is to convert a major source of continuously renewable nonfossil carbon, such as organic wastes and biomass—which consists of all growing organic matter, such as plants, trees, grasses, and algae—to fuels (Anonymous, 1977, 1978a; Kohn, 1978). A renewable, nondepletable raw material that can be converted into substitute fossil fuel would assure a perpetual energy supply. As time passes and the fossil fuel shortage intensifies, renewable energy sources could eventually occupy a position of dominance.

The use of nonfossil renewable carbon as an energy source is not a new concept. Before the booming of the fossil fuels, dried buffalo chips, wood, and wood wastes were important fuel sources. Later, with the onset of the fossil fuel era, an intensive effort was mounted to develop and market a wide range of solid and liquid fuels that almost totally replaced nonfossil energy usage. Now, although the twilight years of fossil fuel supplies have not quite been reached, an effort is being made to close the cycle and return to nonfossil energy sources. The initial concentration has been in the development of waste-to-energy processes, but recently attention has been given to land-based and water-based biomass, because they represent much larger sources of energy and synthetic fuels (synfuels).

The process responsible for the production of biomass is photosynthesis. It was originally responsible for the ultimate deposition of fossil fuels, and it still constitutes the only process by which massive amounts of solar energy are captured and stored today. On a worldwide basis, it has been estimated that $132 \pm 79 \times 10^9$ metric tons ($146 \pm 87 \times 10^9$ tons) of carbon are fixed annually (Brink, 1976). Assuming lignocellulose to be 50% carbon and to have a heat of combustion of 19,800 kJ/kg (8500 Btu/lb) on an ash-free, oven-dried basis, $5.2 \pm 3.2 \times 10^{18}$ kJ ($5 \pm 3 \times 10^{18}$ Btu) are stored annually by photosynthesis. Based on these data, only 2 years are required to photosynthetically produce an equivalent in carbon to the proved and currently available categories of natural gas, natural gas liquid, crude oil, and syncrude from oil, or 8–20 years for an equivalent in fixed carbon to the estimated total remaining recoverable amounts of the four fuel categories (Brink, 1976). Stated in other terms, the amount of carbon fixed per year on a worldwide basis is in the order of 30 times greater than the total world energy production of 15.2×10^{10} kJ (14.4×10^{10} Btu) or 7.0×10^9 metric tons of coal equivalent in 1970.

At present, various species of plant biomass are being investigated as potential feedstocks for conversion to fuels. Potential field crops include sudan grass, napier grass, sweet sorghum, sugarcane, sugar beets, sunflower, and kenaf. Various types of tree crops are also being considered, such as sycamore, poplar, eucalyptus, sweetgum, and alder.

The types of conversion processes available for the production of energy products from wastes and biomass are summarized in Table I.

Figure 1 illustrates the principal routes that can be used to produce gaseous or liquid organic fuels. The most common form of gaseous fuel is synthetic natural gas (SNG), the predominant component of which is methane.

The production of liquid fuels or syncrudes from wastes and biomass is possible by several methods. One of the oldest involves alcoholic yeast fermentation for a few days to afford ethanol. Ethanol can also be converted to a wide range of derivatives. Feeds containing starchy components are suitable for alcohol production. Most processes that would use highly cellulosic biomass as a raw material would have to first hydrolyze the biomass by acids or enzymes to liberate glucose before fermentation. Simple wood distillation was used for many years for methanol production up until the early 1930s.

Another method for converting wastes and biomass to liquid fuels is by direct hydrogenation. This process is similar to hydrogasification except that the liquefaction process is conducted under less vigorous conditions so that all carbon–carbon bonds are not broken and higher yields of intermediate liquid products are formed.

Wastes and biomass can also be converted to liquid fuels by first converting the feed to synthesis gas by pyrolysis, partial oxidation, or steam reforming; the synthesis gas can then be converted to liquid hydrocarbons by the Fischer–Tropsch process. In this process, carbon monoxide is catalytically hydrogenated to a complex mixture of aliphatic hydrocarbons similar to a high-paraffin crude.

Bailie (1976) conducted a technical and economic assessment of processes available for the conversion of agricultural residues (which is applicable to biomass in general) to usable energy. He concluded that at present only

TABLE I
Summary of Nonfossil Carbon-to-Energy Processes and Primary Energy Products[a]

Conversion process	Primary energy products	
Incineration	Energy	Thermal
Separation		Steam
Pyrolysis		Electric
Hydrogenation		
Anaerobic fermentation	Solid fuel	Char
Aerobic fermentation		Combustible
Biophotolysis		
Partial oxidation	Synfuel	Methane (SNG)
Steam reforming		Hydrogen
Chemical hydrolysis		Low-Btu gas
Enzyme hydrolysis		Methanol
Other chemical conversions		Ethanol
		Hydrocarbons
	Energy intensive products	Ammonia
		Steel[b]
		Copper[b]
		Aluminum[b]
		Glass
		Other chemicals

[a] From Kemp and Szego (1976).
[b] Pertain to urban refuse and certain industrial wastes.

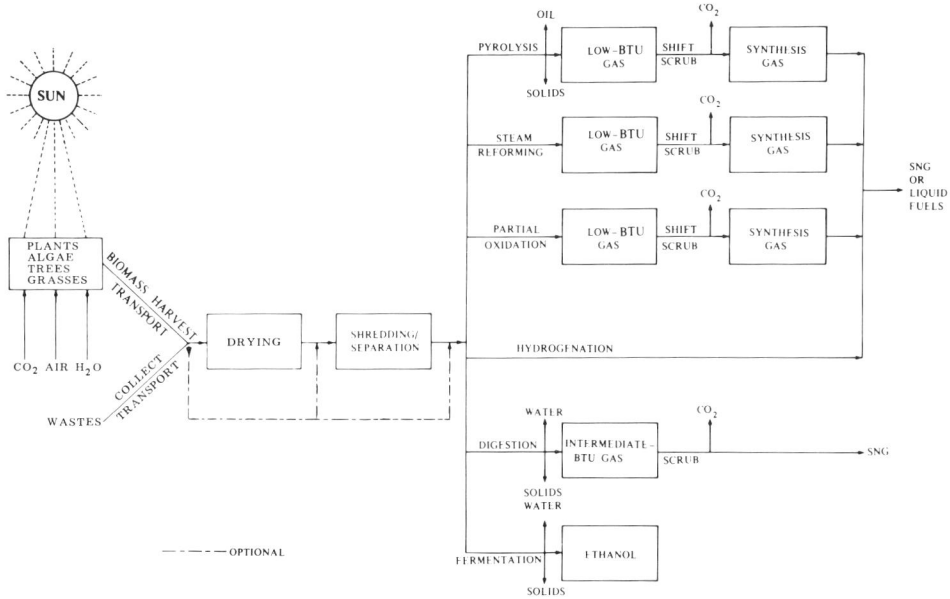

FIG. 1. Production of gaseous and liquid fuels from biomass. From Klass (1976).

anaerobic digestion, hydrolysis and fermentation to ethanol, pyrolysis to low-Btu gas, and pyrolysis to Fischer–Tropsch are technically and economically feasible.

II. Alcohol as a Fuel

The use of alcohol as a motor fuel is not a novel concept, as an ethanol-powered tractor engine was developed as early as 1890. As a motor fuel, the use of both methanol and ethanol grew to be very extensive in Europe during the pre-World War II period (Stone, 1974).

In North America, there has been little widespread use of alcohol as a motor fuel. Alcohol has been used as a special fuel for racing engines and mixtures of water and alcohol were used for injection into high-compression aircraft engines. Relatively low costs and easy availability of petroleum throughout the pre-1970 period made extensive use of alcohol uneconomical.

Today, alcohol technology is again reviving and production of alcohol as a fuel is being given full attention. Alcohol is being considered either as such or in combination with gasoline.

Various test programs are under development, such as the Gasohol Program in Nebraska and the Brazilian National Alcohol Program (Andren *et al.*, 1976; Anonymous, 1978a; Scheller, 1977a). In both programs, a blend of ethanol (up to 20%) and gasoline was tested as a substitute for 100% gasoline. Similar programs were carried out by Volkswagen (Lee *et al.*, 1976) and Shell (Davies, 1976), using methanol–gasoline blends. It was concluded that the alcohol–gasoline blend gave a better gas economy and improved emission characteristics. Tests were also conducted at Stanford University on the use of pure methanol in place of gasoline (McCloskey, 1975). Over 90% reduction in pollutant levels in some of the tests was observed; it is believed similar results will hold for ethanol.

Both ethanol and methanol are probably superior to gasoline as liquid fuels for general automotive propulsion use. They are safer to handle, burn cleaner, and significantly reduce pollution emissions from automative exhaust systems. They have an excellent octane number, around 106, and should not require the addition of tetraethyl lead for satisfactory engine performance. Also, the use of either alcohol should permit greater improvements in engine design, because an explosive mixture of alcohol vapor and air can be used under far higher compression ratios without autoignition than a similar mixture of gasoline and air. However, whereas both alcohols seem to possess certain inherent basic advantages over gasoline, if a choice must be made, ethanol is superior to methanol for a number of reasons (McCloskey, 1975). First, it gives approximately 25% more miles to the gallon than methanol, which is probably because of its higher Btu content of 29,800 kJ/kg (12,800 Btu/lb) as compared with only 22,300 kJ/kg (9600 Btu/lb) for methanol. Further, the solubility of methanol has been determined to be only 13% in regular grade gasoline and as little as 4% in an all-straight run gasoline. In addition, an ethanol content of only 10% in premium fuel at 10°C (50°F) has been found to dry up a tank with less than 0.3% water, whereas a methanol content of 25% would be required to absorb the same amount of water. Based on the above analyses, ethanol is obviously more favorable than methanol as an alternative fuel to fossil fuels.

Besides being a fuel, ethanol is also a very versatile chemical feedstock. A wide variety of chemical products can be derived from ethanol (Prescott and Dunn, 1959). With the increasing shortage of petroleum, which leads to the shortage of petrochemicals as well, ethanol is expected to play a more significant role in the future.

III. Overview of Ethanol Production Processes

Ethanol is produced commercially by chemical synthesis and fermentation. Practically all current industrial ethanol is manufactured synthetically from petroleum and natural gas. All beverage alcohol is produced by fermentation of cereal grains, molasses, and other materials with high starch and sugar contents.

Synthetic ethanol has been produced in the United Stated since 1931. Although ethanol was initially used to prepare ethylene, the wheel of time has turned a full cycle and most of today's ethanol is produced from ethylene obtained from cracked petroleum hydrocarbons. Before World War II, about 72% of ethanol production was by fermentation of molasses. Currently, over 90% of ethanol production is by synthesis from ethylene by the esterification–hydrolysis process or by direct hydration.

One commercial process is the absorption of ethylene by concentrated sulfuric acid to form ethyl sulfuric acid. This intermediate is hydrolyzed to ethanol and sulfuric acid as follows:

$$3C_2H_4 + 2H_2SO_4 \longrightarrow CH_3CH_2OSO_3H + (CH_3CH_2)_2SO_4$$

$$CH_3CH_2OSO_3H + H_2O \longrightarrow CH_3CH_2OH + H_2SO_4$$

Diethyl ether is a valuable by-product and is obtained by the following reaction:

$$CH_3CH_2OH + (CH_3CH_2)_2SO_4 \longrightarrow (CH_3CH_2)_2O + CH_3CH_2OSO_3H$$

Ethanol is also made by direct hydration of ethylene, using phosphoric acid as a catalyst at 68 atmospheric pressures (1000 psi) and 400°C:

$$C_2H_4 + H_2O \xrightarrow{H_3PO_4} CH_3CH_2OH$$

Commercial ethanol contains 95.6% ethanol by weight, which is the composition of a constant boiling point (78.15°C) mixture (an azeotrope) of alcohol and water. Because distillation cannot increase this concentration directly, the addition of benzene forms a lower boiling (64.85°C) tenary mixture of benzene, water, and alcohol, which can remove the water by distillation. Benzene and alcohol produce a low-boiling binary (67.8°C); hence benzene can be separated by distillation. Anhydrous ethanol is obtained commercially in this manner as 200° proof alcohol.

During recent years, production of alcohol by fermentation on a large scale has been of considerable interest. Until the oil crisis of October 1973, only India appeared to appreciate the importance of fermentation alcohol as a strategic material in its economy. Today, projects for fermentation alcohol are moving ahead all over the world.

The most ambitious effort so far is taking place in Brazil. It is probably the largest and most advanced in the world, and it aims at having ethanol from sugarcane account for approximately 20% of Brazil's transportation fuel needs by 1980. By December 1977, the alcohol projects approved under the program numbered 150, with a combined capacity of 2.8 billion liters of ethanol, representing investments of approximately $1 billion. The 1980 goal is a total ethanol capacity of 4.7 billion liters/year, with 4 billion liters/year for fuel and the remainder for chemical feedstocks (Kohn, 1978).

The Philippines are also embarking on a program to produce ethanol for fuel from domestic sugarcane. In Thailand, a project is under consideration to utilize a presently unused 70 million liters/year distillery to produce alcohol for fuel. Also in Europe, ethanol from sugar is gaining its proponents. French sugar beet producers are proponents of such a conversion. They certainly appear to have excess output to redirect toward alcohol production; the Economic Community was expecting a 1.7 million metric ton beet sugar surplus in the 1978 crop year, most of it French (Kohn, 1978).

Fermentation for fuel draws support in the United States as well. Nebraska, for example, has for years been a proponent of blending ethanol into gasoline and is presently involved in an economic feasibility study of the concept (Anderson, 1978; Anonymous, 1978c; Scheller, 1976, 1977a,b; Scheller and Mohr, 1976, 1977). Any raw material containing hexose sugars, or materials that can be transferred into hexose sugars, can be used as fermentation substrates. For example, France and Belgium use sugar beets, Germany utilizes potatoes, other European countries, (i.e., Finland) have converted sulfite liquor and sawdust, whereas corn, sugar beet, and cane molasses are the most popular in the United States. Sawdust and wood flour require the conversion of cellulose to fermentable sugar by hydrolysis, and grains can be converted by the action of malt enzymes. In brief, the major sources for ethanol fermentation are:

1. Sugars from beet, cane, sorghum
2. Grain starch from corn, wheat, barley, and other cereals
3. Cellulosic materials
4. By-product carbohydrates from processing, e.g., whey, fruit wastes.

IV. Ethanol from Sugars

A. Brazil's Alcohol Program

The Brazilian National Alcohol Program is by far the most extensive of its kind in the world. It could provide the necessary blueprint for other countries. The program was initiated in November 1975, its basic strategy being for a total production capability of 3 million m^3 per annum of anhydrous alcohol by the early 1980s. Most of the production will be for gasoline blending, constituting up to 20% by volume of alcohol, based upon projected gasoline demands. During the 1980s, a production of up to 5 million m^3 alcohol per annum is expected, the bulk being from chemical industries. In the initial phase, up to 1980, approximately 120 new distilleries have been scheduled with an average distilling capacity in the order of 120,000 liters/day of anhydrous alcohol. Many distilleries are being projected for 300,000 liters/day, whereas one project based upon both molasses and cane juices, is being planned for a phased development on a modular basis to eventually exceed 1 million liters/day.

Basically three types of distilleries are involved in Brazil. They are (Jackson, 1976):

1. Annexed distillery: attached to an existing or new sugar factory, utilizing factory services and operating on cane and/or cane juice
2. Central distillery: a separate distillery serving a number of sugar factories and sugarcane estates within a specific area and operating on cane molasses and/or cane juice
3. Autonomous distillery: a separate distillery with its own crop estate and operating on sugarcane directly or in other primary materials, such as cassava

At present, sugarcane is considered as the most feasible basic raw material and in order to achieve parity with crystal sugar prices, anhydrous alcohol has been equated at 44 liters per 60 kg of standard crystal sugar. Table II presents data for alcohol produced from annexed distilleries, where the raw material for the alcohol is a heavy viscous final molasses that remains after the maximum amount of sugar has been extracted. This molasses contains 80–90% dissolved solids of which approximately 60% (of total weight) is a mixture of sucrose and invert sugars, dextrose, and levulose. Accompanying

TABLE II
Productivity Factors for Ethanol Production from Sugarcane[a]

	Alcohol indirectly from final molasses		Alcohol directly from sugarcane juice	
Sugarcane yield in 1.5–2 year cycle (south-central region)	63	mt/ha[b]	63	mt/ha
Average sucrose yield (13.2 wt. %)	8.32	mt/ha	8.32	mt/ha
Crystal sugar production	7.0	mt/ha	—	
Final molasses or cane juice production	2.21	mt/ha	66.2	mt/ha
Fermentable sugars, molasses, or juices	1.32	mt/ha	8.73	mt/ha
Alcohol yield at 100% global efficiency	675	kg/ha	4460	kg/ha
Alcohol yield with reasonable 85% global efficiency	11.5	liter/mt cane or	75	liter/mt cane or
	730	liter/ha	4800	liter/ha

[a] From Lindeman and Rocchiccioli (1979).
[b] mt, metric ton; ha, hectare.

data are also presented for independent distilleries, where the raw material is treated juice that enters the distillery directly from the mill house. This green, sticky liquid, slightly more viscous than water, contains an average of 15–16% dissolved solids, of which 85% is sucrose, giving an overall approximately sucrose content of 12–13% of the fresh juice. Fresh cane juice contains very little (0.4–0.8%) invert sugars; however, inversion occurs very easily at the elevated temperatures and low pH incurred in the sugar-producing process and therefore must be carefully controlled to avoid losses.

Figure 2 illustrates in simplified form the steps required to produce ethanol from cane juice–molasses in Brazil. At the distillery, the molasses or clarified juice is weighed and the concentration of solids is adjusted (by dilution in the case of juice) and sent to the fermenting vats.

The fermentation step has changed little in the past 40 years. Fermentation is carried out batchwise in open-air, topless vats of 100–200 m^3 capacity, with the temperature of the wort being controlled manually by simple internal serpentine cooling water tubes. The fermentation is complete in 6–8 hours and cycle times vary between 10 and 12 hours to include filling, emptying, and cleaning the vats.

To increase the efficiency of the fermentation step, the "Melle-Boinot"

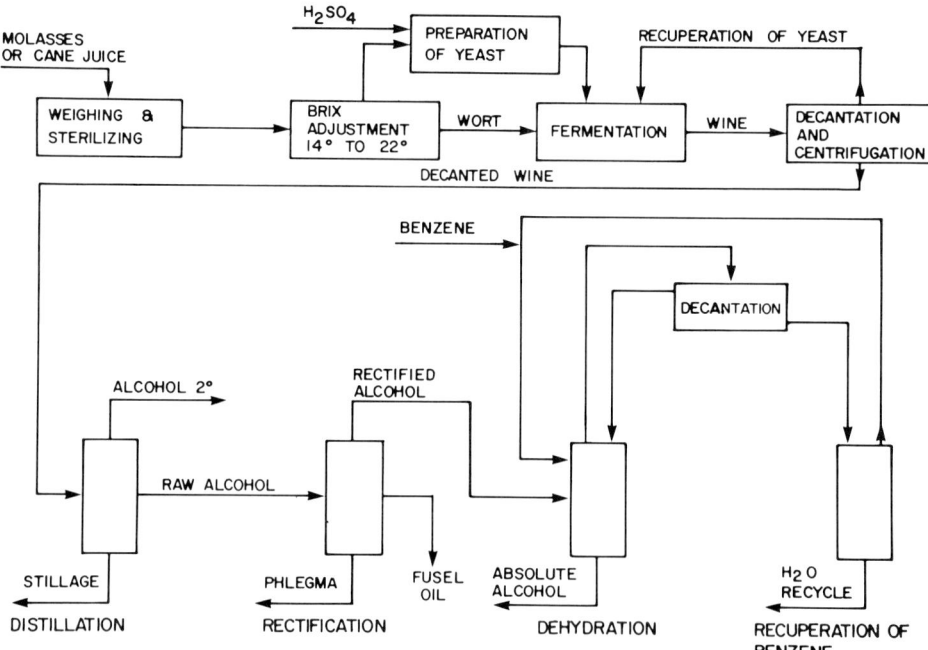

FIG. 2. Block diagram for ethanol distillery. From Lindeman and Rocchiccioli (1979).

process is utilized in most Brazilian distilleries. This process involves centrifugal recuperation of the live yeast from the fermented liquor (normally 10-15% by volume of the total) and reinoculating to other vats. This process reduces considerably the time required for the initial step of yeast cell multiplication.

The fermented liquor containing approximately 8% by volume of ethanol is decanted and stored in a surge vat, which serves to maintain a steady flow to the distillation towers.

Recuperation of ethanol is accomplished using a series of distillation towers, with normally four being required for absolute alcohol production. The first tower serves to separate the solid materials and to drive the aldehydes overhead. The second tower concentrates the fusel oils (higher alcohols) for removal and concentrates the alcohol to the azeotrope. The final two towers are required for alcohol dehydration beyond the azeotrope.

The most popular method to separate the azeotrope utilizes benzene, which has generally replaced glycol as the extracting agent. The last tower is a small stripper, which is needed to recover the benzene from the benzene–water phase formed in the decanter.

Presently, one of the areas receiving great attention in Brazil's alcohol program is the utilization of a continuous fermentation process. This would allow a higher output per unit volume of equipment, produce a higher yield in ethanol, allow for a better control in quality, and allow recuperation of carbon dioxide, if economically justifiable, because the fermentation is carried out in closed vessels. In the area of distillation, it is desirable to produce and evaluate new tower designs that would improve overall thermal efficiency and also reduce the quantity of stillage currently being produced. The predominant bubble-cap-type trayed tower has evolved over the last 30–40 years as being the most simple to operate while providing wide flexibilities in the distillery operation. However, today production demands have increased; consequently more efficient and more complex equipment is needed.

B. Continuous Fermentation

Continuous fermentation of molasses has been studied by Bose and Ghose (1973) on a laboratory scale. The medium for the continuous fermentation has the following composition: molasses, 300 gm; urea, 0.5 gm; o-phosphoric acid, 0.5 ml (84.5% purity); and $MgSO_4 \cdot 7 H_2O$, 0.29/liter. After sterilization at 20 psi for 15 minutes, the sediments were allowed to settle and the clear supernatant was transferred aseptically to the feed storage tanks through the tubes. The supplement solution (3 liters), containing urea (15 gm) and o-phosphoric acid (15 ml), after sterilization was transferred aseptically to the fermenting vessel. Prior to continuous feeding, the yeast was brought into active fermentation stage by growing it batchwise. The medium (3.5 liters) was run into the fermentor and was saturated with CO_2 gas. Yeast grown for 12 hours and followed by 6 hours of fermentation in the same medium was used as the inoculum. After 17 hours of batchwise fermentation, where the total sugar content decreased from 14 to 2.75 w/v, the volume of the mash inside the fermentor was adjusted to 3 liters and continuous feeding of fresh medium along with the supplement solution was started. The rate of addition of supplements (12 ml/hour) and the total volume in the fermentor were kept constant. Four different residence times (17.97, 16.75, 13.5, and 9.62 hours) were studied and the duration of steady states maintained were 8, 7, 6, and 5 hours, respectively. The average yield of alcohol was 43% (w/w), based upon the total sugar input at various residence times; the results are in excellent agreement to the batch yield. Further, it was found that when paddy soak water (prepared by soaking Patnai paddy in water at 70°C for 3 hours and filtering) was used, no addition of supplement (urea and phosphate) was necessary and, indeed, higher yields of alcohol (47% w/w) were observed.

The Danish Distilleries Ltd. of Grenå have commercialized a continuous process for the production of ethanol from molasses (Rosen, 1978). In this process (Fig. 3), the molasses is received by ship and is pumped through a

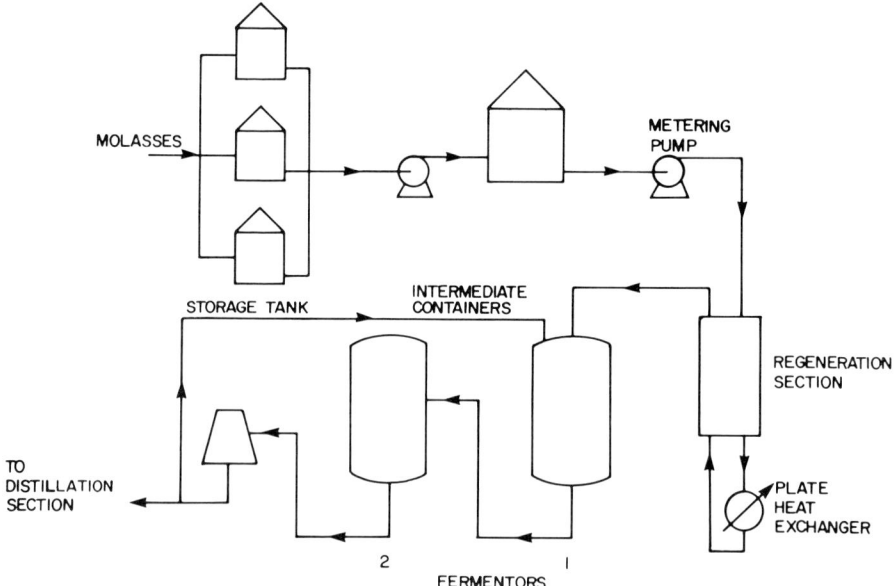

FIG. 3. Continuous production of ethanol by the Danish Distilleries Ltd. From Rosen (1978).

pipeline from the port of Grenå to the factory. The molasses is then stored in 2–3 tanks, each of approximately 1500 m³, and is then pumped to intermediate containers, which rest on electronic weighing cells, in the production room. From there, the raw material is taken to a set of metering pumps, consisting of piston pumps for molasses, water, sulfuric acid, and a solution of diammonium phosphate, to the effect that a diluted molasses solution is conducted from the subsequent blending tank, provided with the necessary chemicals and with pH regulated at approximately 5. This mixture is conveyed through a plate heat exchanger and heated to 100°C. In this way, most of the microorganisms that may be present in the solution are destroyed, resulting in a high degree of sterility. The plate heat exchanger is provided with a regenerative section, in which the solution heated by steam is cooled in countercurrent to the incoming liquid. The outlet temperature is approximately 35°C. This volume is then carried to fermentor 1.

In the fermentation section, two or three fermentors are used, constructed as cylindrical reservoirs with external diameter of 5 m and a height of 8 m. Total volume is about 170 m³. The tanks have conical tops and even, oblique bottoms. Each vessel is provided with in-place cleaning, air spargers, vent pipes for carbon dioxide, and a stirring device, introduced into the cylindrical part and equipped with a three-blade 300-mm propeller driven by a 3-hp electric motor. In addition, the vessels are provided with vacuum and safety valves. Vessels are connected in series as shown in Fig. 3.

The molasses solution carried to fermentor 1 is under constant afflux. There is overflow to the second vessel, and from here the fermented wort undergoes centrifugation prior to distillation. The separated yeast is returned to fermentor 1.

When the fermentation is started, molasses, water, and ordinary bakers' yeast are added to fermentor 1, which is aerated until the yeast has propagated sufficiently to reach the total quantity desired. Approximately 0.02–0.03 liter of air per liter of liquid per minute has to be infused.

Characteristic fermentation data are as shown in the accompanying tabulation.

	Fermentor 1[a]	Fermentor 2
Grams yeast, dry matter/liter	10	10
pH	4.7	4.8
Vol. % alcohol	6.1	8.4
% residual sugar	1.0	0.1
Temperature, °C	35	35

[a] Residence time in each fermentor: 10.5 hours. Afflux: 600 kg molasses diluted in 22,000 liters/hour.

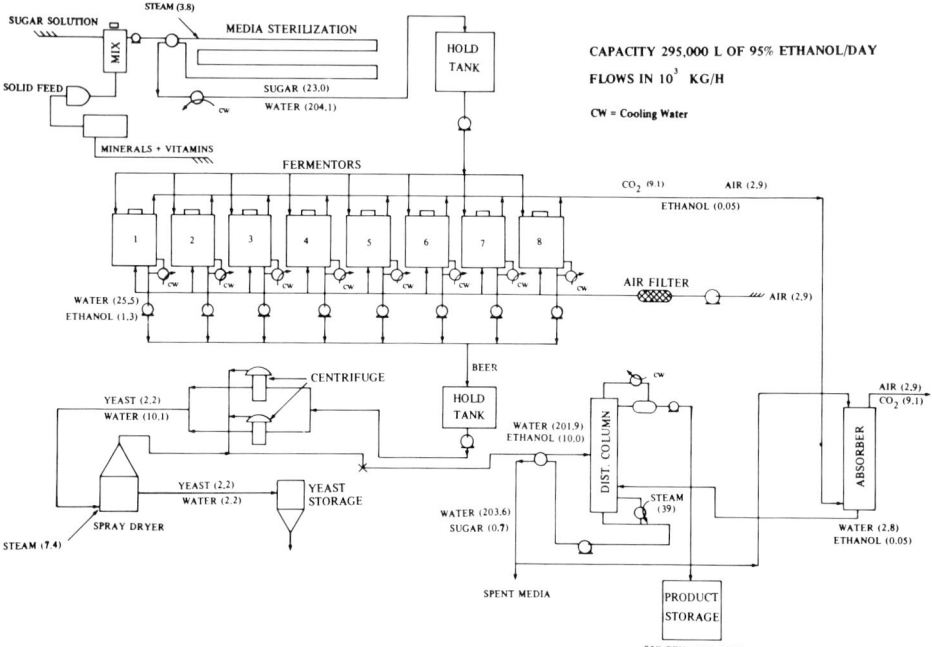

FIG. 4. Flow diagram and mass balance for continuous fermentation. From Cysewski and Wilke (1978).

The yield calculated on molasses amounts to approximately 28.29 liters of alcohol per 100 kg molasses, or a maximum of approximately 65 liters per 100 kg fermentable sugar.

The advantages of this continuous method are that the cubic capacity of the vessel is reduced and instrumentation simplified. The main disadvantage of this process is the loss of flexibility. It is therefore recommended that the two parallel lines of half capacity be used instead of one line of full capacity.

In the United States, Cysewski and Wilke (1977, 1978) designed a continuous process for the production of ethanol from molasses; details of the process are shown in Fig. 4 for a plant with production capacity of 295,000 liters (78,000 U.S. gal) 95% ethanol per day. Their studies showed that a 57% reduction in fixed capital investment is realized by continuous fermentation rather than batch operation. In addition, decreases in required capital investment of 68 and 71% over batch fermentation were obtained for cell recycle and vacuum operation, respectively. However, they pointed out that ethanol production costs were dominated by the cost of molasses (representing over 75% of the total manufacturing cost) and so this will dictate the selling price of fermentation ethanol. Viewed from this point, they concluded fermentation process improvements have a minor effect on total production costs. With the assumption of a reasonable by-product credit, the net production cost for 95% ethanol was estimated at 21.7¢/liter (82.3¢/gal) and 21.3¢/liter (80.6¢/gal) for the cell recycle and vacuum process, respectively.

C. Yeasts Tolerant to High Sugar Concentration

Rose (1976) maintains that if new strains of flocculent yeast which are tolerant to the high sugar concentrations in molasses can be found, the economy of the fermentation process could be improved. The possible economic gains he suggested are improved steam economy in distillation, reduction in mashing time, saving in makeup water, and volumetric reduction of effluent.

Yeast strains were screened from refinery, cane mill, and heavy rum wash samples according to the following criteria:

1. Ability to ferment completely 25% w/v sugars in solutions of molasses
2. Ability to flocculate well to supply clean wash to the still
3. Lack of offensive odors in the wash and distillates.

Four strains were isolated, two each of "budding" and "fission" yeasts: *Saccharomyces carlsbergensis* (*uvarum*) (Cat. No. Sa. 23), *Saccharomyces cerevisiae* (Cat. No. Sa. 28), and *Schizosaccharomyces pombé* (Cat. No. Sz. 10 & Sz. 11). Fermentation experiments were subsequently carried out to study the fermentation characteristics of each isolate. The Sa. 23 and Sz. 10 strains were the most successful, with an alcohol yield of 81.5% and 84%, respectively, from a medium containing 25% w/v sugar. Commercial trials with Sa. 23 were also conducted. The results are shown in Table III.

TABLE III
COMMERCIAL FERMENTATION OF MOLASSES WITH Sa. 23[a]

	Sugars (w/v)	Ethanol (v/v)	Time (hours)	Yield[b] (%)	Initial pH	Final pH
Standard yeast	12.3	6.15	38	77	5.0	4.4
Saccharomyces carlsbergensis, Sa. 23	23.6	12.3	76	80	5.3	4.8

[a] From Rose (1976).
[b] Percentage yield = (ethanol formed/maximum ethanol theoretically possible) × 100.

D. The "ABE" Process

The simultaneous production of acetone, butanol, and ethanol (ABE) from fermentation has been known for a long time. Because of competition from cheaper chemical synthesis, this process has been almost completely displaced since the late 1950s. The only commercial process that has survived the competition is the National Chemical Products (NCP) process of the Republic of South Africa (Spivey, 1978). They have been producing acetone,

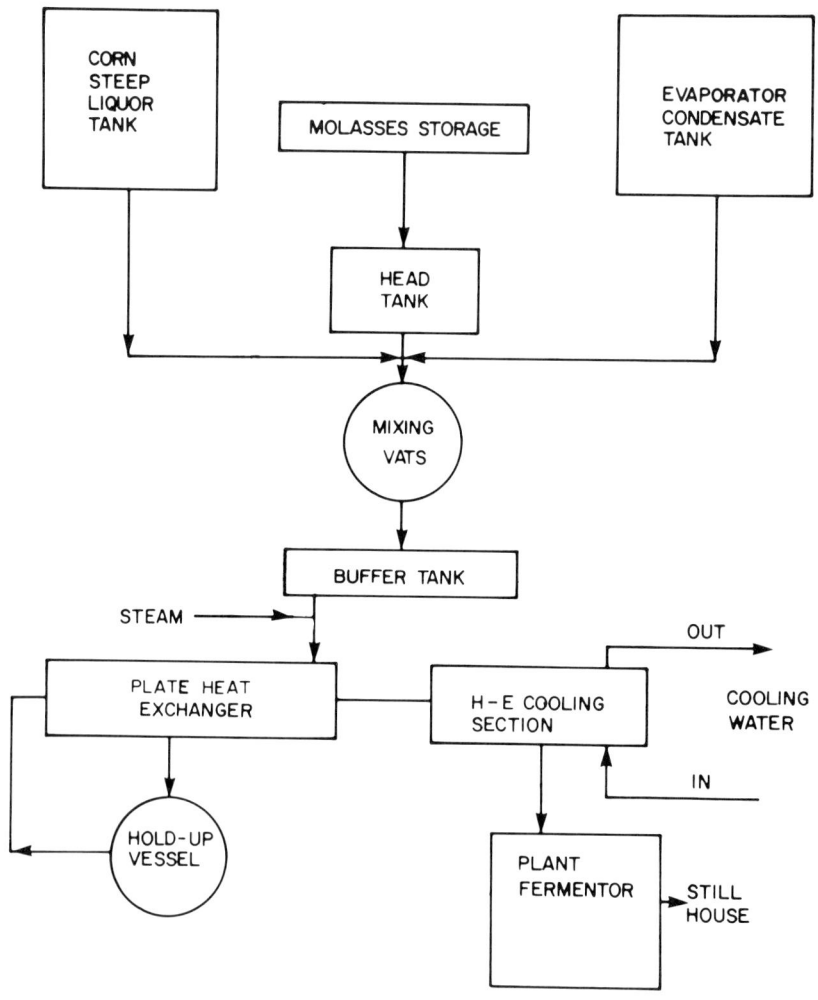

FIG. 5. Flow diagram of the "ABE" fermentation. From Spivey (1978). H-E = Heat exchanger.

butanol, and ethanol from fermentation since 1936. The flow sheet of the process is shown in Fig. 5. The molasses is first weighed and pumped to a mixing vessel, where the other components of the fermentation medium are added and mixed. The mash is sterilized continuously by direct steam injection, using a combination of plate heat exchanger and holding tank to give the required retention time of 4 minutes at 128°C. The sterilized mash is cooled to 34°C in the last section of the heat exchanger and transferred aseptically to a sterile fermentor, which is then flushed with sterile carbon dioxide gas. Once filling of the fermentor is completed, CO_2 continues to agitate the mash for 20 minutes before and after inoculation to ensure good mixing. Finally, if required, the pH is adjusted to 5.8–6.0 with ammonia. The fermentor pressure is allowed to build up to 35 kPa, thus operating the fermentors at a positive pressure at all times which reduces the chance of contamination. The fermentation is complete after about 30–40 hours. The butyl beer is then sent to the stillhouse, where the solvents are recovered. The microorganisms used in this process are strains of *Clostridium acetobutylicum* Weizmann. Strains P265 and P270 are the current cultures in operation at NCP. Typical product composition of this process is given in Table IV.

One of the most serious disadvantages of the ABE fermentation process is the relatively low concentration of products produced. Recovery of the products from a 2% concentration by distillation is energetically unfavorable. Production would be much more economical if some means of increasing the product yields and concentrations in the beer could be developed. Further, the potential of continuous fermentation remains to be exploited. It is also noted that *Cl. acetobutylicum* is one of the few organisms able to ferment pentose sugars as well as hexose. Thus *Cl. acetobutylicum* would be used to convert both hexoses and pentoses, produced by hydrolysis of cellulosic materials, to fuel and chemical feedstocks.

TABLE IV
Products from "ABE" Fermentation[a]

Product	Percentage of sugar fermented	Amount produced (based on 1000 kg of fermentable sugar)
Butanol		180
Acetone	30	90
Ethanol		30
CO_2	50	496
H_2	2	2

[a] From Spivey (1978).

E. Fermentation of Pentoses

In the United States, Batter and Wilke (1977) studied the fermentation of the pentose sugar xylose to ethanol by *Fusarium oxysporum* with an anticipation to utilize by-product xylose from hydrolysis of cellulose. Preliminary results indicated that an 11% higher yield of ethanol can be achieved when *F. oxysporum* was used to ferment sugars produced by acid hydrolysis, as compared to fermentation by *S. cerevisiae* of sugars produced by enzymatic hydrolysis. However, because of the slow growth rate of the organism, future development will be necessary to improve the economic feasibility of the xylose fermentation. As pentoses are predominant in some wood species (particularly high in hardwoods, up to 25%) (according to Dukes *et al.*, 1963), fermentation of these would increase the product yields considerably when cellulose substrates are considered.

F. Other Developments

Elsewhere in the United States, no commercial process has been developed for production of ethanol by direct fermentation of sugar. However, extensive feasibility studies have been carried out. For example, Lipinsky *et al.* (1977b) pointed out that integrating sweet sorghum with sugarcane production could expand the area available for extensive sugar crop production by a factor of 10 or more, thus increasing the availability of sugar for fermentation. They also maintained that the primary key to reducing processing costs lies in increasing the concentration of ethanol in the fermentation mash, not in reduction in fermentation time as practiced in Brazil. Birkett and Polack (1978) examined the economics of an alcohol venture in Louisiana. They concluded at current prices, sucrose is a more valuable product than alcohol, consequently conversion of cane juice to alcohol is not economical. Fermentation of molasses to alcohol could be attractive only if the scale were very large (10,000 tons per day of cane). Moreover, there is insufficient sugarcane available to impact on the total energy requirement of the United States.

A novel fermentation device, the rotorfermentor, has been developed in the United States by Margaritis and Wilke (1972, 1978). This fermentor is designed to achieve high cell concentrations in batch and continuous cultures. It consists basically of a rotating microporous membrane that is enclosed within a stationary fermentor. The metabolic products in the broth are continuously removed by filtration through the rotating microporous membrane while the growing cells are retained inside the fermentor. With this fermentor, Margaritis and Wilke (1978) studied the fermentation of dextrose to ethanol by *S. cerevisiae*. They found that when a dextrose feed concentration of 104 gm/liter was used, almost 100% dextrose utilization was obtained with an ethanol productivity rate of 27.3 gm ethanol per liter per hour. This was about 10 times greater than the ethanol productivity obtained from an ordinary continuous stirred tank (CST) fermentor. In addition, the power consumption and capital cost of the rotorfermentor are claimed to be lower than those of a CST fermentor.

V. Ethanol from Starch

The fermentation of starch from grains has been practiced for a long time. During World War II, industrial grade ethanol was made from grain in water to supply the raw materials needed for synthetic rubber. With the very rapid development of the postwar petrochemical technology, synthesis of ethanol from ethylene replaced fermentation as the economical method for producing industrial alcohol. Following the "energy crisis" of 1973, the price of ethylene has gone from 6.6¢/kg (3¢/lb) to 26.5¢/kg (12¢/lb). Along with the higher fuel costs, this has almost doubled the price of synthetic ethanol. During the same period, grain prices have changed little, making the price of fermentation ethanol more attractive as compared to its synthetic counterpart.

Although various sources of starch are available, corn is by far the most important one, especially in the United States. Cassava root is also an important source of starch; however, its production is mainly restricted to tropical countries. Other sources of starch considered are potato, wheat cereal, and other grains.

A. ETHANOL FROM CORN

Corn is especially attractive for production of fuels because (Lipinsky *et al.*, 1977a):

1. The yield of this crop is relatively high.
2. Corn, like sugarcane, has a C-1,4 photosynthetic mechanism that is inherently quite efficient.
3. The energy output–input ratio for corn is higher than from other major crops (with the exception of sugarcane).
4. Annual production of corn biomass for all purposes probably exceeds 330 billion metric tons (300 billion tons, dry basis) in the United States, about 40% of which is residues that are presently of little commercial value.

There are three alternatives for using corn as an energy feedstock. These include (Lipinsky *et al.*, 1977a):

1. Harvest the entire corn plant for silage, with no separation of grains and plant residues, and use it for energy production.
2. Use corn grain to produce ethanol, with corn residues returned to the soil or used for livestock feed, as currently practiced.
3. Collect semidried corn plant residues, after grain has been harvested, for use in production of furfural, SNG, ammonia, or simple sugars.

The corn-based systems for production of fuels, chemical feedstocks, and coproducts are illustrated in Fig. 6.

According to Lipinsky *et al.* (1977a), the two conversion opportunities that appear promising at present are (1) possible coproduction of SNG and pa-

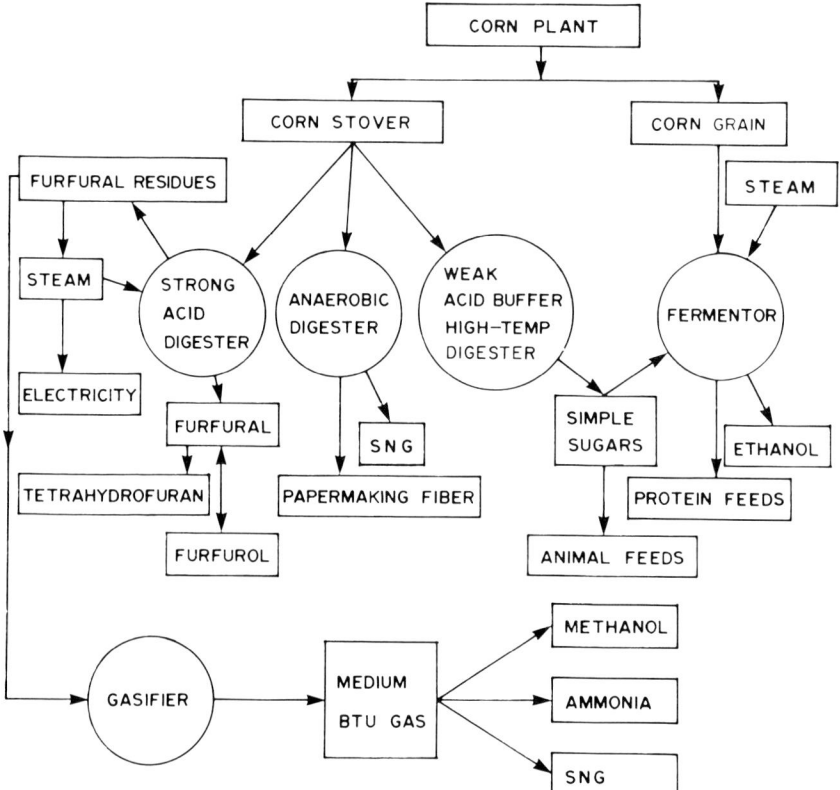

FIG. 6. Corn-based systems for production of fuels, chemical feedstocks, and coproducts. From Lipinsky *et al.* (1977a).

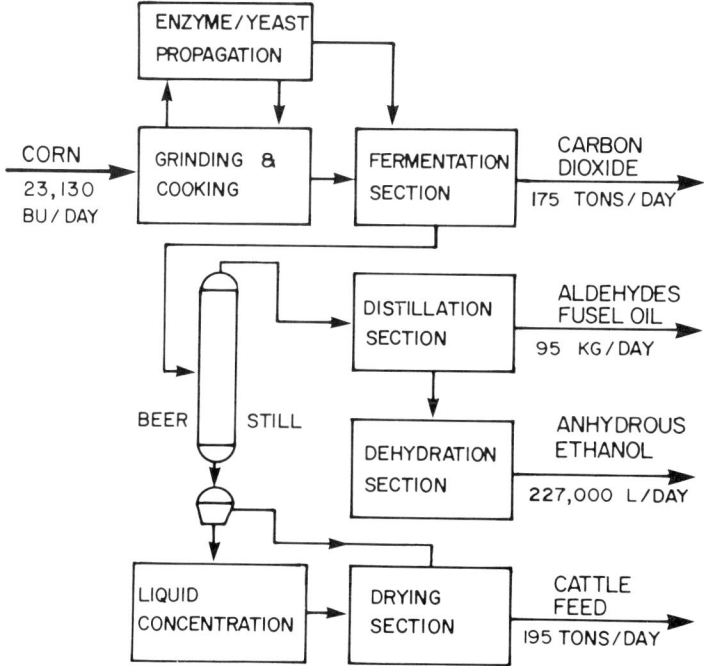

FIG. 7. Production of ethanol from corn. From Scheller (1976).

permaking fiber from corn residues and (2) generation of fermentable sugars from corn stover or silage corn by means of acid hydrolysis catalyzed by specially buffered weak acids. The inexpensive fermentable sugars would then be the raw material for fermentation.

In comparison to sugar as a substrate for fermentation ethanol, corn offers several advantages. As a grain, corn has the advantage over sugarcane in storability for production of ethanol throughout the year. The production and storage of hightest molasses from sugarcane would be relatively inconvenient and expensive. In addition, the coproduction of protein from corn for animal feed markets and corn-based sweetener for food markets can result in cost reduction, making corn-based ethanol somewhat cheaper than sugar-based ethanol (Lipinsky et al., 1977a).

Figure 7 is a block–flow diagram for a conventional fermentation plant producing 76 million liters (20 million gal) per year of anhydrous ethanol from 23,130 bu of corn per operating day.

Corn from storage is fed to a grinder, where the kernel size is reduced to expose the interior portion of the grain. Water is added, pH adjusted, and the ground grain is then cooked to solubilize and gelatinize the starch. After cooking and partial solubilization, fungal amylase, which has been propagated in the plant, is added to convert the starch to dextrose and the mixture is cooled before being fed to the fermentation section of the plant. Yeast, which has also been propagated in the plant, is added and the fermentation is allowed to continue for approximately 48 hours at a temperature of 32°C (90°F). During this time about 90% of the original starch in the grain is converted to alcohol. The fermentation section of the plant consists of a number of vessels and the processing proceeds on a cyclic or batch basis, with some vessels containing fermenting mash while other vessels are being filled, emptied, or sterilized.

Once fermentation is complete the mixture is fed to the beer still, where essentially all of the alcohol is distilled overhead to about 100° proof. The diluted alcohol is purified by further distillation, which removes fermentation by-products (aldehydes, ketones, fusel oils), yielding 190° proof alcohol (95% pure). The distillation process varies somewhat depending on whether beverage grade alcohol or industrial alcohol is being produced. If it is desired to produce anhydrous ethanol, the 190° proof ethanol is fed to a dehydration section consisting of an extractive distillation with benzene to break the azeotrope and remove the residual water.

In the beer still, the remaining water with dissolved and undissolved solids is drawn from the bottom of the still and fed to a centrifuge. The liquid phase from the centrifuge is concentrated to 50% dissolved solids in a multiple-effect evaporator and mixed with the solids from the centrifuge. This mixture is then dried in a fluidized transport-type dryer to 10% moisture and is used as cattle feed. The cattle feed contains all the proteins that were originally present in the grain, plus the additional proteins from the yeast, resulting in a product containing 28% to 36% protein by weight.

In addition to alcohol and cattle feed, the original 23,130 bu/day of corn yield 175 metric tons (193 tons) of carbon dioxide and 95 kg (210 lb) of by-product aldehydes, ketones, and fusel oils. The investment required to build such a plant was estimated to be $21 million in 1975 (Scheller, 1976).

A grain alcohol plant does not need first quality grain for the production of alcohol. Grain that is moldy or has started to sprout is suitable and is known as distressed grain. It exists in grain-producing areas to the extent of about 1% of the annual production. Other low-quality grains below sample grade may raise this level to as high as 5% of the annual grain production.

B. Integration with Processing Corn for Feed

Corn can be converted into food products via two approaches—dry and wet milling. The majority of modern corn dry milling is conducted after the germ has been removed. The corn is usually tempered by being soaked in water for a short time where the germ is broken loose and separated. The endosperm and hull are then ground. The germ is used to make corn oil, and the cake remaining after the oil has been expelled or extracted is sold for animal feed. The ground endosperm can be used for making corn flakes, cornmeal, grits, etc.

In corn wet milling, it is the desire of the processor to separate the starch from the gluten. This is done in a complex process (Fig. 8) beginning with soaking the kernels in hot water with sulfur dioxide present to loosen the kernel components. The grain is then ground quickly to break the germ loose from the kernel and separated. The germ is treated to obtain the oil

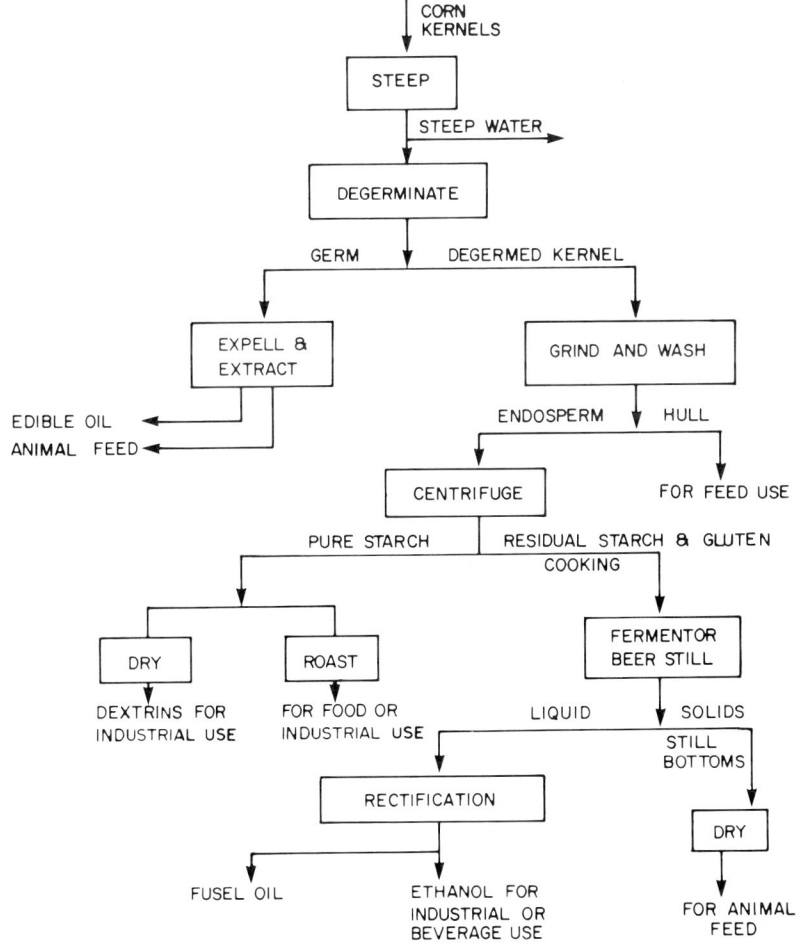

FIG. 8. Integration of ethanol production with wet milling. From Lipinsky et al. (1977a).

either by expelling the oil or extracting it with a solvent. The oil meal is used for animal feed. The degermed kernel is then ground finely and is washed to remove the hull. The remaining material is centrifuged to separate the starch from the gluten. The starch can be dried and sold for industrial or food use, or it can be heated to make specialty products, such as dextrins and specially modified starches. Alternatively, either acid or enzymatic hydrolysis will convert it to corn syrups of varying degrees for conversion to dextrose. Corn syrup can be sold or it can be converted by enzymatic isomerization to a high-fructose syrup, which has a greater sweetening power than the dextrose solution obtained initially as corn syrup. The by-products of wet milling are used as indicated in Fig. 8.

If one wishes to produce ethanol, one can incorporate its production into the wet milling schemes as shown in Fig. 7. There are certain advantages in doing this. The grinding and washing steps and the separation of the starch from the gluten result in high costs and some starch ends up in the gluten and thus in the feed. One can grind less vigorously and then centrifuge in such a way as to get the starch that is easily removed. The residual starch and gluten then can be used in the fermentor with an overall saving in cost. The alcohol produced in this manner, for the most part, is sold for beverage use but it represents a reasonable approach to production of industrial grade ethanol from corn because the value of the starch and by-products would offset much of the cost of the corn. Another process (Lipinsky *et al.*, 1977a), not yet commercialized, involves an enzymatic treatment of starch, whereby dextrose and gluten are used as a fermentation substrate for alcohol production.

A traditional wet milling plant that processes 30,000 bu per day costs approximately $52 million. A wet milling plant of twice that capacity would cost on the order of $80 million (Lipinsky *et al.*, 1977a). These figures are for plants that convert most of the starch into high-fructose syrup. Modification of the design to manufacture ethanol and distillery by-products instead of gluten feeds would decrease the capital cost of the wet milling unit. The overall capital including an ethanol unit should be higher than for a high-fructose corn syrup plant, but detailed cost estimates are not available.

The manufacturing of ethanol by the above processes can be made more competitive in reducing the production cost by using cheaper starting materials and continuous fermentation or other process improvements that would reduce capital investments. Use of corn residues valued at $40/metric ton ($36/ton, 1977 United States price) is an excellent start in the reduction of ethanol cost (Lipinsky *et al.*, 1977a). If an inexpensive hydrolysis process to convert corn stover into a simpler sugar could be developed, the cost of the simple sugars might be reduced to approximately $0.09/kg. If these sugars could be fermented as readily as sugarcane juices, ethanol might cost approximately $0.22/liter ($0.83/gal). If continuous fermentation could replace the batch process now used, the battery limits plant investment might be reduced by approximately 30%.

Chemapec Inc. of New York has disclosed a process for production of anhydrous ethanol from corn using an integrated grain to finished product route. The process is claimed to cost approximately half as much as the conventional route. A pilot plant was demonstrated in Switzerland (Anonymous, 1979a). The process differs from conventional ethanol production and technology in that: (a) much of the nonstarch material in the grain (germ, meal, oils, etc.) is removed before processing; and (b) the mash that is left is not cooked but is digested using "special" enzymes.

A patented waste treatment system is also incorporated, generating methane gas to power the process steam boilers. Net process heat requirements are claimed to be only 4200 kJ/liter (15,000 Btu/gal) of anhydrous ethanol, in contrast to 16,700–23,300 kJ/liter (60,000–80,000 Btu/gal) in conventional fermentation processes.

In this process, the mash (mostly starch, but with some fiber and protein content) is digested and converted to sugar with the aid of "special" enzymes that are active at an unspecified, "but not cooking," temperature. After fermentation, the converted mash proceeds to a vacuum distillation unit, where 199.9° proof ethanol is produced. CO_2 given off is collected for use or sale. Distillation slops go to waste treatment. According to Chemapec, a 38 million to 76 million liter/year (10 million to 20 million gal/year) ethanol plant would be the minimum size feasible, and would cost about $20 million to $25 million. Yield is said to be about 9.8 liter (2.6 gal) of ethanol per bushel of corn.

C. Ethanol from Cassava Roots

To date, sugarcane is the most important substrate for the fermentative production of ethanol in Brazil. Another substrate of great potential is cassava root. The annual world production of cassava root is estimated at about 90 million metric tons (100 million tons), with Brazil (30%), Indonesia (20%), and Zaire (10%) being the largest producers. In Brazil, the world's first commercial plant producing ethanol from cassava root commenced production in December 1977, with an ethanol capacity of 60,000 liters/day (Anonymous, 1978b). It is expected that additional plants will be built in the future using similar processing.

The processing steps used to obtain ethanol from cassava roots are shown in Fig. 9. As the fresh roots are received, they are weighed, washed, peeled, and ground into a mash. Part of this mash, the quantity determined by the plant operating plans, is then side streamed and dried, producing a meal. Unlike sugarcane, which starts to decay and ferment naturally soon after

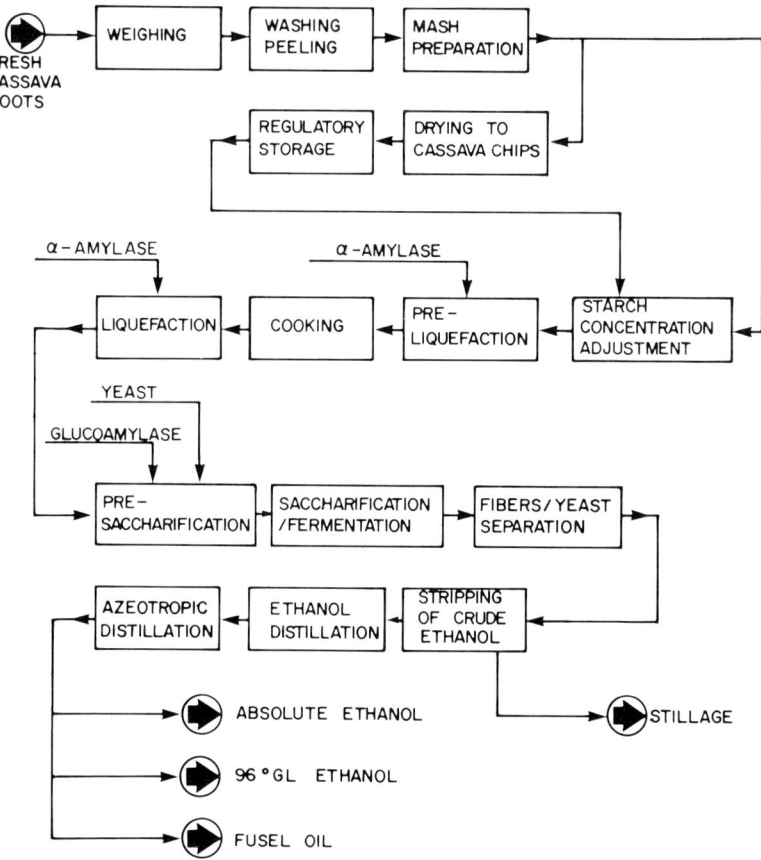

FIG. 9. Production of ethanol from cassava root. From Lindeman and Rocchiccioli (1979).

cutting, dried cassava can be stored for as long as 1 year without a significant loss of starch.

The starch molecules are broken down with the aid of α-amylase which is added in two steps. The preliminary addition is required to decrease the viscosity of the mash and so facilitate the cooking operation. Glucoamylase is then employed to achieve the final conversion of the liquefied starch material into glucose. This is the time limiting step in the overall process.

The saccharification and fermentation processes are carried out simultaneously in the same vat, but because of the limits imposed by the rate of saccharification, the fermentation does not reach completion before 72 hours in some instances.

The final steps in the process of recovering the alcohol from the fermented material are the same as those used for sugarcane juice.

Alcohol yields from cassava are in the region of 165–180 liters/metric ton, which, on a weight for weight basis, is a great improvement over sugarcane (Jackson, 1976). However, as sugarcane harvests can be up to 90 metric ton/ha, the alcohol yield per unit cultivation area is higher for cane under present agricultural methods. Also, the greatest advantage from sugarcane processing is in the use of its fiber (bagasse) as a fuel. A cane stalk contains roughly the same weight of dry fiber as sugars and therefore is more than adequate in providing the energy for alcohol processing. This is not the case with cassava, where the dry fiber content is about 3.5%. Also, because of the necessary conversion of starch to fermentable sugars, a greater energy input is necessary in its processing prior to fermentation.

For batch alcohol manufacture from cassava, the steam requirement is about 1.8 kg steam per kilogram cassava, which must come from an external source. By continuous processing in which efficient heat recovery systems are used, a considerable fuel reduction can be achieved. Therefore the biggest problem in using cassava is in the high energy requirement. Other disadvantages of cassava are (Lewis, 1977) (a) a higher capital investment is required, perhaps as much as 40% over other, same size, independent sugarcane distilleries; and (b) a higher degree of materials handling is involved, and therefore higher operating costs.

There are also a number of factors in favor of cassava (Jackson, 1976; Lewis, 1976):

1. Improvements in future agricultural technology will make available higher yields.

2. There are about 48 hours available before serious deterioration sets in between crop lifting and processing. This is up to five or six times greater than for sugarcane, where the pol (sucrose content) falls off considerably 8 hours after cutting. Therefore, transportation problems to the factory are eased and the radius of estate operation is increased.

3. Although the harvest periods for cassava and sugarcane are of similar duration, cassava chips can be dried (by boiler stack gases) to well below 20% moisture for stable storage during in-crop operations. This will permit maximum production throughout the year using cassava chips during out-of-crop periods.

4. Good conversion yields are attainable by the use of amylolytic enzymes at a reasonable cost within the elements of productions.

5. In bitter cassava varieties, cyanogenetic glucosides do not cause any trouble during processing and the cyanides are flashed off with the steam during the mash cooking. When bitter cassava is used as a foodstuff, these toxic compounds must first be removed by water leaching.

6. It has been found that in mash fermentation, no additions of acids or nutrients are necessary and that a clean spirit, relatively free from con-

generics, results. With a mash pH close to neutral, materials of construction for distillation equipment are not so prone to corrosive attack, allowing a subsequent saving in capital cost.

7. As the living standards in developing countries improve, cassava loses its appeal as the staple "food of the people." Already in many countries, it is being downgraded to a meal for animal feedstuff compounding or processed into a flour with a low nutritional value. Alcohol manufacture, therefore, offers another profitable outlet.

8. With sugarcane, much of the leafy material, apart from the "tops," is burnt prior to cutting, whereas cassava leaves and stems are totally available when the roots are lifted. By a modification of the PROXAN process, the leaves and stems can produce a valuable animal feedstuff. The leaves have a crude protein content of 27% and can provide an extract for monogastrics, whereas the fibrous residues can be used as a forage for cattle.

9. A much lower quality of soil is required than that for sugarcane and therefore vast areas of little used land can be cultivated more profitably.

Menezes (1978) has demonstrated that fungal broths of a basidiomycete and of *Trichoderma viride* increased both the rate of sugar formation and the degree of solubilization, whereas the viscosity of the cassava slurry decreased. The improvement of the fluidal characteristics of the slurry would minimize the sugar and alcohol losses during different steps of the production. Therefore, with addition of cellulase, the use of the whole root becomes possible without losses of substrate.

D. Ethanol from Potatoes

The Danish Distilleries Ltd. of Aalborg has developed a semicontinuous process for the production of ethanol from potatoes and grain (Rosen, 1978). In this process (Fig. 10) the potatoes (or grain) are transported on a conveyor with a variable rate of speed to a belt weigher, which is used partly to register the quantity weighed in, and partly to secure a constant capacity by regulating the velocity of the conveyor. Additionally, the belt weigher is capable of regulating the water supply wanted when grain is used.

Comminution of the potatoes or grain and the blending of grain and water take place in an LS-Gorator. The raw material is pumped from here by means of a monopump for enzyme treatment before gelatinization. The enzyme used is Termamyl 60 prepared by Novo. After enzyme treatment, the potatoes or grain are heated to 90–95°C by coming into contact with flash steam in a condenser. The raw material is then transported from the bottom of this apparatus to a boiler tube, approximately 50 m in length. Heating to a temperature of approximately 150°C is achieved by 10 atm steam using a

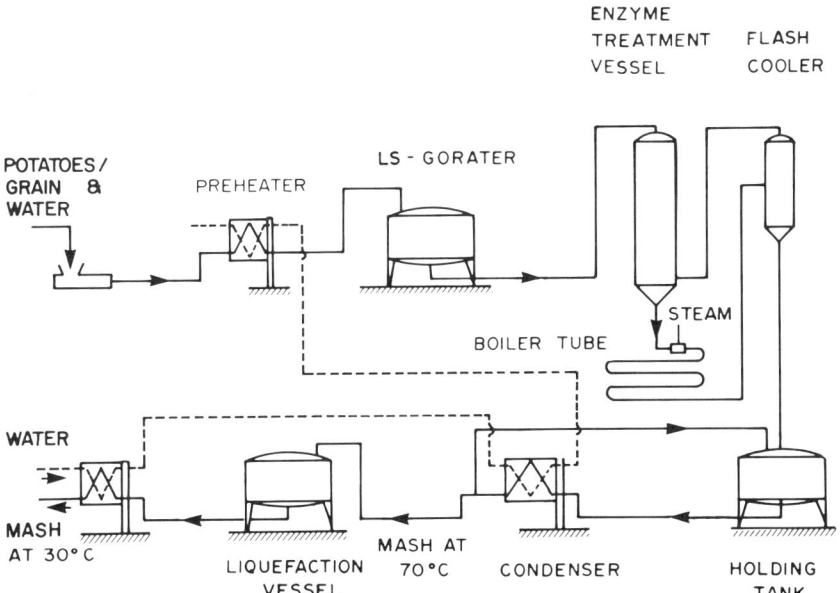

FIG. 10. Danish Distilleries Ltd.'s semicontinuous production of alcohol from potatoes/grains. From Rosen (1978).

Konting steam ejector. The quantity of steam is temperature regulated. Retention time at 150°C is approximately 3 minutes, and the rate of speed in the boiler tube is around 0.3 m/second. The mixture is flashed to atmospheric pressure in a single step, and the steam released is used to preheat the starchy compound. The mash is cooled to 70°C for liquefaction with commercial amylase preparations of bacterial origin. If pH regulation is required it is accomplished with slaked lime. Intermission at 70°C is approximately 25 minutes. The mash is cooled to 30°C and is then pumped to the yeast vessels, where batch fermentation to ethanol is carried out in a conventional manner. All temperature and level regulations are automatic, so that the process can be maintained by one operator.

E. Ethanol from Novel Starch Sources

Tatsumi *et al.* (1977) patented a process for the production of ethanol using a novel starch source—unicellular green algae. Three strains of unicellular green algae that have been found to produce substantial quantities of starch, either intracellularly or extracellularly or both, are *Chlorella vulgaris*, *Chlamydomonas* sp., and *Scenedesmus basilenses*. In this method, agricultural wastes containing acetate are the preferred culture media for the algae. Supplementary nutrients can also be used. The culture can be carried out either batchwise or continuously, in the light or in the dark. In a batch culture, the starch starts to accumulate after the logarithmic growth phase, which usually occurs 24–48 hours after the culture is started, and the starch accumulation reaches its maximum during the stationary phase.

Following the growth phase, the accumulated starch can be recovered or fermented directly using conventional fermentation techniques, and conventional ethanol-producing microorganisms. Before fermentation takes place, the cellular material will be given a pretreatment to rupture the cell walls, thereby releasing the intracellular starch into the fermentation medium. A typical pretreatment comprises a centrifugal separation of the algae (at the stationary phase) for 10 minutes at 3000 rpm. The packed precipitate is then mixed with a small amount of water and sonicated in a 10 kHz oscillator for 10–20 minutes to destroy the algal cells. This treatment increases the starch recovery and facilitates subsequent saccharification and fermentation processes.

VI. Ethanol from Cellulosic Materials

Cellulose is the most abundant organic material that can be used as a source of food, fuels, and chemicals. The net worldwide production of cellulose is estimated at 100 billion tons per year (Spano, 1976). This is approximately 56 kg (150 lb) of cellulose per day for each and every one of the earth's 3.9 billion people. The energy to produce this vast quantity of cellulose comes from the sun and is fixed by photosynthesis. The utilization of this annually replenishable resource is greatly simplified if it is first hydrolyzed into glucose. From glucose, various sources of food consumable by man and animal can be derived. In addition, it can be used as a feedstock to make solvents, plastics, and other chemicals now made from petroleum; it can be converted microbially into single-cell protein; or it can be fermented to a clean-burning fuel, such as ethanol. The various routes for the utilization of cellulose and the possible routes to petrochemicals from cellulose are shown in Fig. 11.

Cellulosic wastes may be agricultural, from forestry, urban, or industrial in origin. Agricultural wastes include crop residues, animal excreta, and crop and animal (meat and animal products) processing wastes. The slashings generated in logging, sawdust, and scrap formed in lumber production and wood product manufacture are the cellulosic wastes produced in forestry originated activities. Residential and light commercial refuse are the primary sources of urban cellulosic wastes. Sewage sludge might also be considered a source of cellulose inasmuch as its cellulose content provides carbon needed

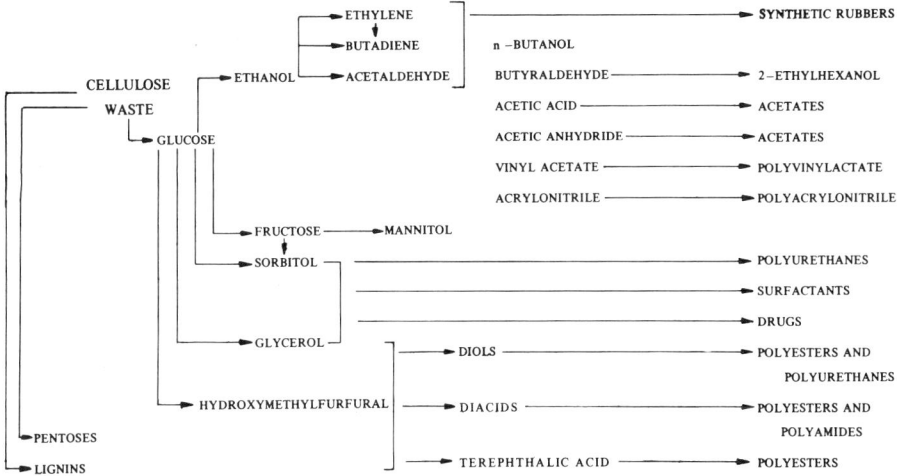

FIG. 11. Possible routes to petrochemicals from cellulosic wastes. From Brenner et al. (1977).

for methane production in the anaerobic digestion sludges. Conversion of cellulose to glucose can be achieved by hydrolysis. Upon hydrolysis, 1 kg of cellulose will theoretically yield 1.1 kg dextrose, which is equivalent to 0.56 kg ethanol. The product of hydrolysis can be put into a variety of processes as a fermentation feedstock. At present, there are two hydrolysis processes available. These are acid and enzyme hydrolysis.

A. Ethanol from Wood via Acid Hydrolysis

One of the earliest commercial hydrolysis processes was a dilute sulfuric acid process developed by M. F. Ewen and G. H. Tomlinson during World War I. Two plants in the United States for producing sugars from wood were operative. This so-called "American process" employed a rotary digester and the hydrolysis was carried out at a comparatively low temperature for prolonged time periods. The end product at both plants was ethanol made by fermenting glucose solution. For economic reasons (the rising price of lumber and low price of blackstrap molasses) both plants were closed at the end of World War I.

Dilute sulfuric acid hydrolysis of wood was reexamined by the Forest Product Laboratory, U.S. Department of Agriculture at the request of the War Production Board in 1943. The pilot plant facilities of the Cliffs-Dow Chemical Company were used for the initial experimental studies. A full-scale, demonstration, wood hydrolysis plant was subsequently designed and built at Springfield, Oregon, but it did not commence operation until after the end of World War II (Hokanson and Katzen, 1978; Azarniouch and Thompson, 1976; Maloney, 1978). A schematic diagram of the ethanol from wood plant is shown in Fig. 12. The process is basically a semicontinuous process in which the hydrolyzate percolates through the "chip" bed, continuously removing the sugars as they are formed. This avoids prolonged retention of the sugars under acid conditions at high temperatures and so reduces degradation. Hydrolyzer operating conditions have a significant effect on the yield of sugar. Optimum conditions are listed in the accompanying tabulation (Hokanson and Katzen, 1978).

	Optimum conditions
Acid concentration in total water	0.53%
Maximum temperature of percolation	196°C
Rate of temperature rise	4°C/min
Percolation time	145–190 min
Ratio total water/oven-dried wood	10
Percolation rate	8.69–14.44 liters/min/m^3

Dilute hydrolysis solution from a previous batch is pumped into the top of the hydrolyzer containing the wood wastes. Sulfuric acid concentration varies during the percolation cycle, averaging 0.53% of the water present in the batch. After the dilute hydrolyzate is charged, hot water and sulfuric acid are added and the temperature is raised to 196°C (15-atm vessel pressure). After 70 minutes of pumping, strong hydrolysis solution starts to flow out of the bottom of the hydrolyzer. Two letdown flash stages are used, the first stage operating at 4.5 atm and the second stage at atmospheric pressure. The condensates from the flash vapor from both stages contain furfural and methanol, which are recovered. The underflow from the two flash stages is the sugar-containing solution. At the end of the percolation cycle the lignin-rich residue is discharged, recovered, and used as a fuel.

The flash condensate passes to a distillation tower for recovery of methanol. From the base of this tower the bottoms pass to a second distillation tower for recovery of the furfural–water azeotrope.

The hot acid hydrolyzate solution is neutralized with a lime slurry and the precipitated calcium sulfate is separated in a clarifier. Calcium sulfate sludge is concentrated to about 50% solids and is trucked to a disposal area.

Neutralized liquor is blended with recovered yeast from previous fermentation and passed to fermentation tanks. From the fermentors, the fermented liquor passes to yeast separators for recovery of the yeast for recycling (*S. cerevisiae*). Ethanol is recovered in a series of distillation towers and is finally rectified to approximately 190° proof.

Bottoms from the beer stripping tower contain pentoses. Instead of dispos-

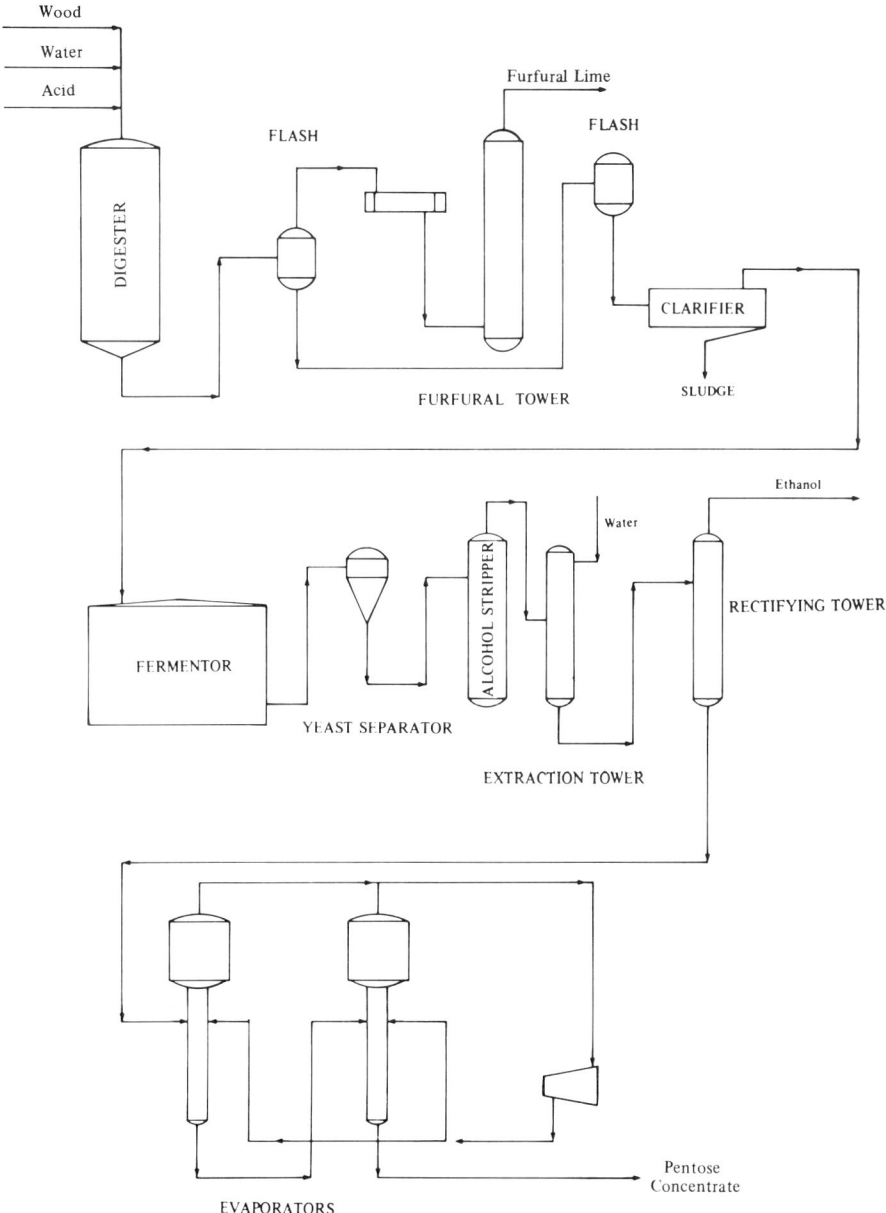

FIG. 12. Ethanol production from wood. From Hokanson and Katzen (1978).

ing of this stream, it is economically feasible to concentrate the sugars to a 65% solution for sale as a feed supplement or for conversion into furfural. The plant was closed for economic reasons after World War I.

More recently, interests in the acid hydrolysis of cellulose have revived on account of two apparently unrelated problems, i.e., the magnitude of the current energy crisis and the increasing problems of solid waste disposal.

B. Acid Hydrolysis of Cellulosic Wastes

A bench-scale production of sugar by acid hydrolysis of municipal wastes has been investigated by the Thayer School of Engineering (Ware, 1976). The process can be accomplished in a continuous flow reactor to yield about 52-54% sugar. The reaction is carried out isothermally at 230°C using 1% (by weight) sulfuric acid and a reaction time of 20 seconds. The yield was found to be a function of reaction time, temperature, concentration of the acid, and solid-to-liquid ratio in the slurry. As the hydrolysis conditions also favor sugar decomposition, the maximum yield expected is 53% conversion of cellulose to sugar.

Fermentation of the hydrolysis products with S. cerevisiae in an 800 ml fermentor was subsequently carried out giving a yield of ethanol from 85-93% after approximately 20 hours. It is also found that the rate of fermentation increased with higher sugar concentration (from 4 gm/100 ml to 12 gm/100 ml), but not the yield.

In Britain, Porteous (1972) has proposed a process to produce ethanol from domestic refuse. A flow diagram is presented in Fig. 13 for an input of 250 metric tons/day. The refuse is pulverized and discharged into a floatation separator (dry separation may also be possible) or a special pulper to allow separation of the refuse into a dense and a light fraction. The pulped fraction, mainly cellulosic material, is then fed to a fine particle and plastic removal section and then to a reactor for hydrolysis at 230°C with 0.4% H_2SO_4 at an optimum residence time of 1.2 minutes to obtain a maximum conversion to fermentable sugars. This is followed by flash cooling using the process feed water as coolant, neutralization with $CaCO_3$, and filtration. Fermentation is then carried out for roughly 20 hours at 40°C and the resulting 1.7% aqueous ethanol solution is distilled or rectified to yield 95% ethanol.

The disadvantages of the acid hydrolysis process have been summarized by Ghose (1978):

1. Expensive corrosion-proof equipment is needed for the acid treatment, resulting in a high capital cost.

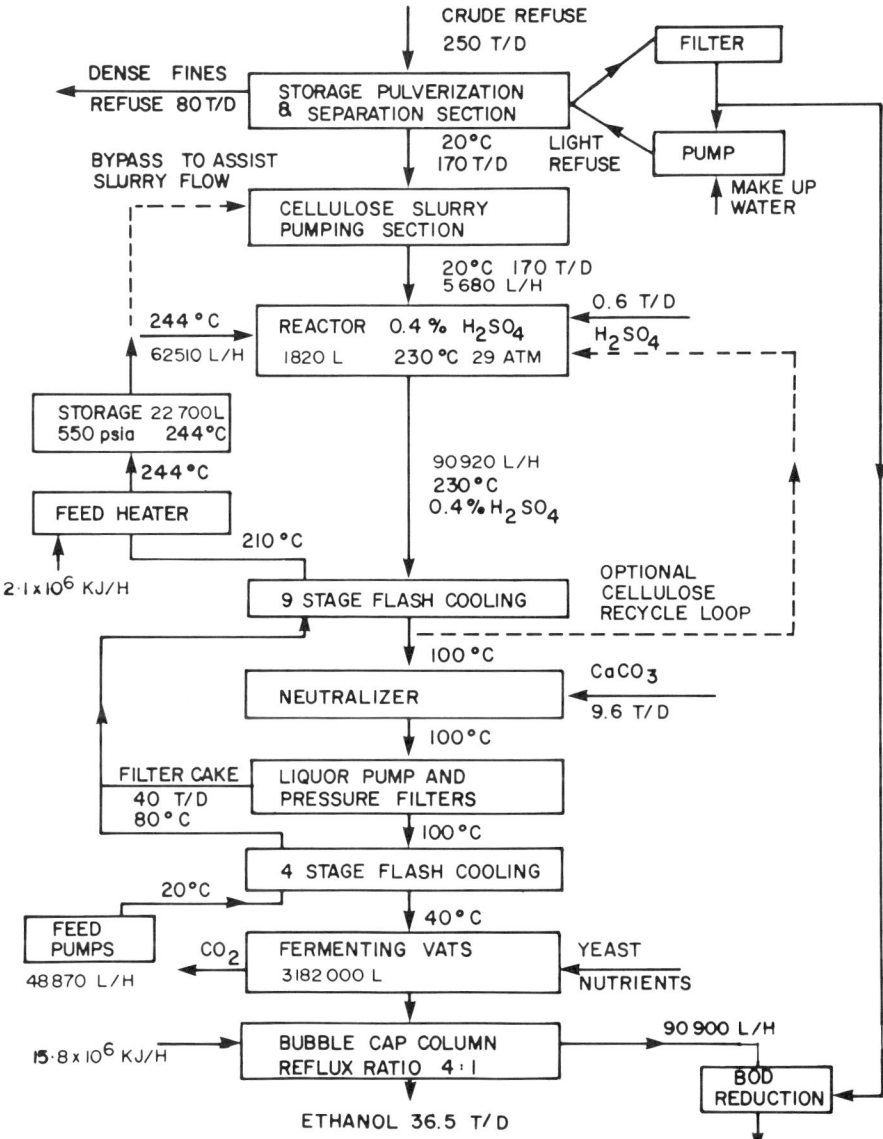

FIG. 13. Flow diagram for continuous production of ethanol from refuse with 60% content. T/D, tons per day; L/H, liters per hour. From Porteous (1972).

2. The crystalline structure of cellulose makes it very acid resistant so that high temperature and high acid concentration are needed for hydrolysis. Consequently, the resulting sugar is partially decomposed.

3. Waste cellulose usually contains impurities that will also react with acid, producing undesirable by-products and conversion compounds.

Because of these disadvantages, the acid hydrolysis has been deemphasized in recent years and efforts have been directed toward enzymatic hydrolysis.

C. Enzymatic Hydrolysis of Cellulose

Although many microorganisms are cellulolytic, it has been seen that all active developments of practical processes for saccharification are being considered on the basis of the *Trichoderma* enzyme system. Based on filter paper as an assay substrate very few organisms appear to produce more than 0.3 IU/ml in broths. Wild strains of the fungus *Trichoderma*, such as QM 6a, produce 0.5 IU/ml, and a recently obtained mutant strain of QM 9414 is reported to have produced up to 8.2 IU/ml. A comprehensive list of other cellulolytic microorganisms is presented by Mandels and Andreotti (1978).

In order to obtain more active strains of cellulolytic fungi, which reach better yields and show faster rates of hydrolysis, Reese (1975) suggested a mutation program as well as the optimization of the culturing conditions. Both aspects are currently being extensively studied.

Elder (1976) recognized the importance of microorganisms in the rapid decay of fallen trees in tropical areas, and suggested looking in such a location for cellulases with high activities. Herr *et al.* (1978) followed this suggestion and examined five strains of fungi for cellulolytic activities. The five fungi examined were: *Aspergillus niger, Lenzites trabea, Myrothecium verrucaria, Trichoderma komingii*, and *Trichoderma lignorum*. Two of them (*A. niger* and *T. lignorum*) were isolated from soil samples coming from rain forests of Zaire, Africa. They found that *Trichoderma, Aspergillus*, and *Myrothecium* were among the most active strains, especially *A. niger* and *T. lignorum*, which exhibited considerable cellulolytic activities. Further research in this area is needed.

D. *Trichoderma* Cellulase

The *Trichoderma* cellulase fermentation has some unusual problem areas that relate to the insolubility of the substrate and the multiplicity of the enzyme system. Ghose (1978) and Mandels and Andreotti (1978) have presented the following:

1. Cellulase is an induced enzyme in *Trichoderma* and is produced only when the fungus is grown on cellulose or on other glucosides containing β-1,4 linkages, such as cellobiose, lactose, or sepharose. This means that for practical purposes the fungus must be grown on cellulose, which is a slowly metabolized, insoluble substrate. This creates challenging fermentation problems.

2. Cellulose and insoluble impurities, such as lignin, absorb cellulase, thus reducing the amount of cellulase available for hydrolysis.

3. The synthesis of enzyme is strongly repressed by soluble sugars or other easily metabolizable substrates.

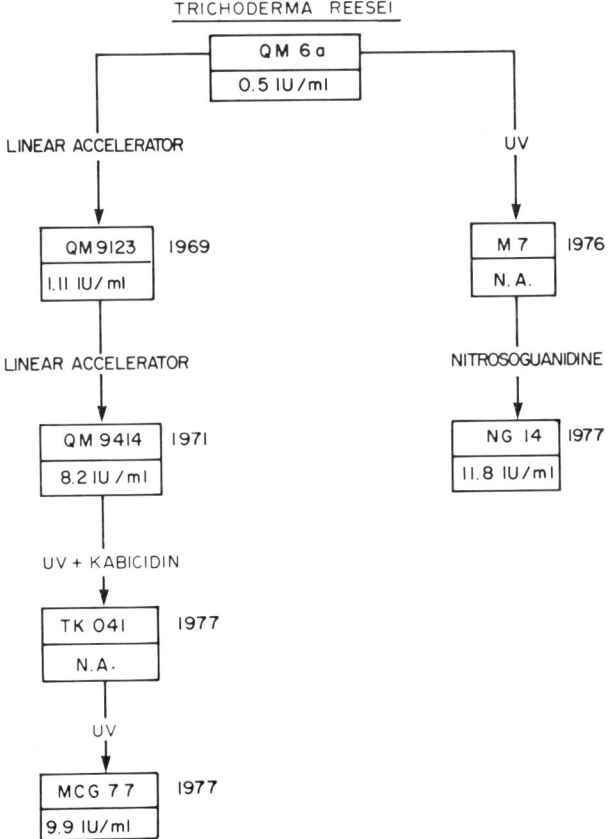

FIG. 14. *Trichoderma reesei* strain improvement by mutation. UV, ultraviolet; N.A., not analyzed. From Ghose (1978).

4. Control of pH is critical. If pH is not controlled, enzymes will be inactivated by severely acidic conditions. However, the yield of cellulase is reduced at higher pH.

5. Cellulase has a low specific activity on crystalline cellulose, which is because of its insolubility and recalcitrance. Several alternative routes are proposed: (a) increase of cellulose concentration, (b) addition of glucose to the cellulose media, (c) use of continuous cultures, (d) use of two-stage fermentation, and (e) recycling of cells. Among these alternatives, Mandels and Andreotti (1978) considered cell recycling as the more promising area for further research.

6. Because of low bulk density of cellulose, higher concentrations result in thick slurries, which leads to severe foaming.

7. Temperature profiling and pH cycling are essential inputs for productivity enhancement.

Mutagenic improvement in *Trichoderma* strains has resulted in 16-fold enhancement in filter paper (FP) activity over the parent strain QM 6a (Fig. 14). Continuous developments in the mutation work have made it possible to achieve 11.8 IU/ml FP activity with a productivity of 56 IU/ml with a new mutant *Tr* NG14. Some of the significant improvements made over the past 3 years are listed in Table V.

Submerged fermentation processes for cellulase production have been proposed and were economically evaluated (Wilke *et al.*, 1976a). Cost of the enzyme is estimated at 2.7¢/liter and is very likely to be cut down by 80% with further developments in process improvements and mutant strains. Practically all of the commercial cellulases available today are produced in Japan by solid culture fermentation (Table VI) (Ghose, 1978; Toyama, 1976). This fermentation is of special interest in developing countries because the technology is fairly simple. The drawbacks of solid culture for enzyme production are (a) difficulties of monitoring and controlling the fermentation, and that (b) some of the cellulase components tend to remain adsorbed on the solid residue.

According to Ghose (1978) and Spano (1976) the following future research is recommended:

1. The wild-type strain *Trichoderma* (*Tr* QM 6a) has already been mutated through four improved strains to a promising level; its further improvement requires extensive studies.

2. Optimization of the hydrolysis reaction to achieve a higher sugar production rate.

3. Development of a cost-effective pretreatment for cellulosic materials to make them readily susceptible to enzymatic hydrolysis.

TABLE V
IMPROVEMENTS IN CELLULASE PRODUCTION PROCESSES

Organism	Inducer[a]	Productivity (IU/liter/hour)	Residence time (QR) in CR (hours) or fermentation time (t) in BR (hours)[a]	Nature of process	Enzyme activity (IU/ml FP)	References
Tr QM 9414 (second generation mutant of Tr QM 6a)	0.5% cellulose	18.0	QR = 50	Continuous with cell recycle in 14-liter fermentor	0.91	Wilke et al., 1976a
	0.5% VP cellulose	21.2	QR = 50	Continuous process in 5-liter fermentor	1.01	Sahai and Ghose, 1977
	5.0% NIB 40 cellulose	42.0	t = 120	Batch process in 5-liter fermentor	4.60	Nyström and DiLuca, 1977
	0.5% VP cellulose	30.0	QR = 50	Continuous with cell recycle in 5-liter fermentor	1.20	Ghose and Sahai, 1979
	6.0% cellulose	39.0	t = 212	Batch process in 10-liter fermentor	8.20	Benedict et al., 1978
Tr MCG 77	6.0% cellulose	47.0	t = 212	Batch process in 10-liter fermentor	9.90	Benedict et al., 1978
Tr MCG 14	6.0% cellulose	56.0	t = 212	Batch process in 10-liter fermentor	11.80	Benedict et al., 1978

[a] CR, continuous reactor; BR, batch reactor; VP, purified powdered cellulose of native cotton from V. P. Chest Institute, Delhi; NIB 40, partially delignified unbleached fibrous hardwood pulp from Brown and Co. (Berlin, New Hampshire).

TABLE VI
COMMERCIAL CELLULASE PREPARATIONS AVAILABLE FROM JAPAN[a]

Commercial name	Remarks
Cellulase Onozuka CUY	Commercial grade
Cellulase Onozuka CLA 820-1	Commercial grade
Meicelase CMB-233-1	Commercial grade
Meicelase CEP-233	Original crude

[a] From Toyama (1976).

4. Development of adequate cellulosic feedstocks, be they from urban/industrial waste, agricultural residues, or biomass from land and water farming.

5. Pilot plant investigation should be made for a realistic economic evaluation of the cellulase production process.

6. More quantitative efforts should be directed toward the development of solid culture technique.

Of particular interest is also the study on mixed enzymes (Ghose, 1978). If any particular component is in short supply in a cellulase system from one source it can be fortified with the same component from another source, richer in that component. As an example, *Trichoderma viride* (Tv) enzyme is deficient in xylanase and can be fortified with *Aspergillus wenti* (Aw) cellulase which is comparatively rich in this component. The results of such a tailored mixture are presented in Table VII.

E. Pretreatment of Cellulose

In order to enhance the susceptibility of cellulose for enzymatic hydrolysis, various techniques to pretreat the cellulosic material have been developed. Table VIII gives a summary of the pretreatment processes.

TABLE VII
EFFECT OF MIXED CELLULASE ENZYME ON HYDROLYSIS[a]

Substrate	Enzyme: Hydrolysis in first hour (milligram RS[b] produced per milliliter)				
	Tv[b]	Aw[b]	Tv:Aw 2:1	Tv:Aw 1:2	Tv:Aw 1:1
Cellulose (5%)	3.2	0.9	2.7	1.9	2.5
Bagasse (5%)	1.8	1.6	2.8	2.6	3.4

[a] From Ghose (1978).
[b] RS, reducing sugars; Tv, *Trichoderma viride*; Aw, *Aspergillus wenti*.

TABLE VIII
METHODS FOR PRETREATMENT OF CELLULOSE[a]

Physical	Chemical	Combination
Ball milling	Sodium hydroxide	Hot ball milling
Hammer milling	Ammonia (liquid)	NaOH and ball milling
Weathering	Ammonia (gas)	NO_3^- and irradiation
Boiling	Hydrochloric acid	
High-pressure steam	Acetic acid	
Electron irradiation	Sulfuric acid	
Photooxidation	Sodium sulfide	
Wetting	Sulfur dioxide	
γ-Irradiation	Nitrogen dioxide	
	Potassium hydroxide	
	Phosphoric acid	

[a] From Dunlap et al. (1976).

Ball milling and γ-irradiation are among the more effective physical methods, whereas sodium hydroxide treatment and sulfur dioxide gas treatment are the more viable chemical methods.

1. Ball Milling

Milling cellulosic material is an often used and popular method for increasing cellulose digestibility. The material to be treated is simply placed on a ball mill and subjected to shearing and compressive forces generated by the mill for a specific period of time. The most obvious changes in cellulose physical properties include a reduction in crystallinity, a reduction in mean degree of polymerization, an increase in the fraction of the material that is water soluble, and a marked decrease in particle size. It is reported that the digestibility of the milled cellulose is directly related to the time of milling. The cost of ball milling cellulose on a large scale is not very well defined. Estimates of costs ranging from about $33 to $110/metric ton ($30 to $100/ton) of cellulose reduced to less than 200 mesh size have been reported (Dunlap et al., 1976).

2. γ-Irradiation

γ-Irradiation of cellulosic materials has provided increased digestibilities resulting in a lower degree of polymerization, lower crystallinity, and higher moisture absorption capacity. Studies have shown that a level of about 5×10^7 rad on wheat straw and various hardwoods is needed before a significant increase in digestibility is observed (Dunlap et al., 1976). The cost of this level of irradiation is about $138–$165/metric ton ($125–$150/ton). It is known that addition of certain nitrate and nitrite salts to cellulosic materials prior to irradiation seems to aid in catalyzing the degradation reactions, but costs for these processes have not yet been reported.

3. Sodium Hydroxide Treatment

Sodium hydroxide treatment of cellulose is probably the oldest and best known method of increasing digestibility. The application of the alkali results in disruption of lignin structure, hydration and swelling of the cellulose, and decrease in cellulose crystallinity. As long as the treated cellulose remains moist these modifications persist, but drying often irreversibly negates them. Maximum digestibility was found to be reached at an alkali level between 0.1 and 0.15 gm NaOH per gram solids. The cost of sodium hydroxide to treat 1 metric ton of cellulosic material was estimated to be about $22–$33, and the cost of handling, mixing, neutralizing excess alkali, and washing the treated material would probably add an additional $44–$66/metric ton. Therefore, the treatment cost would be about $66–$99/metric ton of cellulose material fed to the process (Dunlap et al., 1976).

4. Sulfur Dioxide Treatment

Sulfur dioxide gas treatment of cellulose is probably the most recent innovation in cellulose treatment techniques and may be one of the most promising. Early work at the U.S. Forest Products Laboratory has shown that reaction of gaseous sulfur dioxide with moist cellulosic materials at 120°C for about 2–3 hours resulted in an increase of several times in digestibility. The possibility of using a gaseous agent to treat cellulose is most interesting for several reasons: The gas could be easily removed from the cellulose after treatment and presumably recovered for total reuse; the treated material

TABLE IX
COMPARISON OF PHYSICAL AND CHEMICAL PRETREATMENT OF CELLULOSE[a]

	Physical	Chemical
Bulk density	Increases (8–30%) but at increased cost (10–15% slurry can be handled easily)	Decreases considerably; even 7–10% slurry becomes difficult to handle
Reactor space	Increased reactor volume required because of lignin	With delignification less reactor volume is required
Extent of hydrolysis	Nearly 50% when 10% cellulose is hydrolyzed	Nearly the same extent as in mechanical treatment
Waste outflow	None	Causes problems for wash water which can be recycled

[a] From Ghose (1978).

would be in a solid form, and solubilized materials could be removed if desired by a small volume of wash liquid; and penetration of the agent into the cellulose structure would be more rapid and complete with a small molecule the size of sulfur dioxide or the resulting anion in an aqueous solution.

A comparison of some features of physical and chemical pretreatment methods is given in Table IX.

F. Enzymatic Hydrolysis Processes

So far enzymatic hydrolysis of cellulose has been carried out at most on a small pilot plant scale. A schematic diagram of an enzymatic hydrolysis process is given in Fig. 15. As the figure shows, the process is divided into several steps and includes two basic inputs, namely, nutrients for the fungus and the cellulosic materials to be hydrolyzed. The nutrients supply nitrogen and other requirements not satisfied by the cellulosic materials. Consequently, the first step involved nutrient "feed" preparation. The nutrients are dissolved in water and the nutrient medium is then heat sterilized. While these two steps are in progress, the cellulosic wastes are heated to 200°C and then size reduced to a particle size of about 50 μm. This finely milled cellulose would give a sugar recovery of about 70%, whereas in the case of unmilled cellulose only 6-7% sugar can be recovered. Of the various milling processes available, only ball milling results in an extensive size reduction, greatest bulk density, and maximum susceptibility. Hammer mill-

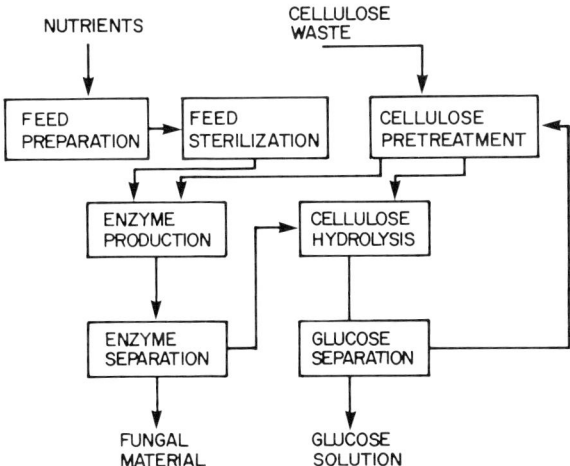

Fig. 15. A schematic diagram of a process to enzymatically saccharify cellulosic waste. From Golueke (1977).

ing, fluid energy milling, colloid milling, and alkali treatment proved to be much less satisfactory (Golueke, 1977).

The enzyme production step, which is the major one in the process, involves microbial growth and subsequent enzyme production. The fungus is propagated as a submerged culture in a fermentor unit equipped for mixing and aerating the growth medium. Here, the attainment of high levels of cellulase occurs simultaneously with the establishment of a cellulose-degrading enzyme system that is induced in the presence of cellulose. Rosenbluth and Wilke (1970) estimated that it would take four fermentor units of 197,000 liters (52,000 gal) each to process 9 metric tons (10 tons) of pretreated cellulose per day. Aeration is accomplished by injection and is kept at the maximum speed to maintain an adequate oxygen transfer. The separation of the enzyme solution from the fungal mycelial mass can be accomplished by means of continuous filtration or by centrifugation. The filter cake, consisting mainly of cell material, is sent to a dryer and the enzyme-rich filtrate is fed continuously into the hydrolysis vessel.

The cellulose hydrolysis or saccharification step is the most important one in the process. Here, the enzyme solution produced in the previous step comes into contact with the pretreated cellulosic materials. The enzyme solution catalyzes a hydrolysis of the solid cellulose to the glucose product. The product stream (about 5% glucose) is continuously withdrawn from the unit. Further growth of the fungus in this step is prevented by elevating the temperature of the mixture to 50°C.

Finally the glucose solution is separated from the unhydrolyzed cellulose by means of filtration. This glucose solution can be used for fermentation to ethanol. As for the residue, it is dried, reground, and returned to the hydrolysis vessel. The recycled cellulosic material must be ground repeatedly to render the residual cellulose more susceptible to hydrolysis. Cell material filtered from the culture broth (enzyme solution) is passed through a dryer.

Research has been carried out at the Natick Development Center (NDC) and a California university (Berkeley) to design and evaluate a process for the enzymatic hydrolysis of newsprint. Newspaper was chosen as the substrate because it has a high cellulose content, its composition is consistent, and it is readily available.

The process designed by the NDC is shown in Fig. 16. Process description is given by Allen (1976), Andren *et al.* (1976), Bailie (1976), Spano (1976), and Ware (1976). Economic analysis of the process indicated that a major part of the overall process cost is the production of enzymes. With the current process, much of the enzyme is lost by adsorption on the noncellulosic fraction of the substrate. Little work, however, has been reported on enzyme recovery. Another major part of the overall process cost is substrate pretreatment. Ball milling has been found to be the most effective physical

PILOT PLANT PROCESS FOR NEWSPAPER HYDROLYSIS

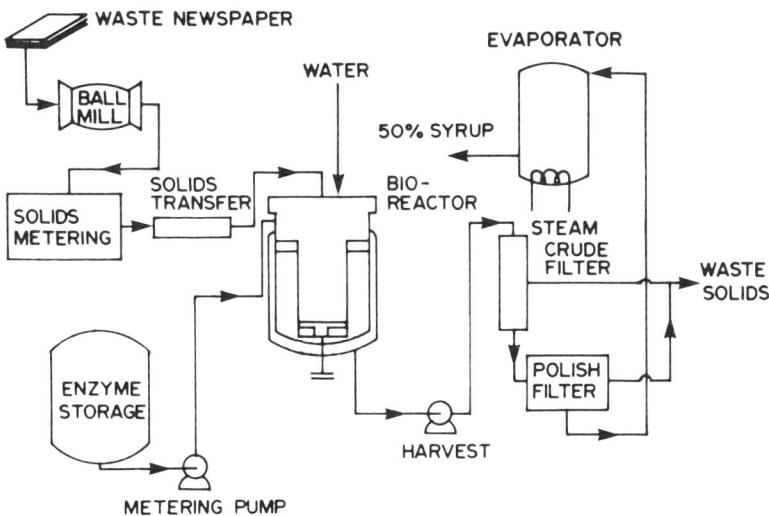

PILOT PLANT PROCESS FOR CELLULASE PRODUCTION

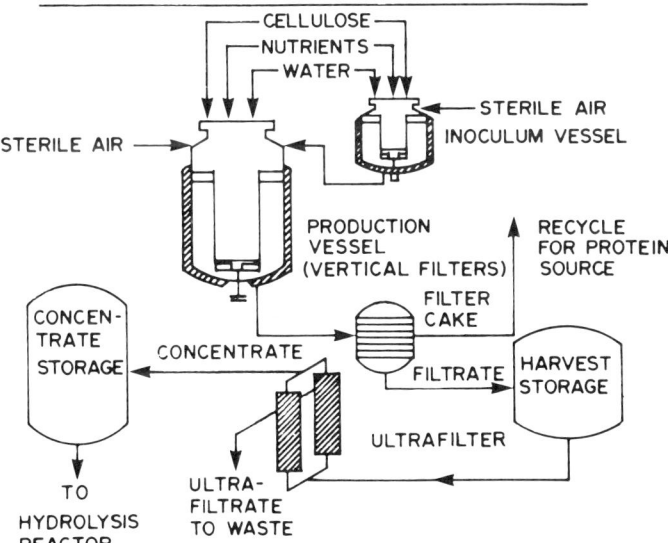

FIG. 16. Enzymatic hydrolysis of newsprint by Natick Development Center (NDC). From Spano (1976).

method of making the maximum amount of cellulose in the substrate available for enzymatic hydrolysis, but it is expensive. Andren and Nyström (1976) reported that wet chemical pulps are by far the most susceptible substrates for enzymatic hydrolysis. This suggests that either a chemical wood pulp be produced specifically for use as a substrate for enzymatic hydrolysis or that chemical pulping technology be applied as a pretreatment method for other potential cellulosic substrate. In the former case, cost would be lowered by using cheaper previously wasted or unused wood and less costly physical treatment techniques because maintenance of fiber strength for use in the paper industry is no longer a criteria. Also, fillers and other additives would no longer be needed.

The process developed at the University of California is shown in Fig. 17. The primary feed consists of 803 metric tons/day (885 tons/day) of newsprint containing 6% moisture. By means of moderate shredding and milling the feed is reduced to approximately 20 mesh. The size reduction is not critical as long as the material will form an aqueous suspension that can be pumped, agitated, and filtered. An additional 0.54 metric tons/day (0.6 tons/day) of feed material is diverted to the first enzyme induction fermentor after sterili-

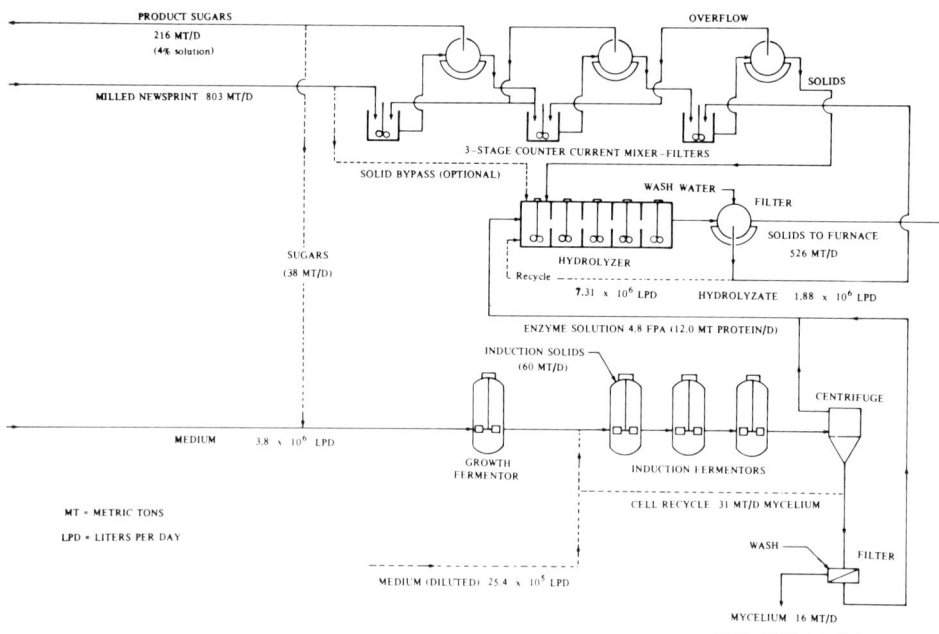

FIG. 17. Enzymatic hydrolysis of newsprint (Wilke et al., 1976b).

zation with steam. The product sugar stream from the hydrolysis is contacted countercurrently in three mixer-filter stages with feed solids for enzyme recovery. Each mixer-filter stage consists of a mixing tank to provide 30 minutes of contact time and a horizontal belt vacuum filter to separate the solids from the liquid. A total enzyme recovery of 90% is theoretically predicted based on adsorption studies.

Hydrolysis is conducted over 40 hours at 45°C at a solid/liquid ratio of 1:20 w/w based on inputs to the hydrolyzer. The latter consists of five agitated, cylindrical, concrete digesters of the type used for solid waste treatment in sanitary engineering. Cellulose conversion of 50% is assumed, at an overall enzyme strength equivalent to 3.5 filter paper activity (FPA) units in the hydrolyzer. Provision is made for the recycling of a portion of the product sugar solution (plus enzyme) back to the hydrolysis vessel. A sugar concentration of 40% is obtained. A range of sugar levels is possible depending on the mode of operation and amount of sugar recycling employed.

Makeup enzyme is produced in a two-stage fermentation system, employing the fungus *Trichoderma viride* QM 9414. Cell growth is obtained in the first stage at the dilution rate of 0.2/hour, employing a medium containing 1% product sugars plus minerals and protein nutrients. The induction system is operated at an overall dilution rate of 0.017/hour excluding the cell recycle stream. Both stages employed stainless steel vessels with agitators operated at 30°C with aeration rates of 0.15 and 0.015 vol/vol/min in the growth and induction stages, respectively. The growth stage feed is sterilized in a heat exchanger system. The induction section effluent is passed through a centrifuge, from which a portion of the underflow is fed back to the first induction stage. Ten induction stages in series are employed. A cell recycling fraction of 0.65 is used. The recycling fraction, is the fraction of cells leaving the last induction stage which is returned to the first stage. In this case, the use of recycling will maintain the cell density in the induction system at 7 gm/liter, assuming there is negligible growth in the induction system when newsprint is employed. The resultant enzyme production is sufficient to provide an enzyme concentration of 3.5 FPA in the hydrolyzer. A portion of the centrifuge underflow is filtered and the cells are discarded to maintain adequate cell viability. The centrifuge overflow will contain a small concentration of cells. Removal of these cells prior to hydrolysis is assumed unnecessary because *T. viride* will not grow at the hydrolysis temperature. Spent solids from the hydrolyzer following filtration are fed to a furnace and steam power plant to provide process steam and electricity for the process. A substantial excess of energy is available in the spent solids, sufficient to operate an alcohol fermentation plant and to produce some additional by-product power.

G. Fermentation of Hydrolyzate to Ethanol

Fermentation of the cellulose hydrolyzate was studied by Cysewski and Wilke (1976). They designed an industrial size fermentation plant capable of producing 89,700 liters/day (24,000 gal/day) of 95% ethanol using the sugars produced from the enzymatic hydrolysis of cellulosic wastes. The hydrolysis product was found to be 70% fermentable by *Saccharomyces cerevisiae*, thus requiring a 14.3% solution of hydrolyzate sugars to obtain the optimum feed of 10% fermentable sugars. Preliminary cost analysis indicated that it is economically favorable to concentrate the sugar to 14.3%. It is claimed that the concentration costs of 0.71¢/liter (2.7¢/gal) of ethanol produced more than offset the savings in fermentation and distillation cost.

Figure 18 shows a schematic flow diagram of the ethanol fermentation process. After the hydrolyzate sugars have been evaporatively concentrated from a 4.0% to a 14.3% solution, protein and mineral supplements are mixed with the sugars. Sterilized by steam injection, the fermentation broth is distributed to five continuous fermentors, each operating at a dilution rate of

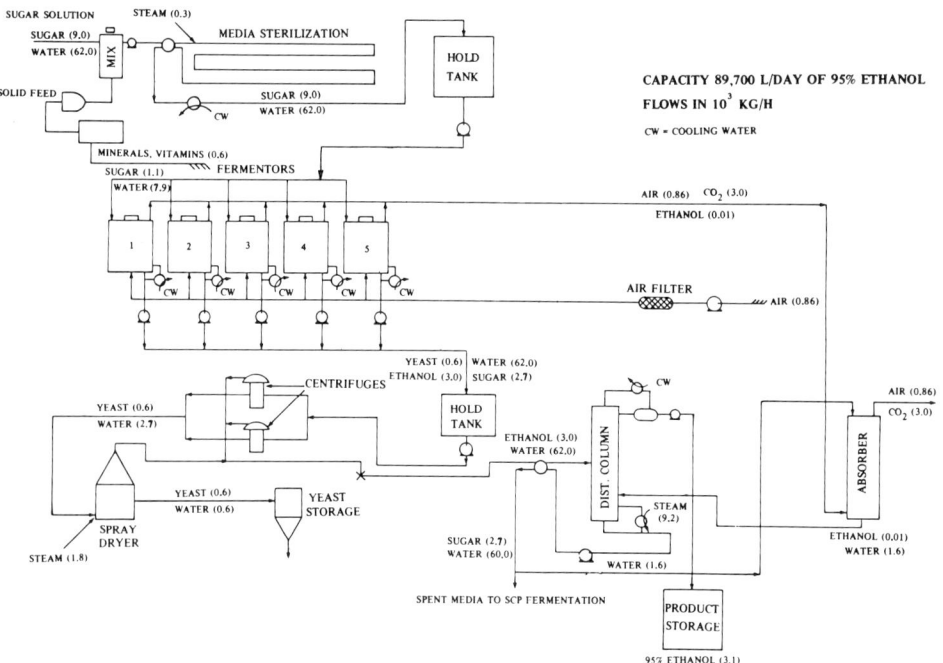

FIG. 18. Flow diagram for ethanol production from cellulose hydrolyzate. From Cysewski and Wilke (1976).

0.17/hour. A low flow of air (8.0×10^{-4} vol/vol/min) is sparged through the fermentors to maintain the oxygen at the optimum level of 0.07 torr. The fermented liquor then passes to two continuous centrifuges and the yeast is removed. The yeast is subsequently dried and stored for sale as a protein feed supplement. The clarified "beer" from the centrifuge is next distilled to concentrate the ethanol to 95% by weight. An absorber, using the distillate bottoms as the absorbing liquid, is employed to recover ethanol lost in the exit gases (air and CO_2) from the fermentors. The ethanol-rich stream from the absorber is also fed to the main distillation unit for final ethanol recovery.

Saccharomyces cerevisiae used in the ethanol fermentation will ferment only 70% of the reducing sugars in the hydrolyzate. The remaining 30% of the sugars (xylose and cellobiose) are fed to an aerobic fermentation process to produce single-cell protein from a torula yeast. Although torula yeasts will utilize the remaining sugars, it is a facultative aerobe and does not produce ethanol.

The cost of production is about $0.28/liter ($1.05/gal) for 95% ethanol. Therefore, improvement of the process, both in hydrolysis of cellulose and fermentation of the hydrolyzate, is needed before the process is commercialized.

H. Enzymatic Hydrolysis of Cellulose Waste from the Food Processing Industry

The potential for utilizing solid wastes resulting from fruit and vegetable processing operations as a source of cellulose for enzymatic conversion has been studied by Cooper (1976). He concluded that at this time, cellulose waste from food processing industries does not have a great potential for the following reasons:

1. Most processors have already found markets for a large part of their residues.
2. On a ton for ton basis, most food processing residues are not as good sources of cellulose as are many other potential sources. Of the 2,867,000 metric tons (3,160,000 tons) of dry residues generated yearly in the United States, only 499,000 metric tons (550,000 tons) are crude fiber. Yet they contained about 1,786,000 metric tons (1,969,000 tons) of extractable carbohydrates, primarily soluble sugars and carbohydrate materials easily hydrolyzed.
3. All food processing residues are highly putrescible in their natural form and, therefore, cannot be stored long.
4. Most food processing residues are seasonal in nature.

I. ONE-STEP PROCESS FOR PRODUCTION OF ETHANOL FROM CELLULOSIC WASTES

Cellulose hydrolysis and ethanol fermentation as a one-step process have been studied by Takagi et al. (1977). This process is operated aerobically at 30°-40°C. They observed an accelerated production rate of ethanol compared to a two-step process of saccharification and fermentation. Inhibition by ethanol and cellulase was reduced to less than half compared to glucose.

A similar one-step process is described in Hoge's patent (1977). The process is essentially comprised of three stages, which are:

1. Sterilization. The cellulosic material is subjected to a steam treatment. Thereafter the resultant sterilized mass is adjusted to the proper solid content, temperature, and pH for effective fermentation reactions.

2. Concurrent digestion and fermentation. The sterilized cellulosic material is hydrolyzed to fermentable sugars by inoculation and fermentation of the mass with yeast and enzymes to produce alcohol.

3. Alcohol removal. The alcohol formed is removed by stripping, e.g., vacuum distillation; the yeast and enzymes are then recycled for reuse.

Improvement features of this one-step process include, for example:

1. The use of nonproteolytic yeasts to ferment sugars, in conjunction with the removal of carbon dioxide and vacuum distillation of ethanol in the same reaction vat as the enzymatic hydrolysis of cellulose to sugars. By this, the following desirable results are obtained:

    a. Elimination of the large volume of cooling water, such as is normally required to control the temperature of fermentation. In conventional processes, 190–300 liters (50–80 gallons) of cooling water per gallon of alcohol are needed, whereas practically none is required in this process.

    b. The removal of the reaction products may be carried out in a continuous or intermittent manner in order to force the reactions to completion. Sugar formed by the enzymatic hydrolysis is immediately removed by fermentation by the yeasts. The alcohol formed by the fermentation of the sugars is likewise continuously removed by the vacuum distillation.

    c. The bottoms or solutions that remain after the sugar and alcohol productions are completed will still retain active enzymes and yeasts. This inoculum solution, therefore, may be recycled to dilute new quantities of incoming sterilized cellulosic fibers.

2. Addition of mineral acid to the cellulosic materials having a high solid content (e.g., 25–40% by weight) prior to the steam sterilization. This would result in a partial breakdown of cellulose, which is then more accessible to subsequent enzymatic hydrolysis. Cellulosic materials sterilized in the presence of an acid can be slurried at a higher solid content than is possible with

fibers sterilized with steam alone. The use of higher cellulosic concentrations will help by reducing the tank sizes as well as speeding up the reactions.

3. Combination of sterilization with partial acid hydrolysis to give the maximum benefit with respect to the cost of this treatment step.

J. Comparison of Acid and Enzymatic Hydrolysis

Some important features of acid and enzymatic hydrolysis of cellulosic wastes to produce sugars for alcohol fermentation are compared in Table X.

K. Future Developments

At present, the production of industrial alcohols from cellulosic materials is technically feasible but commercially unattractive. The technical barriers that must be lowered or completely removed to achieve a viable economical process include (Spano, 1976):

1. Development of very effective substrate pretreatment that significantly increases the susceptibility of crystalline cellulose to enzymatic hydrolysis.

2. Increase of cellulase productivity of the fungus through mutation. Today, the cellulase productivity of *Trichoderma* sp. has been quadrupled by

TABLE X
Comparison of Acid and Enzymatic Hydrolysis[a]

	Acid	Enzymatic
Pretreatment	May be necessary	Necessary
Rate of hydrolysis	Fast (minutes)	Slow (hours)
Temperature	High (230°C) (Battelle Res. Institute, Geneva, is engaged in developing a process)	Low (50°C)
Pressure	Above atmospheric	Atmospheric
Yield	Varies depending on material and process conditions	Varies depending on substrate, enzyme, and process conditions
Formation of interfering by-products	Probably formed	Not likely
Commercial processes	Yes, in USSR	No, not even an adequate pilot plant

[a] From Ghose (1978).

mutation of the parent and subsequent mutant strains. A tenfold increase over the original wild strain appears possible through more extensive mutation studies.

3. Optimization of the submerged fermentation process to increase the production rate of cellulase and reduce enzyme production costs.

4. Optimization of the cellulosic hydrolysis step to achieve no less than 50% saccharification of the cellulose substrate in 24 hours or less.

5. Development of commercially applicable techniques for harvesting and processing syrups for fermentation.

6. Development of commercially applicable techniques for processing residues for recycling or other applications.

A discovery at Stanford Research Institute (SRI) International (Menlo Park, California) of a high-temperature cellulase enzyme complex may help to cut down the cost of cellulase production (Anonymous, 1978b). The source of this enzyme complex is *Thielatin terrestris,* a thermophilic soil fungus. It may be cultured at high temperatures, between 60° and 90°C; contaminating organisms therefore are purged from the culture automatically. In addition, the tolerance of this enzyme complex to high temperatures enables the reaction to be operated at a higher temperature, which results in faster reaction rates. This could translate into important savings for users. At present, the SRI is looking into producing this cellulase commercially.

Wilke and Blanch (1977) are currently investigating various alternatives to improve the efficiency as well as the economics of the cellulase enzymatic process. For example, at present countercurrent hydrolysis is under study. It is expected that more economical procedures could be developed in the near future to make the enzymatic process more competitive.

L. Cellulose vs. Corn

Ethanol can be produced from the fermentation of sugar (or molasses) directly, or from sugars derived from cellulose saccharification and hydrolysis of starch, such as that found in corn. At present, cellulose saccharification is only in the developmental stage, whereas the corn sugar industry is a well-established one. However, it is expected that cellulose saccharification will become competitive with the corn sugar industry in the near future.

Table XI presents certain factors that affect the competitive situation of each starting material. Yield per acre makes the base cost for cellulose clearly superior to corn. The removal of hemicelluloses and lignin may be equated, costwise, to the physical removal of the hull germ of the corn kernel. However, starch has the big advantage of solubility and, therefore, less costly concentration. As far as sugar quality is concerned, if adequate removal of

TABLE XI
COMPARISON OF CORN AND CELLULOSE AS SOURCES FOR SUGAR PRODUCTION[a]

	Corn		Cellulose	
Supply	Broad based, large quantities, single, defined source, 2 ton/acre		Broad based, large quantities many potential sources, 10-15 ton/acre	
Feedstock	Amylopectin, amylose, some protein		Cellulose, hemicellulose, lignins	
Solubility	35%–40% (after thinning)		5–7.5% without excessive power required (needs concentration)	
Sugar stream quality	95% glucose, 5% trimers, isomaltose, maltose at 30–35% concentration		95% glucose, 5% other	
Costs	Corn starch[b]	9¢/lb	Cellulose	2–5¢/lb
	Enzyme & processing	2¢/lb	Enzyme & processing	7–10¢/lb
	Total	11¢/lb	Total	9–15¢/lb
	in 30% concentration		in 5–10% concentration	

[a] From Seeley (1976).
[b] 1975 value.

lignin and hemicelluloses is assumed, the quality of the sugar stream from a two-step enzyme process should be about the same.

VII. Ethanol from Other Wastes

Purification of wastewater is necessary to most industries, which means increased costs. One alternative to the conventional biologic purification is to use the wastewater in a process resulting in a product the value of which may cover the costs of purification wholly or partially. An alcohol product is considered as one of the viable alternatives.

A. Cheese Whey

During recent years much attention has been paid to utilization of the increasing amount of whey all over the world. Only a few countries succeeded in utilizing their total whey production (Reesen, 1978; Reesen and Strube, 1978). In the United States, only 60% is utilized. Although much has been done to solve this problem, the percentage utilization seems to be stagnant because of the increasing cheese production.

Reesen and Strube (1978) studied various alternatives for whey utilization. They concluded that from an economic point of view, one of the more promising ways seemed to be the production of alcohol, with the possibility of combining it with an ultrafiltration to separate and utilize the whey proteins. In this process, a selected yeast strain, *Kluyveromyces fragilis*, is cultured on molasses in a two-stage batch operation and is used as fermentor inoculum.

A plant is presently under construction in Denmark for the continuous fermentation of whey into alcohol for beverages. The whey is first concentrated by reverse osmosis and then by ultrafiltration. A by-product of the process, methane gas, will help power the operation (Anonymous, 1978a; Elias, 1979).

The ultimate economic impetus for alcohol production from whey is the search for alternatives to crude oil, as both an energy source and as a raw material for the production of chemicals. Considering the whey permeate, a raw material with no available energy, a total energy balance for alcohol production gives a net energy profit of 30%; i.e., the energy released by the alcohol is 30% greater than the energy required to produce it. The yield from the process corresponds to approximately 80% of what is theoretically obtainable, calculated on the basis of lactose content. About 42 liters of whey containing 4.4% lactose is required for production of 1 liter of 100% alcohol. A strain of *K. fragilis* will be utilized. The schematic diagram of the processing operations is shown in Fig. 19.

Whey permeate is pumped from the ultrafiltration plant to the storage tank, where the pH is adjusted. The permeate is then transferred through a sterilizer to the first fermentor. The yeast is cultured on molasses in a two-stage batch process. All of the culture from the second stage is pumped to the first fermentor, which has already been charged with permeate. The continuous addition of permeate starts after a couple of hours. When the first fermentor has been filled, the pump between the fermentors is started. When the second fermentor is filled, the fermented whey is pumped to the separator. The system is kept at a steady state and throughout the process, chemicals and antifoam agents are added. The yeast cream is removed by the separator and is pumped to a storage tank, where it is inactivated. The fermented whey is pumped via a surge tank to a distillation plant, where the crude alcohol is distilled.

Economic analysis of this process has shown that it can compete with synthetic alcohol produced from ethylene.

Because glucose and galactose are more universally fermentable sugars than is lactose, it is suggested that β-galactosidase-treated whey would make a better substrate for industrial fermentation. This aspect has been investigated by O'Leary *et al.* (1977a). They studied the production of ethanol by *K. fragilis* and *S. cerevisiae* using cottage cheese whey, in which 80–90% of the lactose had been prehydrolyzed to glucose and galactose. The results showed that the fermentation time was increased over that of unhy-

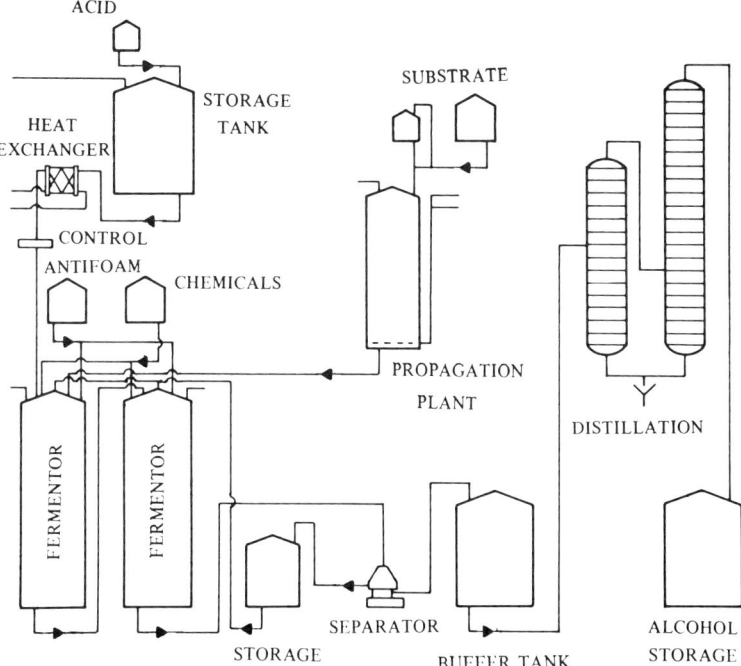

FIG. 19. Continuous production of ethanol from whey. From Reesen (1978).

drolyzed whey (120 hours vs. 72 hours by *K. fragilis* at 30 °C and an ethanol yield of 2%) because of a diauxic pattern of fermentation. However, they suggested that even though the hydrolyzed whey requires a longer fermentation time, it could lend itself to the preparation of concentrates with relatively high solid contents because of the greater solubility of glucose and galactose. This may result in higher alcohol yields than is usually possible in normal whey, because high-alcohol-producing yeast strains that are unable to ferment lactose can be used for the fermentation. In a subsequent study Reese (1975) found that an ethanol yield of 6.5% could be obtained using *S. cerevisiae* with a lactase-hydrolyzed acid whey permeate containing 30–35% total solids.

B. Coffee Waste

Use of coffee waste as a substrate for ethanol fermentation has been studied by Bhat and Singh (1975). In this process, freshly pulped coffee beans with mucilage and fruit skins were used. Fermentation was carried out using *S. cerevisiae* at 28°C over a period of 1 week. It was found that about 85% of sugar present had been utilized. Alcohol obtained was only 3–4% in concentration. The process does not seem to be economical at the present time.

C. POTATO PEEL WASTE

The use of peel waste from dry caustic peeling of potatoes for the production of alcohol was studied by Bloch et al. (1973). In this process, the peel waste is first mixed and acidified to a pH of 6 before it is autoclaved. The peel waste is then cooled to 67°C and mixed with unfiltered mold (*Aspergillus factidus*) broth in order to hydrolyze the starch into fermentable sugars. The mixture of peel waste and mold broth is agitated at 62–67 °C for 20 minutes before it is cooled to 30°C. Bakers' yeast is then added. Fermentations are carried out over a period of 68–72 hours. An alcohol yield of up to 90% was observed. However, no economic analysis of the process has been reported.

D. PINEAPPLE WASTES

Tan and Lau (1975) studied the production of alcohol from the mill juice of pineapple cannery wastes on a bench scale. Because of the low sugar content of the mill juice, it was necessary to enrich it with other sugar sources. Cooked sago (a local agricultural product of Singapore) was chosen in the study to enrich the mill juice to a level of 8 w/w %. Fermentation was carried out with 0.7 w/w % of rice cake (a very commonly used yeast for the fermentation of cooked rice to rice wine in Asia) at 38°C for 72 hours. They found that 13% by weight of mill juice or 9.4% by weight of the cannery wastes could be recovered as industrial alcohol. Presaccharification of the sago by acid hydrolysis is not necessary if rice cake is used as the source of yeast for fermentation. The use of rice cake not only simplifies the process and reduces the production cost, but also gives a better alcohol yield. This is because of the higher alcohol tolerance limit of the yeast contained in the rice cake than that of the conventional *S. cerevisiae*. Aside from the higher alcohol yield rice cake is cheap and easily available in Singapore.

E. SPENT SULFITE LIQUOR

The sulfite process, involving the delignification of wood with acid bisulfite, is widely used by mills in Europe and North America. The digestion of wood in an aqueous solution containing alkaline-earth bisulfites and an excess of sulfur dioxide solubilizes the lignin by combination with SO_2 or HSO_3^-. The wood cellulose largely remains undegraded, whereas the less resistant hemicelluloses are hydrolyzed to monosaccharides.

About 9180 liters (2430 U.S. gal) of spent sulfite liquor (SSL) are recovered per metric ton of pulp produced (Ericsson, 1947). This SSL contains 11–14% solids, depending upon the kind of cook and ratio of cooking liquor to wood. These solids represent up to 50% of the wood raw material. They consist essentially of lignin sulfonates (65–70%), hexoses and pentoses (20–30%), polysaccharides, galacturonic, and acetic acids, some resin, unconsumed bilsulfite, and ash (6–10%).

Spent sulfite liquor creates a major pollution problem when discharged into receiving water because its constituents—wood sugar and lignosulfonic acid—exert a large biochemical and chemical oxygen demand (BOD and COD), color the water, and may poison aquatic life. The readily fermentable sugars are the primary cause of the high BOD of the SSL, which ranges from 25,000 to 40,000 ppm (Ericsson, 1947). The lignin fraction is very resistant to microbial attack, and therefore constitutes little to the BOD but contributes to the COD (about 100,000 ppm).

Fermentative production of ethanol from SSL, dates back to 1907 when the first experimental plant was built in Sweden. The process found wide acceptance, especially in Europe during the war periods; for example, by 1948 there were 34 plants in operation in Sweden alone, with enough capacity to process some 80% of the spent sulfite liquor produced in that country. Today, only 14 plants are left in operation in Sweden. This reduction reflects the shift from sulfite pulping to pulping by Kraft process (Shreve and Brink, 1977), and the fierce competition with the petroleum industry for the alcohol market.

Ericsson (1947) has described the production of ethanol from SSL at the Bellingham plant (Fig. 20). Spent sulfite liquor from the digesters is discharged, together with wood pulp fibers, into blow pits that have perforated

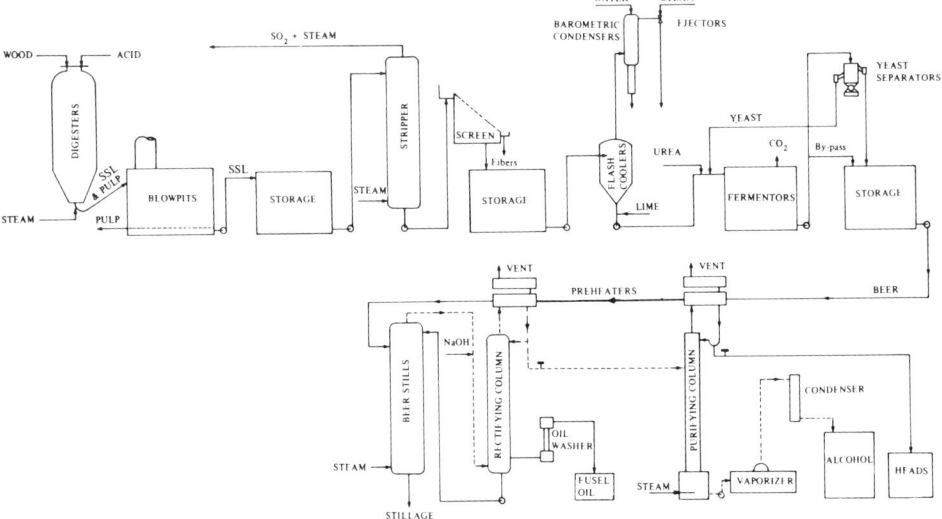

FIG. 20. Process flow diagram for the production of ethanol from spent sulfite liquor. From Ericsson (1947).

bottoms or plates of stainless steel. The sulfite liquor is drained off and stored in a tank at a temperature of about 90°C. From the storage tank, the liquor is pumped to the top of a column and flows downward over stainless steel perforated plates, while steam introduced at the bottom of the column and flowing upward removes sulfur dioxide and other volatile materials. The sulfur dioxide is reused. The hot sulfite liquor is pumped through screens that remove wood pulp fibers. It flows by gravity to a storage tank. Passage of the liquor through flash coolers reduces its temperature to 30°C and concentrates it by about 10–12%. A slurry of lime is added to adjust the pH to 4.5, and urea is added to the liquor as a source of nitrogen. The conditioned sulfite liquor is pumped into the first of a series of seven interconnected fermentation tanks. Yeast from a previous cycle is mixed with the sulfite liquor as it flows continuously to the fermentor. Fermentation is carried out at 30°C. After the fermentation is over, in about 20 hours, the fermented liquor is pumped to yeast separators. The required amount of yeast is returned to the first fermentor for reuse. The clarified liquor (beer) is passed by gravity to a storage tank and distilled. Finally, the alcohol is rectified to 190° proof or higher, purified, and warehoused.

A batch process developed by Les Usines de Melle involves the "reuse of yeast" (Ericsson, 1947; Prescott and Dunn, 1959). Essentially the process is comprised of the centrifugal separation of yeast cream from the fermented liquor and the reuse of this recovered yeast as an inoculum for batch fermentation of the new liquor. The advantages of this method are increased alcohol yield because of better utilization of the sugars, faster fermentation rates because of higher yeast concentration, and better acclimatization of the yeast to the liquor.

Avela et al. (1969) investigated the effect of concentration of SSL on the yield of alcoholic fermentation. They found that an 18% laboratory concentrate of a mill spruce spent liquor gave an 8% higher yield of alcohol than the original unevaporated spent liquor. The unevaporated SSL contained 9.4% dry matter and gave 1.5–15 kg of alcohol per ton of dry matter.

VIII. Economic Analysis of Ethanol Production Processes[2]

A. ETHANOL FROM CHEMICAL SYNTHESIS

Practically all synthetic ethanol is now produced from ethylene, which is obtained from petroleum and natural gas. Commercial experience has shown that 1 kg of ethylene will yield about 2.08 liters (0.55 gal) of 190° proof

---

[2]All cost figures are given in U.S. dollars.

TABLE XII
CONVERSION COST OF 190° AND 200° PROOF ETHYL ALCOHOL FROM ETHYLENE
(EXCLUSIVE OF COST OF ETHYLENE) (MAY 1974)[a]

	Cost	
Alcohol	¢/liter	¢/gal
190° Proof		
Base conversion cost	4.0	15.1
Depreciation	1.7	6.4
($3.2 million/year, 10 year, 190 million liters)		
Net	5.7	21.5
200° Proof		
Alcohol	5.9	22.5
(3.967 liters at 5.7¢/liter)		
Cost of dehydration	0.7	2.6
Total cost (exclusive of ethylene, profit, packaging, and sales expenses)	6.6	25.1

[a] From Miller (1975).

alcohol. Most new synthetic plants have a minimum yearly alcohol capacity of the order of 190 million liters (50 million gal) (Miller, 1975). In 1974, such a plant would cost approximately $32 million. By means of the chemical engineering (CE) plant cost indices, which were 165.4 in 1974 and 226.0 in 1978 (Anonymous, 1979a), the cost of the plant at present would be approxi-

TABLE XIII
EFFECT OF ETHYLENE COST ON ETHYL ALCOHOL COST (MAY 1974)[a]

Alcohol	Cost in ¢/liter (¢/gal) at an ethylene cost in ¢/kg (¢/lb) of:							
	11.0	(15.0)	13.2	(16.0)	15.4	(17.0)	17.6	(18.0)
190° Proof								
Ethylene	5.3	(20.0)	6.3	(24.0)	7.4	(28.0)	8.5	(32.0)
Conversion	5.7	(21.5)	5.7	(21.5)	5.7	(21.5)	5.7	(21.5)
Total manufacturing cost/liter (exclusive of profit, packaging, and sales expenses)	11.0	(41.5)	12.0	(45.5)	13.1	(49.5)	14.2	(53.5)
200° Proof								
Ethylene	5.5	(21.0)	6.7	(25.2)	7.7	(29.3)	8.8	(33.5)
Conversion	6.6	(25.1)	6.6	(25.1)	6.6	(25.1)	6.6	(25.1)
Total manufacturing cost/liter (exclusive of profit, packaging, and sales expenses)	12.1	(46.1)	13.3	(50.3)	14.3	(54.4)	15.4	(58.6)

[a] From Miller (1975).

mately $44 million.[3] The ethanol conversion cost, excluding the cost of ethylene, was about 5.7¢/liter (21.5¢/gal) in 1974, as shown in Table XII. Ethylene costs vary, depending on location and included expenses. The effect of ethylene price on alcohol cost is shown in Table XIII. At present, the ethylene cost is about 31¢/kg (14¢/lb) and the price of 190° proof ethanol is $4.33/liter ($1.25/gal) (Anonymous, 1979b).

B. Ethanol from Sugar and Cassava Root

The processes for the production of ethanol from sugar and cassava are currently widely used in Brazil. Table XIV shows the economics of alcohol produced from cassava and sugarcane, considering independent distilleries of the same daily capacity (150 m³/day). The difference in annual capacities results from the limited sugarcane harvesting season.

The cassava distillery requires a fixed capital investment that is 20% higher than that of an equivalent independent sugarcane distillery. Extra equipment for liquefaction and saccharification accounts for the difference. Higher working capital for the sugarcane distillery is a result of the fuel alcohol legislation in Brazil, which requires larger alcohol inventory for the independent sugarcane distillery (Yang and Trindade, 1979).

The alcohol (from cassava and sugarcane distillation) selling prices were computed including a return on investment of 12%/year, discounted-cash flow (DCF) method (Yang and Trindade, 1979). Economics of cassava alcohol and sugarcane alcohol on the basis of selling prices differ by less than 10%. The administered price of fuel alcohol currently, close to $0.31/liter ($1.17/gal), is lower than the calculated selling prices for cassava and sugarcane alcohols, which are in the range of $0.36–$0.38/liter ($1.36–$1.44/gal) freight on board (FOB) distillery. Subsidized financing from government is therefore necessary before the fermentation alcohol can be marketed commercially.

Closer examination of Table XIV reveals that agricultural feedstock accounts for the larger fraction of the estimated selling prices. Hence, high sensitivities of alcohol economics to agricultural and distillery yield increases are expected (Yang and Trindade, 1979). By-product credit in both cases appears small according to Table XIV. Note, however, that the raw stillage was valued mainly as a mineral fertilizer substitute. If the proper value of the organics in the stillage is assigned, or other more efficient stillage upgrading methods are considered, it may be possible to generate a higher net by-product credit. Moreover, excess bagasse in independent distilleries,

---

[3]Present day cost = cost in 1974 × (CE index in 1978/CE index in 1974).

TABLE XIV

ECONOMICS OF ANHYDROUS ETHANOL PRODUCTION FROM CASSAVA AND SUGARCANE[a,b]

Cost item	Cassava hypothetical distillery		Sugarcane distillery	
Investment				
Fixed investment ($10⁶)	15.8		13.3	
Working capital ($10⁶)	1.2		2.2	
Total	17.0		15.5	
Composition of selling price	$/m³	(%)	$/m³	(%)
Feedstock:				
Cassava Roots at $33.3/ton[c]	228	60.1	—	—
Sugarcane at $13.6/ton[c]	—	—	204	57.0
Enzymes, chemicals, and utilities	60	15.8	6	1.7
By-products[d]	(16)	(4.2)	(15)	(4.2)
Labor	13	3.4	15	4.2
Maintenance, materials, operating supplies, insurance, and administrative expenses	18	4.8	25	7.0
Taxes[e]	15	4.0	24	6.7
Depreciation	32	8.4	49	13.6
Net operating profit[f]	29	7.7	50	14.0
Calculated selling price as fuel, F.O.B. distillery	379	100.0	358	100.0

[a] From Yang and Trindade (1979).
[b] Basis: 150 m³/day cassava distillery operating 330 days/year (49,500 m³/year); 150 m³/day sugarcane distillery operating 180 days/year (27,000 m³/year). Exchange rate: Cr. $18,000/U.S. $1.00
[c] Includes value added tax (sugarcane) and social tax (sugarcane and cassava).
[d] Difference between the cost of direct application of stillage as fertilizer and the credit from sales of 96°GL ethanol and fusel oil.
[e] Includes income tax and social tax.
[f] Return on investment of 12%/year, DCF, based on the annual sum of depreciation and net operating profit, and 15-year operational life for the distillery.

available above steam raising needs, may be a source of additional by-product revenues if proper pricing and market conditions are found.

C. ETHANOL FROM CORN

Economics of the processes for production of ethanol from corn were studied by Lipinsky *et al.* (1977a), Miller (1975), and Scheller (1977c).

The cost of ethanol production depends on many conditions, such as locations, regional labor rates, and raw materials used. Miller (1975) made an estimate of the conversion costs in a practical sized grain fermentation plant

TABLE XV
FERMENTATIVE CONVERSION COST OF 190° AND 200° PROOF ETHYL ALCOHOL FROM CORN (EXCLUSIVE OF COST OF CORN) (MAY 1974)[a]

Alcohol	¢/liter	¢/gal
190° Proof (0.67 liter/bu)		
Base conversion cost	9.5	35.9
Depreciation	2.3	8.7
($1.54 million/year, 10 years, 67.0 million liters)		
Total	11.8	44.6
By-product feed credit	9.0	34.0
(0.81 kg/liter alcohol at $110/metric ton)		
Net	2.8	10.6
200° Proof (10.2 liters/bu)		
Alcohol	2.9	11.1
(3.967 gal at 2.8¢/liter)		
Cost of dehydration	0.7	2.6
Total cost (exclusive of corn, profit, packaging, and sales expenses)	3.6	13.7

[a] From Miller (1975).

processing 20,000 bu of corn per day. The result is given in Table XV. A by-product feed credit has been taken in the 2.8¢/liter (10.6¢/gal) of alcohol conversion costs. This optimum sized plant could produce 67 million liters (17.7 million gal) of 190° proof alcohol annually. The plant investment in 1974 was approximately $15.4 million. In terms of the present day value, the plant cost is about $21 million. The effects of corn costs on ethanol cost are shown in Table XVI. Recovered by-product feed (distillers' grains), which is a valuable protein animal feed, has a significant effect on the alcohol cost. The other major by-product is carbon dioxide, which could be recovered. However, only in selected metropolitan regions would it be economical to do so. In the economic analysis reported above, no credit for carbon dioxide recovery is included.

Scheller (1977c) presented another economic evaluation for the production of 76 million liters (20 million gal) per year of anhydrous ethanol from milo, as shown in Table XVII. Milo, a feed grain and a member of the maize family, was considered because it is less expensive than corn. The cost of milo at $7.70 per hundred kilograms ($3.50 per hundred pounds), as shown in Table XVII, was higher than the prevailing price of $5.50 per hundred kilograms ($2.50 per hundred pounds) in 1977. The conversion cost has been estimated to be 8¢/liter (30¢/gal) ethanol produced, based on design studies using coal as a fuel source. On the income side, a total of $22 million/year can be collected from the sale of 76 million liters/year of ethanol at $0.29/liter ($1.10/gal). The value of $0.29/liter was somewhat lower than the price

TABLE XVI
Effect of Corn Cost on Ethyl Alcohol Cost
(Basis: 10.2 liters 200° Proof Alcohol per Bushel) (May 1974)[a]

Corn price/bushel ($)	Alcohol cost					
	Corn		Conversion[b]		Total base cost[c]	
	¢/liter	¢/gal	¢/liter	¢/gal	¢/liter	¢/gal
1.00	9.8	37.0	3.6	13.7	13.4	50.7
1.25	12.2	46.3	3.6	13.7	15.8	60.0
1.50	14.7	55.5	3.6	13.7	18.3	69.2
1.75	17.1	64.8	3.6	13.7	20.7	78.5
2.00	19.5	74.0	3.6	13.7	23.2	87.7
2.25	22.0	83.3	3.6	13.7	25.6	97.0
2.50	24.5	92.6	3.6	13.7	28.1	106.3

[a] From Miller (1975).
[b] These costs do not include profits, packaging, and sales expenses.
[c] By-product grains credited at $110/metric ton in conversion cost.

TABLE XVII
Material Balance and Economics[a,b]

	Conventional plant
Investment:	
Plant investment	$23,000,000
Working capital	4,000,000
Total investment	$27,000,000
Income:	
Ethanol, 76 million liters/year	
at $0.29/liter ($1.10/gal)	$22,000,000
Distillers' dried grains	
232 metric tons (210 tons)/day at $132/metric ton ($120/ton)	9,200,000
Carbon dioxide	
192 metric tons (174 tons)/day at $2.2/metric ton ($2/ton)	130,000
Total income	$31,330,000
Expenses:	
Milo, 21,490 bu/day	
at 7.7¢/kg ($3.50/cwt)	$15,370,000
Conversion cost	
at 7.9¢/liter (30¢/gal) ethanol	6,000,000
Total expenses	$21,370,000
Depreciation: (10% straight line)	$ 2,300,000
Taxes: (50%)	$ 3,830,000
Net Cash Flow:	$ 6,130,000
As percent of investment	22.7

[a] From Scheller (1977c).
[b] Plant producing 76 million liters/year of 200° proof ethanol from milo.

quoted for synthetic anhydrous ethanol in 1977, which was $0.35/liter ($1.32/gal) F.O.B. This is because Nebraska, where the study was carried out, is somewhat removed from the principal chemical centers. It is necessary to lower the price in order to make the delivered price to the customer competitive (Scheller, 1977c). It should be noted from Table XVI that the income from ethanol is only slightly higher than the total expense of operations. This means that the profit for the operation will depend on the income received for the by-products. In Table XVII, distillers dried grains plus solubles were priced at $132.2/metric ton ($120/ton) which was somewhat below the average 1975 price of $150.8/metric ton ($136.80/ton). The distillers dried grains plus solubles produced an income of about $9,200,000, whereas the carbon dioxide, if sold at $2.2/metric ton ($2/ton) at atmospheric pressure, would bring an additional $130,000. The profit before taxes and depreciation is the difference between the income and expenses, or $9,960,000 (Scheller, 1977c).

The total capital investment for the 76 million liter/year grain alcohol plant was estimated to be $27 million in 1977, including $4 million for the working capital. The 10% straight line depreciation on the plant would be $2.3 million/year. It was also assumed that the federal and local corporation taxes would amount to 50% of the profit before taxes, or $3.83 million after the deduction of depreciation, leaving a net profit of $3.83 million. The net cash flow is $6.13 million/year or 22.7% of the total investment.

Scheller also prepared an economic evaluation of plants with capacities of 190 million and 380 million liters (50 million and 100 million gal) per year, as

TABLE XVIII
COST OF PRODUCING ETHANOL FROM CORN AND MILO[a]

	190 million liters ethanol (50 million gal)/year				380 million liters ethanol (100 million gal)/year			
	Milo in ¢/kg		Corn in $/bu		Milo in ¢/kg		Corn in $/bu	
	5.5	8.8	1.60	2.60	4.0	8.0	1.60	2.60
Grain cost[b]	14.9	23.8	16.20	26.40	14.8	23.8	16.20	26.40
By-product credit[c]	−9.7	−14.4	−8.60	−12.90	−9.7	−14.4	−8.60	−12.90
Net grain cost	5.2	9.4	7.60	13.50	5.1	9.4	7.60	13.50
Conversion cost	6.3	6.3	6.30	6.30	5.4	5.4	5.30	5.30
Depreciation (20 years)	1.1	1.1	1.00	1.00	0.8	0.8	0.80	0.80
Ethanol cost[d]	12.6	16.8	14.90	20.80	11.3	15.6	13.70	19.60

[a] From Scheller (1977c).

[b] All costs are in cents per liter of ethanol.

[c] Distillers' grains are priced at $99.2/metric ton ($90.00/ton) for a milo price of 5.5¢/kg ($2.50/cwt) and a corn price of $1.60/bu and at $149/metric ton ($135/ton) in the other cases.

[d] To this ethanol cost must be added profit and taxes to obtain an ethanol price.

shown in Table XVIII using corn or milo as feed. Ethanol costs, excluding profit and taxes, range from 11.3¢/liter (43.2¢/gal) to 20.8¢/liter (78.3¢/gal) depending on plant size and grain type.

Lipinsky et al. (1977a) prepared yet another economic evaluation for the production of 259.3 million liters (68.5 million gals) per year of 190° proof ethanol from 24 million bushels of corn. The total capital investment for the plant is shown in Table XIX. The estimated product costs and annual operating costs are given in Table XX and XXI, respectively. The operating costs were calculated using $2.50/bu of corn and a by-product feed credit of $116/metric ton ($105/ton). The final cost of ethanol production ranged from $0.31 to $0.34/liter ($1.18 to $1.29/gal) in 1976, depending on whether high- or low-debt economics are used in the analysis. The costs estimated were in

TABLE XIX
ESTIMATED CAPITAL COST FOR ETHANOL FROM CORN[a,b]

Scale:	Feed (corn grain), 24.29 million bu/year [617,000 metric tons (680,000 tons)] (15.5% moisture)		
Product:	Ethanol, 259.2 million liters (68.5 million gal)/year, 190° proof		
Basis:	330 stream days/year		

Cost element	In millions of 1976 dollars	
Battery limits plant (BLP), including land and site preparation, utilities, and services	69.23	
Interest during construction at 12%	8.31	
Estimated total fixed capital		77.54
Startup at 20% of gross annual operating costs		18.20
Working capital:		
Inventory		
Corn	2.34	
Chemicals	0.29	
Ethanol	6.23	
Fusel Oil	0.01	
Supplies	0.06	
Spare Parts	0.78	
By-product feed	—	
Accounts receivable	8.46	
Miscellaneous	2.83	
Total		21.00
Total estimated capital cost		116.74
Annualized capital charge	21.91[c]	14.04[d]

[a] From Lipinsky et al. (1977a).
[b] Note: These are tentative results. In any case, they are intended to be only plus 50% or minus 30% accurate.
[c] Capital charges using low-debt economics.
[d] Capital charges using high-debt economics.

TABLE XX
ESTIMATED PRODUCT COSTS—ETHANOL FACILITY[a,b]

	Millions of 1976 dollars	
	Low-debt economics	High-debt economics
Corn grain valued at $2.50/bu		
Raw material (grain)	60.73	60.73
Conversion costs	30.28	30.28
By-product feed credit	−24.46	−24.46
Annualized capital charge	21.92	14.04
Total annual cost	88.47	80.59
Cost per liter (gallon) in dollars	$0.34 ($1.29)	$0.31 ($1.18)

[a] From Lipinsky et al. (1977a).
[b] Note: These are tentative results. In any case, they are intended to be only plus 50% or minus 30% accurate.

close agreement with those obtained by Miller and Scheller, as discussed previously. Lipinsky et al. (1977a) noted that the product costs quoted are at the plant gate; excluding the cost of warehousing, distribution, and marketing. If the latter are included, about 10% must be added to the cost. This would bring the cost of ethanol to $0.34–$0.38/liter ($1.30–$1.42/gal).

D. ETHANOL FROM WOOD

The Madison process, as discussed above, was chosen by Inter Group Consulting Economists Ltd. (1978) for economic evaluation of ethanol production from wood. They based their estimates on the report prepared by Raphael Katzen Associates (1975). In addition, they separated the wood al-

TABLE XXI
ESTIMATED NET ANNUAL OPERATING COSTS—ETHANOL FACILITY[a,b]

Cost element	In millions of 1976 dollars	
Raw material (corn grain) at $2.50/bu	60.72	
Conversion costs (at $0.442/gal or $0.117/liter)	30.28	
Gross annual operating costs		91.00
By-product feed credit at $105/ton	−24.5	
Net annual operating costs		66.5

[a] From Lipinsky et al. (1977a).
[b] Note: These are tentative results. In any case, they are intended to be only plus 50% or minus 30% accurate.

cohol plant into two subunits; the acid hydrolysis plant and the fermentation plant.

On the investment side, in 1977 $65 million was estimated as the cost of a Canadian acid hydrolysis plant of a scale designed to serve a 227 million liter (60 million U.S. gal) per year fermentation facility (i.e., 382,832 tons of hexose from cellulose). This estimate did not include additional investments required for the conversion of pentoses of xylitol or furfural, recovery of hexoses from the prehydrolysis of hemicellulose, or conversion of lignin to potential chemical by-products. For the fermentation plant, it would require an investment of about $60 million. This investment level included concentrators and evaporators required to bring the glucose solution to the required level of 12%, fermentation facilities, storage, and shipping facilities, as well as utilities and services. The product output is 100% anhydrous ethanol. The investment level did not include requirements for drying or treating the spent liquors for by-product recovery. Table XXII summarizes the investment requirements for the overall facility.

The nonfeedstock processing costs for the hydrolysis and fermentation subprocesses are summarized in Table XXIII. The major elements in the processing costs are capital recovery and return, and the cost of fuel. The capital recovery and return element will vary up or down by about 35% as the return on capital desired is varied to 5% or 15%. The major fuel requirements are for process steam for use in the hydrolysis units, distilling towers, and for direct combustion in the glucose evaporation units. Inter Group Consulting Economists Ltd. (1978) also noted that coal is purchased for the process under consideration. In addition, processing costs excluded any consideration of credits for the by-products produced.

E. Ethanol from Cellulosic Wastes

Wilke *et al.* (1976b) have made an economic assessment for an integrated processing scheme, whereby a cellulosic waste (newsprint) is first converted

TABLE XXII
Investment Requirements for a 227 Million Liter/Year
Ethanol Facility Utilizing Acid Hydrolysis (1977 dollars)[a]

	$-Million
Acid hydrolysis unit	37.39
Fermentation unit	32.40
Off-site facilities and utilities	55.22
Total	125.01

[a] From Inter Group Consulting Economist Ltd. (1978).

TABLE XXIII
Nonfeedstock Processing Costs for a 227 Million Liter/Year
Ethanol Facility Utilizing Acid Hydrolysis (assumed wood feedstock)[a]

	Cost/ton of ethanol (EtOH) (1977 $/ton)	
	Hydrolysis	Fermentation
Noncapital related		
Fuel[b]	19.84	38.58
Chemicals	31.93	34.17
Power	2.94	1.93
Water	0.39	—
Operating labor	32.03	5.47
Supervision and administrative labor	12.81	2.18
Plant overhead	22.42	3.83
Capital related		
Maintenance	21.50	13.76
Local taxes and insurance	10.75	9.92
Capital recovery and return[c]	42.10	38.87
Subtotal: Cost per ton of EtOH	196.71[d]	148.71
Total: Cost per ton of EtOH (net of feedstock costs)	345.42[e]	

[a] From Inter Group Consulting Economist Ltd. (1978).
[b] Coal fuel requirements are in the range of 30.26–32.58 GJ per ton of ethanol produced; coal price of about $1.89/GJ is assumed for this table.
[c] Cost of fermentable sugar is 9.9¢/kg on a sugar yielded basis.
[d] Assumed 10% real before tax rate of return on capital over 20-year plant life.
[e] Table XXIII does not consider any credits for by-products produced.

TABLE XXIV
Product Summary and Overall Cost Analysis for
Ethanol Production from Newsprint[a]

	Metric tons/day	Unit	Production cost (¢/unit)	Assumed market value (¢/unit)	By-product cost credit per liter ethanol (¢)
Ethanol (95%)	73.9	liter	27.5	—	—
Carbon dioxide	102.5	—	—	—	—
Yeast cake	15.1	kg	—	66	88
Torula yeast	29.8	kg	66	66	—
Electricity	—	kWh	1	2	4

[a] From Wilke et al. (1976b).

TABLE XXV
CAPITAL INVESTMENT SUMMARY OF ETHANOL
PRODUCTION FROM NEWSPRINT[a]

Hydrolysis	$23.4 × 10^6
Alcohol and single-cell protein fermentation	5.4 × 10^6
Power plant	4.6 × 10^6
Total	$33.4 × 10^6

[a] From Wilke et al. (1976b).

to sugars by enzymatic hydrolysis, and then to ethanol by yeast fermentation (as described earlier). Table XXIV summarizes the products of the process, production costs, and estimated market value of the by-products. The production costs for ethanol and for torula yeast were estimated from a previous study (Cysewski and Wilke, 1976). Yeast cake and torula yeast values were based on the prevailing quotation for distiller's yeast. A credit of 1¢/kWh was taken assuming that the surplus power can be sold for 2¢/kWh at the process site. No credit was taken for the torula yeast because the market value is just equal to the estimated production cost.

On the basis of the assumed by-product credits, the analysis indicates that 95% ethanol might be produced F.O.B. the plant for about 16¢/liter (61¢/gal) assuming zero cost for the cellulosic feed. Table XXV summarizes the estimated capital investment.

F. By-product Credits

As noted above, by-products generated at various points in each stage of the hydrolysis–fermentation process constitute a critical element within overall ethanol production economics. These credits will clearly reduce net processing costs for ethanol; the only uncertainty pertains to the magnitude of the potential credits (Inter Group Consulting Economists Ltd., 1978).

Table XXVI indicates the range of by-products that can be attained from wood and grain feedstocks. Although relevant data are not available, by-products obtained from potatoes, straw, and processed municipal waste would probably tend to be more similar to grain by-products than to wood by-products. It is apparent that significant differences in by-product formation are associated with different feedstocks.

By-product credits have been estimated by Inter Group Consulting Economists Ltd. (1978) and are summarized in Table XXVII. As seen from this table, grain feedstocks yield by-product credits that are less sensitive to energy values than is the case for wood feedstocks; the spent grain is as-

TABLE XXVI
BY-PRODUCT SOURCE[a]

Feedstock	Lignin	Hemicellulose	Hydrolysis vapors		Spent grain	Spent yeast	Distillation	
			Methanol	Furfural			Fusel oils	Bottoms
Wood	Fuel Chemicals	Fuel Fodder Chemicals (xylitol, furfural)	Fuel Chemicals	Fuel Chemicals	—	Fuel Fodder	Fuel	Fertilizer
Grain	—	—	—	—	Fodder	Fuel Fodder	Fuel	Fertilizer

[a] From Inter Group Consulting Economist Ltd. (1978).

sumed to be sold to feed producers with little or no additional cost of processing. By-product credits for wood feedstocks, however, could be significantly greater if higher values could be realized for the lignin. In all cases, fertilizer markets for distillation bottoms would also act to increase by-product credits.

G. COMPARISON OF VARIOUS PROCESSES

Inter Group Consulting Economists Ltd. (1978) compared the capital investments and processing costs for the production of ethanol from various sources of biomass. The result is shown in Table XXVIII. Also included in this table is the cost of ethanol production from sugarcane. It is obvious that the most economical processes use sugar or starch (from corn, potatoes, etc.) as the raw material. They require the least capital investment as well as a lower processing cost. Cellulose from wood or municipal wastes is not economical at present.

H. POTENTIAL OF ETHANOL AS A FUEL

The cost of ethanol produced by fermentation as reported by various workers (Yang and Trindade, 1979; Lipinsky et al., 1977a; Miller, 1975; Scheller, 1977c; Inter Group Consulting Economists Ltd., 1978; Wilke et al., 1976b) differs widely. This is probably because of the use of different feedstocks as well as different processes. In fact, as noted by Quittenton (1979), there is currently a fierce argument in the United States about the cost of ethanol produced from grain by fermentation. Those in the business of actually producing ethanol from grain in the United States maintain they cannot do it for under $0.33/liter ($1.25/gal as of January 1979). They projected that by 1985 the cost of producing ethanol will be about $0.42/liter

TABLE XXVII
CALCULATION OF NET BY-PRODUCT CREDITS FROM ETHANOL PRODUCTION FOR WOOD AND GRAIN FEEDSTOCKS (1977 DOLLARS)[a]

	Volume/ metric ton (ton)	Market	Net by-product credit[b] ($/metric ton of EtOH)			
			1985	1990	2000	2010+
Wood feedstock by-products[c]						
Lignin[d]	0.78	Fuel	34.39	39.55	44.71	48.15
Hemicellulose (pentose sugars)[e]	0.84	Fuel	27.56	27.56	27.56	27.56
Methanol	0.05	Fuel	7.00	7.00	7.00	7.00
Furfural	0.06	Fuel	5.99	7.18	8.38	9.57
Yeast	0.25	Fodder	27.56	27.56	27.56	27.56
Fusel oils	0.03	Fuel	2.85	3.42	3.99	4.55
Distillation bottoms	0.91	Fertilizer	—	—	—	—
Total/metric ton of EtOH			105.35	112.27	119.20	124.39
Grain feedstock by-products						
Grain residues	0.25	Fodder	27.56	27.56	27.56	27.56
Yeast	0.25	Fodder	27.56	27.56	27.56	27.56
Fusel oils	0.06	Fuel	5.70	6.83	7.98	9.12
Distillation bottoms	0.91	Fertilizer	—	—	—	—
Total/metric ton of EtOH			60.82	61.95	63.10	63.24

[a] From Inter Group Consulting Economists Ltd. (1978).
[b] Net by-product credits are in most cases related to commodity values relative to oil products.
[c] Based on aspen.
[d] Credit in this table assumes coal price value of $1.89/GJ 1985 as per Table XXIII with gradual escalation.
[e] Excludes potentially large credits for fodder residue.

($1.60/gal) as compared to the predicted cost of gasoline production at $0.59/liter ($2.25/gal). It has also been estimated (Quittenton, 1979) that the price of crude oil on the world market might reach $21/barrel by 1985. However, the recent increases by the Organization of Petroleum Exporting Countries (OPEC) have brought crude oil prices at the present time (Anonymous, 1979c) to $21/barrel; this is up 50% since January 1, 1979 and 1000% from the $1.80 price per barrel at the start of the 1970s. The narrowing price gap between gasoline and ethanol leaves little doubt as to ethanol's potential as an economically viable fuel source.

TABLE XXVIII
COMPARISON OF ALTERNATIVE SOURCES OF BIOMASS
FEEDSTOCK FOR THE PRODUCTION OF ETHANOL[a]

Feedstock	Yield/metric ton (liters/metric ton)	1977 dollars per liter of ethanol	
		Investment[b] requirement	Processing[c] costs
Wood[d]	200.4–255.6	0.55	0.27–0.30
Grain (acid)[e]	441.0	0.53	0.27
Potatoes	105.2	0.33	0.11
Municipal solid waste[f]	250.6	0.55	0.27
Straw[f]	265.6	0.55	0.27
Grains (enzyme)	436.0	0.35	0.11
Beans	300.7	na[g]	na[g]
Sugarcane	75.0	0.31	0.03

[a] From Inter Group Consulting Economists Ltd. (1978).

[b] Investment required per liter of annual capacity.

[c] Net of feedstock costs—includes capital recovery and a 10% return before taxes. By-product credits are excluded from consideration.

[d] Range indicates potential variation with different types of wood (oven dry tons); higher yields reflect aspen species. Coal price of about $1.89/GJ is assumed.

[e] Yield assumes soft wheat with starch content of 69% and a yield of 0.511 kg of alcohol per kilogram of starch. Cost differential between grain and wood based on DREE analyses.

[f] Very immature technology—yields and costs assumed on the basis of a cellulose composition similar to wood.

[g] na, not available.

IX. Energy Considerations

Ethanol production from biomass involves the conversion of stored solar energy from a bulky form to usable energy with advantages in storage and transportation. The efficiency of conversion from one form to the other is a question of some interest during periods of scarce energy supplies.

Inter Group Consulting Economists Ltd. (1978) pointed out that evaluation of the level of conversion efficiency requires a careful identification of the system boundaries for comparison purposes. Several analyses have presented comparisons of energy input and energy of ethanol produced from plantation and fermentation operations. One analysis (de Carvalho et al., 1977) showed a net release of solar energy (crops and firewood) equal to five times the energy consumed in biomass harvesting and processing (excluding sustenance of labor). Another analysis by Schmidt-Holthausen and Engelbart (1977) carefully differentiates between biomass utilized for fermentation (manioc) and the by-products used for internal energy production. In this case a net yield of released solar energy of 50% is estimated (excluding an allowance for labor). A similar energy analysis was reported by Scheller and Mohr (1976) for the production of ethanol from corn. The result is summarized in Table XXIX, which is an overall energy balance for farming and grain alcohol production operations. The total energy production from ethanol, aldehydes, fusel oils, corn stalks, cobs, and husks is 67,578 kJ/liter (242,494 Btu/gal) of ethanol produced. This energy does not include the energy associated with the production of distiller's dried grains and solubles, because it is not directly related to the production of ethanol but rather to a by-product recovery operation carried out for economic reasons. The net energy loss associated with the by-product grain production is 5060 kJ/liter (18,170 Btu/gal) of ethanol, as shown in Table XXIX. Scheller (1977b) estimated a 30% net gain in solar energy by fermenting grain to ethanol when considering farming and processing energy requirements. In comparison, the production of ethanol from ethane and other gases via ethylene hydration, incurs a net loss of energy equal to about 30% of the ethanol produced.

A positive energy balance can usually be demonstrated for ethanol production if plant fuel needs are met from by-products and/or additional primary feedstock consumption (e.g., wood, straw). An energy deficit will tend to emerge, however, if a separate fuel (e.g., coal) is "imported" in order to meet the plant's fuel needs. The ultimate desirability and feasibility of any option depends, in reality, on comparative economic analysis of the optimum use for each feedstock—net energy flows can be misleading in that they implicitly assume equal value per unit of energy regardless of feedstock.

Lewis (1977) analyzed net energy production using a different approach.

TABLE XXIX
Overall Energy Balance for Grain Alcohol Production from Corn[a]

Energy production	kJ/bu Corn	Btu/bu Corn	kJ/liter	Btu/gal EtOH
Ethanol	206,734	196,109	21,054	75,600
Aldehydes, fusel oil	2,954	2,802	301	1,080
Stalks, cobs, husks	453,431	430,127	46,177	165,814
Total	663,119	629,038	67,532	242,494
Energy consumption				
Farming operation	125,752	119,289	12,805	45,986
Transportation of stalks, etc.	4,266	4,047	434	1,560
Alcohol plant	295,128	279,960	30,054	107,920
Total	425,146	403,296	43,293	155,466
Net energy production	237,973	225,742	24,239	87,028
Net energy loss				
By-product grain production	49,673	47,120	5,060	18,170

[a] From Scheller and Mohr (1976).

He adopted the concept of gross energy equipment (GER) in his analysis. The GER value is the amount of energy resource needed to make the particular input available and is normally expressed in terms of megajoules (MJ) per kilogram. Table XXX presents data in terms of energy requirements, net energy gains and losses, and land area equivalents for a number of relevant conversion and production systems. The energy figures for methane and ethanol production include energy gain or loss through the growth of the substrate in each case. It can be seen that only the sugarcane–ethanol route is currently a net energy producer (the corn–ethanol route is excluded in Lewis' analysis), although all the processes remain to be fully optimized. For cellulosic raw materials, many technological innovations would be required to derive a favorable energy balance because of the highly energy intensive pretreatment procedures required.

X. Summary and Conclusions

Recent oil shortages and escalating oil prices have indicated a priority need to develop alternative energy sources to replace petroleum. Synthetic liquid fuels, alcohol in particular, have been considered as one of the most suitable substitutes.

Alcohol fuels do offer certain attractive features as petroleum substitutes:

1. Alcohols are readily available liquids that can be produced and utilized within existing technologies. No further processing of alcohols for their use as fuel is required, which is advantageous as compared to petroleum fuels.

2. Alcohols can be produced from a number of renewable resources and have a potential not only as fuel, but also as a chemical feedstock for production of a number of today's "petroleum" products. In this respect, technologies are also available and most of today's petrochemicals could more or less be produced from alcohol instead of petroleum.

3. Alcohol fuels burn cleaner than petroleum, which is environmentally more acceptable. Also, alcohol is a "natural" substrate for microbial growth, which means that it is readily biodegradable, which is quite important in case of spills and uncontrolled discharge into the environment.

4. As alcohols can be produced from a variety of natural resources, their production could be developed to best suit a particular region. This allows a utilization of a wide spectrum of resource areas that are economically and strategically superior.

5. Alcohols produced from waste and inedible agricultural products could be efficiently used as substrates for controlled fermentations to produce food chemicals, drugs, enzymes, etc. In this way, edible carbohydrates may be directly consumed as food, whereas rather inedible commodities could be diverted to industrial use.

TABLE XXX

ENERGY REQUIREMENTS, NET ENERGY GAINS AND LOSSES AND LAND AREA EQUIVALENTS FOR A NUMBER OF CONVERSION AND PRODUCTION SYSTEMS[a]

Principal substrate	Product	GER product (GJ/ton)	Net energy GJ/ton	Net energy GJ/ha/year	Key inputs
CO_2	"Energy crops"	1.26	+16	+1090	Waste,[b] fertilizer, fossil fuel
Raw sewage	Algae[c]	57	−34	−850	Fossil fuel
Raw sewage	Algae[d]	18	−5	+125	Fossil fuel, flocculants
Algae	Methane[d]	168	−112	−627	Fossil fuel
Livestock waste (U.K.)	Methane	144	−88	−0.88	Fossil fuel
Sugarcane	Ethanol	24	+3	+51	N, P, fossil fuel
Cassava	Ethanol	61	−34	−71	N, P, fossil fuel
Timber	Ethanol[e]	239	−212	−74[f]	N, P, fossil fuel
Timber	Ethanol[g]	98	−71	−18[f]	N, P, H_2SO_4 fossil fuel
Straw	Ethanol	222	−195	−138	N, P, fossil fuel

[a] From Lewis (1976).
[b] Water availability is assumed to be sufficient, although in many locations this is not the case, which results in a higher energy requirement to grow the crops or else a reduction in productivity.
[c] The figures relate to current methods adopted.
[d] The figures are estimates of what should be possible at present.
[e] Cellulose hydrolyzed to fermentable sugars by fungal enzymes.
[f] Figures expressed on basis of land area requirement to annually replenish the quantity of wood substrate used.
[g] Cellulose hydrolyzed to fermentable sugars by acids. Also requires 470% manpower increase over enzyme route.

The technology for large-scale production of ethanol is presently available. However, conversion costs display wide variations depending upon the feedstocks and the process adopted.

There are four major sources of feedback available for ethanol production from biomass. They are:

1. Sugars from cane, beet, and sorghum
2. Grain starch from corn, wheat, barley, and other cereals
3. Cellulosic materials from wood and wastes
4. By-product carbohydrates from processing, such as waste sulfite liquor, whey, or food industry wastes.

In tropical countries, such as Brazil, sugarcane and molasses are readily available and inexpensive. They are the most common feedstocks used in ethanol production. The conversion of sugars to ethanol is carried out by batch fermentation and subsequently the product is separated in a bubble-capped distillation column. This approach has been considered inefficient and costly. As a result, the cost of production is high, and government subsidy is needed to make the product competitive. Future developments call for study of continuous fermentation and the design of more efficient distillation towers. In addition, it is also noted that the economy of the fermentation process can be improved by using yeasts tolerant to high sugar concentration.

The ABE process, which is only practiced in one plant commercially, needs further refinement before it can compete with other routes. The main problem is the high cost involved in separation of the low concentration product from other fermentation metabolites.

Among the various sources of starch available, corn is by far the most important. It is generally agreed that corn is the least costly substrate for ethanol production from renewable resources at the present time. Further research and development of the process is still needed to make the process more competitive. Suggested areas of study include integration of facilities, such as a wet milling plant for corn and an ethanol production facility. This would allow more effective utilization of resources and would reduce the overall investment and cost of production. Improvement of by-product recovery or upgrading the value of by-products can also help to reduce the effective cost of production.

For carbohydrates to be metabolized by the selected yeast, an effective method of hydrolysis to monosaccharides is needed. The release of monomers from the complex polysaccharide is therefore the key step in its utilization as a fermentation substrate for alcohol production. Therefore, processes that yield improvements in hydrolysis technology are needed. Research in this area is necessary to establish optimum conditions for the hydrolysis reaction and optimum equipment design for industrial production.

In the fermentation itself, improvements are needed in continuous fermentation and in maintenance of a high cell mass concentration of yeast by proper recycling. Also, development of improved strains for specific fermentation is imperative. The expanding field of genetic engineering may find a farreaching application in the high-yield alcohol production. Other routes aimed at improvement of alcohol production are in the blocking of product inhibition by continuous removal of the product as it is produced, e.g., by vacuum distillation. Fermentation under vacuum with boiling permits cost reductions and also a better utilization of the fermentation heat, which is otherwise not recovered. More applied research in this area is required.

The study of ethanol production from cellulosic materials has received much attention recently. It generally involves two stages of reaction: hydrolysis of cellulose to sugar and fermentation of sugar to ethanol. The technology for the latter is readily available. However, much controversy still exists between acid and enzymatic hydrolysis of cellulose to sugars. Acid hydrolysis suffers from the requirement for corrosion-proof equipment because of the need for high acid concentration and high temperature to achieve hydrolysis. In addition, acid hydrolysis also produced many undesirable by-products. Enzymatic hydrolysis is more specific in nature, however, and requires milder reaction conditions, but the rate of reaction is slow and the production of cellulase is expensive and requires costly pretreatment of the raw material. Much effort has been directed toward the development of a less expensive pretreatment scheme, the improvement of cellulase production, and the optimization of the hydrolysis reaction.

A one-step process for ethanol production from cellulose has been investigated. This process combines acid and enzymatic hydrolysis reactions in a single step and attempts to take the advantage of each individual process. The discovery of a high-temperature cellulase enzyme may help cut down the cost of cellulase production as well as enable faster reaction rates.

Among the various industrial wastes available for ethanol production, only cheese whey and spent sulfite liquor have been commercialized. The high cost of waste collection and a generally low concentration of sugars in the wastes appear to be the main barriers for the commercialization of the waste-based process. However, a coupling of a controlled fermentation process with industrial waste treatment would prove advantageous. These wastes have to be treated before their release into the environment, which treatment is always at no profit. A recovery of a valuable product may considerably offset the treatment cost and possibly either break even or make the whole operation economical. The amount of industrial wastes that may be available for this route is often underestimated. For example, spent sulfite liquors from the pulp and paper industry are available in enormous quantities (about 1 billion gal/day in Canada). These liquors contain fermentable sugars but also have a high concentration of other chemicals that inhibit or otherwise affect alcohol production by yeast. Research directed to the solution of this problem is also needed.

In summary, alcohol can be produced in large quantities from a variety of renewable resources. Improvement of processes, economics, and yield are the key parameters to be solved in order to produce large quantities of ethanol. We shall still depend on liquid petroleum fuels for decades to come, but ethanol is a very logical alternative to pursue and develop. With the recent increases in the price of crude oil by the Organization of Petroleum Exporting Countries (OPEC), ethanol is undoubtedly becoming a more competitive alternate to gasoline.

REFERENCES

Allen, A. L. (1976). *AIChE Symp. Ser.* **72** (158), 115.
Anderson, E. V. (1978). *Chem. Eng. News* **56** (31), 8, 15.
Andren, R. K., and Nyström, J. M. (1976). *AIChE Symp. Ser.* **72** (158), 91.
Andren, R. K., Erickson, R. J., and Medeiros, J. E. (1976). *In* "Enzymatic Conversion of Cellulosic Materials: Technology and Applications" (E. L. Gaden, M. H. Mandels, E. T. Reese, and L. A. Spano, eds.), pp. 177–203. Wiley, New York.
Anonymous (1977). *Chem. Eng. News* **55** (28), 32.
Anonymous (1978a). *Chem. Eng. News* **56** (17), 21.
Anonymous (1978b). *Chem. Eng. News* **56** (32), 22.
Anonymous (1978c). *Promon Tech. Cent. Newsl.* **3** (1), 1.
Anonymous (1979a). *Chem. Eng.* **86** (7), 103.
Anonymous (1979b). *Chem. Mark. Rep.*, April 30.
Anonymous (1979c). *Time* July 9, p. 6.
Avela, E., Hase, A., and Soila, R. (1969). *Pap. Puu* **51** (7), 565.
Azarniouch, M. K., and Thompson, K. M. (1976). *Symposium on Alcohols as Alternative Fuel for Ontario*, Pulp and Paper Research Inst. of Canada, Pointe Claire, Quebec, Nov. 19.
Bailie, R. C. (1976). "Technical and Economic Assessment of Methods for Direct Conversion of Agricultural Residue to Usable Energy." Report to U.S.D.A., No. 12-14-1001-598, Oct. 20, p. 12.
Batter, T. R., and Wilke, C. R. (1977). "A Study of the Fermentation of Xylose to Ethanol by *Fusarium oxysporum.*" Lawrence Berkeley Lab., University of California, p. 160.
Benedict, J. G., Andreotti, R., and Roche, C. (1978). *Symp. Biotechnol. Energy Prod. Conversion.*
Bhat, P. K., and Singh, M. B. D. (1975). *J. Coffee Res.* **5** (3–4), 71.
Birkett, H. S., and Polack, J. A. (1978). *Sugar J.* **5**, 9.
Bloch, F., Brown, G. E., and Farkas, D. F. (1973). *Am. Potato J.* **50** (10), 357.
Bose, K., and Ghose, T. K. (1973). *Process Biochem.* **8** (2), 23, 33.

Brenner, W., Rugg, B., and Rogers, C. (1977). *Symp. Clean Fuels Biomass, Sewage, Urban Refuse, Agric. Wastes*, pp. 25-28.
Brink, D. L. (1976). *Appl. Polym. Symp.* **28**, 1377-1391.
Cherimisinoff, P. N., and Morresi, A. C. (1976). *Pollut. Eng. Technol.* **1**, 1-505.
Cooper, J. L. (1976). *In* "Enzymatic Conversion of Cellulosic Materials: Technology and Applications" (E. L. Gaden, M. H. Mandel, E. T. Reese, and L. A. Spano, eds.), pp. 251-271. Wiley, New York.
Cysewski, G. R., and Wilke, C. R. (1976). *Biotechnol. Bioeng.* **18**, 1297.
Cysewski, G. R., and Wilke, C. R. (1977). *Biotechnol. Bioeng.* **19**, 1125.
Cysewski, G. R., and Wilke, C. R. (1978). *Biotechnol. Bioeng.* **20**, 1421.
Davies, R. G. (1976). *Symposium on Alcohols as Alternative Fuel for Ontario*, Pulp and Paper Research Inst. of Canada, Pointe Claire, Quebec, Nov. 19.
de Carvalho, A. V., Milfont, W. N., Jr., Yang, V., and Trindade, S. C. (1977). *Int. Symp. Alcohol Fuel Technol.—Methanol Ethanol*, 6-1(1) to 6-1(9).
Dukes, E. P., Holzer, L., Klingsberg, A., and Wronker, C. C., eds. (1963). "Encyclopedia of Chemical Technology," Vol. 22, p. 364. Wiley (Interscience), New York.
Dunlap, C. E., Thomson, J., and Chiang, L. C. (1976). *AIChE Symp. Ser.* **72** (158), 58.
Elder, A. L. (1976). *Biotechnol. Bioeng. Symp.* **6**, 275.
Elias, S. (1979). *Food Eng.* **51** (1), 99.
Ericsson, E. O. (1947). *Chem. Eng. Prog.* **43** (4), 165.
Ghose, T. K. (1978). "Microbial Technology in the Provision of Energy and Chemicals from Renewable Resources" (submitted to the Fermentation Commission IUPAC, RO 171978).
Ghose, T. K., and Sahai, V. (1979). *Biotechnol. Bioeng.* **21** (2), 283.
Golueke, C. G., ed. (1977). "Biological Reclamation of Solid Wastes," pp. 189-205. Rodale Press, Emmaus, Pennsylvania.
Herr, D., Luck, G., and Dellweg, H. (1978). *J. Ferment. Technol.* **56** (4), 273.
Hoge, W. H. (1977). U.S. Patent 4,009,075.
Hokanson, A. E., and Katzen, R. (1978). *Chem. Eng. Progr.* **74** (1), 67.
Inter Group Consulting Economists Ltd. (1978). "Liquid Fuels from Renewable Resources: Feasibility Study." A report prepared for the Government of Canada, Interdepartmental Steering Committee on Canadian Renewable Liquid Fuels, Ottawa, Canada.
Jackson, E. A. (1976). *Process Biochem.* **11** (5), 28.
Kemp, C. C., and Szego, G. C. (1976). *AIChE Symp. Ser.* **72** (158), 1.
Klass, D. L. (1976). *Symp. Clean Fuels Biomass, Sewage, Urban Refuse, Agric. Wastes*, pp. 21-58.
Kohn, P. M. (1978). *Chem. Eng.* **85** (5), 58.
Lee, W., Konig, A., and Bernhardt, W. (1976). *Symposium on Alcohols as Alternative Fuel for Ontario*, Pulp and Paper Research Inst. of Canada, Pointe Claire, Quebec, Nov. 19.
Lewis, C. W. (1976). *Process Biochem.* **11** (9), 29.
Lewis, C. W. (1977). *Energy* **2**, 241.
Lindeman, L. R., and Rocchiccioli, C. (1979). *Biotechnol. Bioeng.* **2**, 1107.
Lipinsky, E. S., Sheppard, W. J., Otis, J. L., Helper, E. W., McClure, T. A., and Scantland, D. A. (1977a). "Systems Study of Fuels from Sugarcane, Sweet Sorghum, Sugar Beets and Corn," Vol. 5. A final report prepared for the ERDA, Battelle, Columbus, Ohio.
Lipinsky, E. S., Kresovich, S., McClure, T. A., Helper, E. W., and Lawhon, W. T. (1977b). "Fuels from Sugar Crops," 2nd quarterly report to DOE, Battelle, Columbus, Ohio.
McCloskey, J. R. (1975). *Energy Sources* **2** (1), 53.
Maloney, G. T. (1978). Chemicals from Pulp and Wood Waste: Production and Applications. *Chem. Technol. Rev.* No. 101. Noyes Data Corporaton, Park Ridge, New Jersey.

Mandels, M., and Andreotti, R. E. (1978). *Process Biochem.* **13** (5), 6.
Margaritis, A., and Wilke, C. R. (1972). *Dev. Ind. Microbiol.* **13** (4), 159.
Margaritis, A., and Wilke, C. R. (1978). *Biotechnol. Bioeng.* **20**, 709.
Menezes, T. J. B. (1978). *Process Biochem.* **13** (9), 24.
Miller, D. L. (1975). *Biotechnol. Bioeng. Symp.* **5**, 345–352.
Nyström, J. M., and DiLuca, P. H. (1977). *In* "Bioconversion of Cellulosic Substances into Energy, Chemicals and Microbial Protein," (T. K. Ghose, ed.), p. 293. IIT, New Delhi, India.
O'Leary, V. S., Green, R., Sullivan, B. C., and Holsinger, V. H. (1977a). *Biotechnol. Bioeng.* **19**, 1019.
O'Leary, V. S., Sutton, C., Bencivengo, M., Sullivan, B. C., and Holsinger, V. H. (1977b). *Biotechnol. Bioeng.* **19**, 1689.
Porteous, A. (1972). *Pap. Trade J.* **156** (6), 30.
Prescott, S. C., and Dunn, C. G., eds. (1959). "Industrial Microbiology," 3rd ed., pp. 102–128. McGraw-Hill, New York.
Quittenton, R. C. (1979). *Eng. Dig.* **25** (2), 17.
Raphael Katzen Associates (1975). "Chemicals from Wood Wastes." A report prepared for the USDA, Madison, Wisconsin.
Reese, E. T. (1975). *Biotechnol. Bioeng. Symp.* **5**, 77–80.
Reesen, L. (1978). *Dairy Ind. Int.* **43** (1), 9.
Reesen, L., and Strube, R. (1978). *Process Biochem.* **13** (11), 21.
Rose, D. (1976). *Process Biochem.* **11** (2), 10, 36.
Rosen, K. (1978). *Process Biochem.* **13** (5), 25.
Rosenbluth, R. R., and Wilke, C. R. (1970). S.E.R.L. Report 70-9. College of Engineering and School of Public Health, N. California, Berkeley.
Sahai, V., and Ghose, T. K. (1977). *In* "Bioconversion of Cellulosic Substances into Energy, Chemicals and Microbial Protein," (T. K. Ghose, ed.), p. 269. IIT, New Delhi, India.
Scheller, W. A. (1976). *Fermentation in Cereal Processing*, 61st Natl. Meet. Am. Assoc. Cereal Chem.
Scheller, W. A. (1977a). *Symp. Clean Fuels Biomass, Sewage, Urban Refuse, Agric. Wastes*, pp. 185–200.
Scheller, W. A. (1977b). "Stone and Webster International Biochemical Symposium." Toronto, Ontario.
Scheller, W. A. (1977c). *Int. Symp. Alcohol Fuel Technol.—Methanol Ethanol*, 5–1(1) to 5–1(4).
Scheller, W. A., and Mohr, B. J. (1976). *Am. Chem. Soc., Div. Fuel Chem. Prep.* **21** (2), 28.
Scheller, W. A., and Mohr, B. J. (1977). *Chemtech* **6** (10), 616.
Schmidt-Holthausen, H., and Engelbart, W. (1977). *Int. Symp. Alcohol Fuel Technol.—Methanol Ethanol*, 5–2(1) to 5–2(4).
Seeley, D. B. (1976). *In* "Enzymatic Conversion of Cellulosic Materials: Technology and Applications" (E. L. Gaden, M. H. Mandels, E. T. Reese, and L. A. Spano, eds.), pp. 285–292. Wiley, New York.
Shreve, R. N., and Brink, J., eds. (1977). "Chemical Process Industries," 4th ed., pp. 558–561. McGraw-Hill, New York.
Spano, L. A. (1976). *Symp. Clean Fuels Biomass, Sewage, Urban Refuse, Agric. Wastes*, pp. 325–348.
Spivey, M. J. (1978). *Process Biochem.* **13** (11), 2, 25.
Stone, J. (1974). "Survey of Alcohol Fuel Technology." An interim report to Mitre Corp., McLean, Virginia.
Takagi, M., Abe, S., Suzuki, S., Emert, G. H., and Yata, N. (1977). *In* "Bioconversion of Cellulosic Substances into Energy, Chemicals and Microbial Proteins," (T. K. Ghose, ed.), pp. 551–571. IIT, New Delhi, India.

Tan, T. C., and Lau, C. M. (1975). *J. Singapore Natl. Acad. Sci.* **4** (3), 152.
Tatsumi, C., Ogaki, M., and Tanabe, I. (1977). British Patent 1,493,480.
Toyama, N. (1976). *In* "Enzymatic Conversion of Cellulosic Materials: Technology and Applications" (E. L. Gaden, M. H. Mandels, E. T. Reese, and L. A. Spano, eds.). pp. 207–219. Wiley, New York.
Ware, S. A. (1976). *U.S.N.T.I.S.*, *PB Rep.* **PB-258** 499, p. 78.
Wilke, C. R. and Blanch, H. W. (1977). "Pilot Plant Studies of the Bioconversion of Cellulose and Production of Ethanol." Lawrence Berkeley Lab., University of California.
Wilke, C. R., von Stockar, U., and Yang, R. D. (1976a). *AIChE Symp. Ser.* **72** (158), 104.
Wilke, C. R., Yang, R. D., and von Stockar, U. (1976b). *In* "Enzymatic Conversion of Cellulosic Materials: Technology and Applications" (E. L. Gaden, M. H. Mandels, E. T. Reese, and L. A. Spano, eds.), pp. 155–175. Wiley, New York.
Wilke, C. R., Cysewski, G. R., Yang, R. D., and von Stockar, U. (1976c). *Biotechnol. Bioeng.* **18**, 1315.
Yang, V., and Trindade, S. C. (1979). The Brazilian Gasohol Program, Centro de Technologia Promon, Rio de Janeiro, Brazil.

Surface-Active Compounds from Microorganisms

D. G. COOPER AND J. E. ZAJIC[1]

Chemical and Biochemical Engineering, Faculty of Engineering Science,
The University of Western Ontario,
London, Ontario, Canada

I.	Introduction	229
	A. Importance of Biosurfactants and Scope of This Review	229
	B. Microorganisms and Biosurfactants	230
	C. Uses for Biosurfactants	231
II.	Carbohydrate-Containing Surfactants	231
	A. Trehalose Lipids	231
	B. Rhamnolipids	234
	C. Sophorose Lipids	236
	D. Diglycosyl Diglycerides	238
	E. Polysaccharide Lipid Complexes	238
III.	Amino Acid-Containing Surfactants	239
	A. Lipopeptides	239
	B. Ornithine Lipids	241
	C. Protein	242
IV.	Phospholipids	242
V.	Fatty Acids and Neutral Lipids	245
	A. Introduction	245
	B. Carboxylic Acids	245
	C. Surface Properties of Carboxylic Acids	246
	D. Neutral Lipids and Mixtures with Fatty Acids	249
VI.	Conclusions	250
	References	250

I. Introduction

A. IMPORTANCE OF BIOSURFACTANTS AND SCOPE OF THIS REVIEW

Industry uses at least 10^6 tons/year of synthetic surfactants (Gerson and Zajic, 1978a). Many different types of surfactants are being used, but it is important to develop even more to broaden the spectrum of properties available. No surfactant is suitable for all the potential applications.

Biosurfactants are important because they present a much broader range of surfactant types and properties than the available synthetic surfactants. Furthermore, they are usually biodegradable, which reduces the potential of pollution (Zajic *et al.*, 1977a,b; Margaritis *et al.*, 1979).

This review considers metabolites produced by microorganisms which

---

[1]Present address: College of Science, University of Texas at El Paso, El Paso, Texas 79968.

demonstrate surface activity. Particular emphasis is given to systems in which the metabolite has been characterized and shown to be surface active. In some cases the surfactant has not been identified and, in many others, a compound has been isolated that is analogous to a known surfactant but no attempt has been made to evaluate surface activity. The review also includes examples of biosurfactant modification—usually by substrate manipulation.

B. Microorganisms and Biosurfactants

As will be shown in this review, surfactants produced by microorganisms are usually lipids. Their surfactant properties result from a combination of polar and apolar moieties in a single molecule. The apolar or hydrophobic portion is normally a hydrocarbon. The most common example is the hydrocarbon chain of a fatty acid. The polar or hydrophilic groups include a wide range of possibilities. Examples are the ester and alcohol functional groups of neutral lipids, the phosphate-containing portions of phospholipids, and the sugars of glycolipids. Recent review articles by Helenius and Simons (1975) and by Small (1968) both include well-written introductions to surfactants with an emphasis on biosurfactants.

This work is not intended to be a general review of microbial lipids. Review articles of areas of this field of research include those by Asselineau (1966), Lederer (1967), Shaw (1970, 1974), Komura et al. (1975a,b), and Steck et al. (1978).

When monitoring a microbial culture for surfactant production, it is necessary to have a method to determine surface activity. The most useful method is to measure the surface tension of the whole broth or a dilution of the whole broth. Distilled water has a surface tension of about 73 dyn/cm. An effective biosurfactant can lower this to under 30 dyn/cm. Furthermore, if an excess of surfactant is produced, diluting the whole broth until the surface tension increases at the critical micelle concentration gives an estimate of the amount of surfactant present (Cooper et al., 1979a).

Other measurements indicate surface activity. One is the lowering of the interfacial tension between an aqueous phase and an emissible organic phase. Stabilization of an oil and water emulsion is a commonly used indicator. Foaming indicates surface activity. Lysis of protoplasts may be caused by surfactants (Arima et al., 1968; Bernheimer and Avigad, 1970).

Oil and water emulsification is an important property for many microbes. These are the organisms capable of using hydrocarbons as their only carbon source. The potential hydrocarbon substrates include straight chain, branched, and cyclic alkanes, alkenes, and aromatics (Einsele and Fiechter, 1971; Klug and Markovitz, 1971; McKenna and Kallio, 1965; Humphrey, 1967). It is generally concluded that organisms growing on water-insoluble hydrocarbon substrates benefit from the presence of a surfactant that can emulsify the oil into the water (Gutierrez and Erickson, 1977; Nakahara et al., 1977; Hug et al., 1974; Mallee and Blanch, 1977). Emulsification results in small oil drops in the water, which increases the surface area between the two phases. These surfaces are the most efficient for microbial growth. These conclusions are supported by evidence that the addition of surfactants to oil and water culture media stimulates microbial growth (Whitworth et al., 1973; Humphrey and Erickson, 1972; Hisatsuka et al., 1971).

C. Uses for Biosurfactants

A use of biosurfactants that shows particular promise is in the cleanup of oil spills, both on water and on land (Jobson et al., 1972; Cundell and Traxler, 1973; Mulkins-Phillips and Stewart, 1974; Walker and Colwell, 1974). The oil can be inoculated with an appropriate organism that will use it as a substrate. The organism will produce biosurfactants to improve the accessibility of the substrate. This allows further biodegradation both by the original organism and by others, which may each show different preferences for different fractions of the oil. If the spill is on water, the biosurfactant may also aid in dispersing the oil. A related use for biosurfactants is the release of bitumen from mixtures such as the Athabasca Tar Sands (Gerson and Zajic, 1977, 1978a,b).

An indirect importance of biosurfactants is that they facilitate the growth of microorganisms on hydrocarbons and a large number of useful products can be obtained from these substrates (Abbott and Gledhill, 1971). An important product is single-cell protein (Mateles et al., 1968) as well as a large number of individual amino acids. Other useful metabolites include alkane transformation products, sugars and polysaccharides, nucleic acids, vitamins, and pigments (Abbot and Gledhill, 1971). Flocculating agents have been produced from kerosene (Zajic and Knettig, 1971; Knettig and Zajic, 1972).

Biosurfactants can lead to problems as well. Fuel tanks can be contaminated by microorganisms and their bioemulsifiers can cause water dispersion into the oil and corrosion (Odier, 1976; Hedrick et al., 1968).

II. Carbohydrate-Containing Surfactants

A. Trehalose Lipids

The most commonly isolated microbial surfactants, or emulsifying agents, are glycolipids. Glycolipids containing the disaccharide trehalose are common in the literature of the extracellular lipids of coryneform and related bacteria.

Suzuki et al. (1969) isolated a trehalose lipid with strong emulsifying properties from *Arthrobacter paraffineus* KY4303 grown on n-paraffins. Each molecule contained trehalose and two β-hydroxy α-branched fatty acids (corynomycolic acids) esterified to undetermined sites on the sugar. The fatty acids had the general formula R^1—CH(OH)—CHR^2—COOH (Fig. 1), where R^1 varied from 18 to 23 carbons and R^2 from 7 to 12 carbons long. Trehalose containing glycolipids have also been isolated from several strains of *Arthrobacter, Mycobacterium, Brevibacterium, Corynebacterium*, and *Nocardia* growing on hydrocarbons (Suzuki et al., 1969; Rapp and Wagner, 1976; Anderson and Neuman, 1933). These have also been cited as examples of bacterial production of emulsifiers (Gutierrez and Erickson, 1977), however, none has been completely characterized.

The most extensively studied trehalose lipids are the cord factors, originally isolated from mycobacteria but eventually found in a wide range of related organisms (Lederer, 1967; Asselineau, 1966). Several of these have been completely characterized and they all have the same general structure (Fig. 2). The cord factors from mycobacteria are 6,6'-dimycolates of trehalose. Mycolic acids are α-branched β-hydroxycarboxylic acids having from 60 to 90 carbon atoms. Normally the R^2 substituent in R^1—CH(OH)—CHR^2—COOH is a saturated alkane, $C_{22}H_{45}$ or $C_{24}H_{49}$ (Lederer, 1967). However, the R^1 chain has a much larger variation in length and often contains other functional groups. Some examples are: $C_{62}H_{122}O_3$, with one unsaturation (Etémadi, 1966) and $C_{77}H_{154}O_3$ with two unsaturations (Etémadi et al., 1964a) both from *Mycobacterium smegmatis*; $C_{80}H_{156}O_3$, with two cyclopropane groups, from *Mycobacterium kansasii* (Etémadi et al., 1964b); $C_{85}H_{168}O_4$, with a methoxyl function and a cyclopropane group (Etémadi, 1966) and $C_{87}H_{160}O_4$, with a keto function and a cyclopropane group (Lederer, 1967) both from *Mycobacterium tuberculosis*; and $C_{56}H_{108}O_5$, with one unsaturation and a second carboxylic group, from *Mycobacterium phlei* (Markovits et al., 1966). Cord factors containing smaller corynomycolic acids have been isolated from *Corynebacterium diphtheriae* (Ioneda et al., 1963) and *Nocardia asteroides* (Ioneda et al., 1970). These contained 32 or 34 carbon atoms, and some of them were unsaturated. Finally, there is a report of a cord factor from *Mycobacterium fortuitum* that contains palmitic and stearic acid instead of the hydroxyl-containing acids (Vilkas and Rojas, 1964).

The surfactant characteristics of most of these trehalose lipids or cord

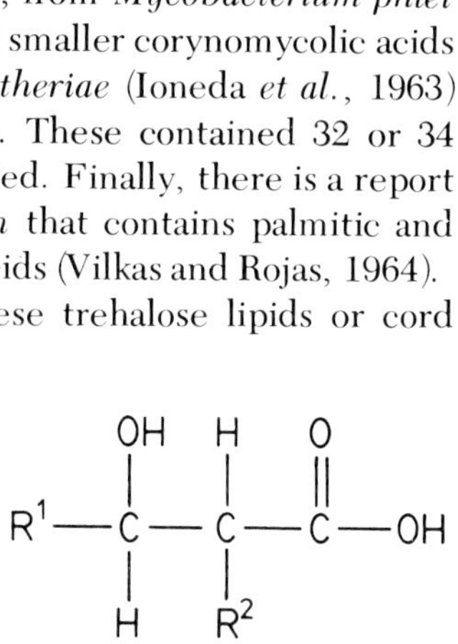

FIG. 1. The structure of a corynomycolic acid: a β-hydroxy α-branched carboxylic acid in which R^1 and R^2 are alkyl substituents having a total of fewer than 40 carbon atoms. The related mycolic acids have longer substituents (up to 90 carbon atoms) and may carry various functional groups.

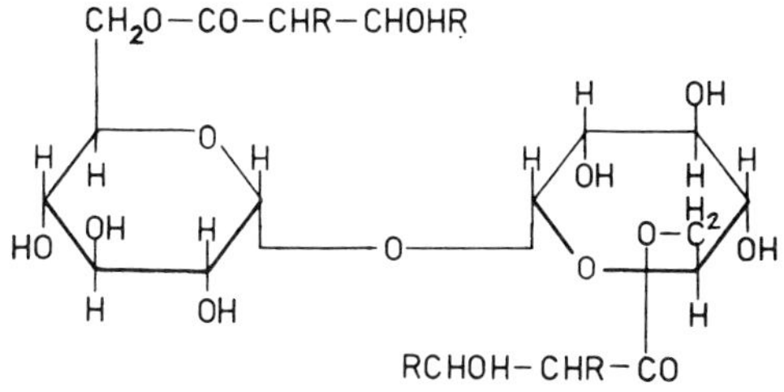

FIG. 2. Structure of cord factor: a diester of trehalose and two mycolic (or corynomycolic) acids.

factors have not been published. The few measurements in the literature indicate that they are effective emulsifiers. Suzuki et al. (1969) concluded that the trehalose lipid isolated from *A. paraffineus* grown on alkanes was responsible for the observed emulsion of the hydrocarbon and aqueous phases. From a *Nocardia* species grown on alkanes Rapp and Wagner (1976) isolated a trehalose lipid that was present in only trace amounts when the organism was grown on glycerol. This lipid emulsified water and hydrocarbon mixtures and stimulated growth of the *Nocardia* on hydrocarbons. The addition of small amounts of the purified lipid to shake flasks before inoculation enhanced the dry weight of samples taken with 24 hours (Table I). There was a direct relationship between the amount of lipid added and the biomass measurement.

Most of the trehalose lipids characterized have had the same general structure of two α-branched β-hydroxy fatty acids esterified to a trehalose molecule. One acid is esterified to each of the two glucose units of the dimer. It is possible to obtain modified glycolipids by altering the substrate used to grow the bacteria. When species of *Arthrobacter*, *Corynebacterium*, and *Nocardia* were grown on sucrose (Suzuki et al., 1974) or fructose (Itoh and Suzuki, 1974) the trehalose portion of the lipids was substituted. When

TABLE I
EFFECT OF TREHALOSE LIPIDS ON GROWTH OF
Nocardia ON n-ALKANES[a]

Lipid added (%)	Dry wt. (gm/liter)[b]
0	1.9
0.0001	2.10
0.001	2.70
0.01	3.87

[a] From Rapp and Wagner (1976).
[b] Sample taken 23 hours after inoculation.

sucrose was the substrate, two sucrose glycolipids were isolated—having one or two corynomycolic acids. Similarly, when fructose was the substrate, fructose 6-corynomycolate and fructose 1,6-dicorynomycolate were isolated. Unfortunately, these modified glycolipids were not tested for surfactant properties. However, this is an example of the possibilities for altering biosurfactants to modify surfactant parameters.

B. Rhamnolipids

Glycolipids that contain the sugar rhamnose and β-hydroxycarboxylic acids have been isolated from several strains of *Pseudomonas aeruginosa* (Edwards and Hayashi, 1965; Itoh *et al.*, 1971). The first rhamnolipid to be characterized (R-1) consisted of two molecules of rhamnose and two molecules of β-hydroxydecanoic acid (Edwards and Hayashi, 1965). The structure of this compound, 2-O-α-L-rhamnopyranosyl-α-L-rhamnopyranosyl-β-hydroxydecanoyl-β-hydroxydecanoate, is illustrated in Fig. 3.

Unlike the trehalose lipids, the fatty acids are not bound to the sugar by acyl bonds. The hydroxyl function of one acid is condensed with a carbohydrate of the second acid. This results in a glycolipid with a free carboxyl group. Itoh *et al.* (1971) later identified a second extracellular rhamnolipid (R-2), which was similar to R-1 but with only one rhamnose unit (Fig. 4). It was postulated that R-2 was the precursor of R-1.

Hisatsuka *et al.* (1971) measured the surfactant properties of the R-1 lipid. The minimum surface tension of a solution of R-1 in 0.1 M $NaHCO_3$ buffer was under 40 dyn/cm. The critical micelle concentration was about 0.006%. This lipid was also found to be an excellent emulsion stabilizer. It produced very stable emulsions more efficiently than two commercial surfactants used for a comparison (Tween 20 and Noigen EA 141). From these properties, it was postulated that the function of the extracellular rhamnolipid was to act as

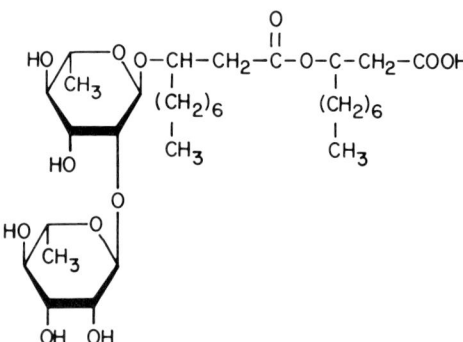

Fig. 3. The R-1 rhamnolipid isolation from *Arthrobacter paraffineus*. Adapted from Cooper *et al.*, 1979c, by permission of the U.S. Environmental Protection Agency.

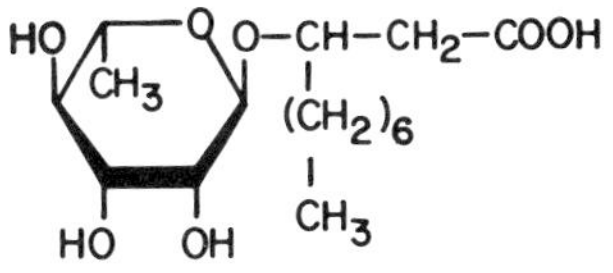

FIG. 4. The R-2 rhamnolipid isolated from *Arthrobacter paraffineus*. Adapted from Cooper *et al.*, 1979c, by permission of the U.S. Environmental Protection Agency.

an emulsifying agent for the hydrocarbon substrate of *P. aeruginosa*. It has been shown that strains of *P. aeruginosa* produced up to an order of magnitude more rhamnolipid when growing of hydrocarbon as compared to a glucose substrate (Itoh *et al.*, 1971). Hisatsuka *et al.* (1971) isolated and purified the rhamnolipid and then measured the effect of adding this compound to fresh media before inoculation. They observed dramatic increases in the rate of growth of *P. aeruginosa* as the amount of lipid was increased up to a maximum effect at a concentration of about 0.0025%. Commercial surfactants also caused significant stimulation of the rate of growth. However, none of these agents had any effect on the rate of growth on glucose. This R-1 rhamnolipid also stimulated the growth of other *P. aeruginosa* strains on *n*-hexadecane but not of other species of bacteria. Itoh and Suzuki (1972) later demonstrated that the R-2 or the R-1 rhamnolipids were essential for the growth of *P. aeruginosa* on hydrocarbons. They isolated a mutant strain that could not grow on hydrocarbons and did not produce either rhamnolipid. When R-1 or R-2 rhamnolipids isolated from other strains were added to the media of the mutant strain there was a good rate of growth on hydrocarbons (Table II). Trehalose lipids isolated from *Nocardia* or

TABLE II
EFFECT OF GLYCOLIPIDS ON THE GROWTH ON HYDROCARBON OF A
P. aeruginosa MUTANT INCAPABLE OF PRODUCING RHAMNOLIPIDS[a,b]

Glycolipid	Concentration (μg/ml)	O.D. at 660 μm after 50 hours
None		0.030
R-1[c]	10	0.140
R-1[c]	50	0.645
R-2[c]	10	0.105
R-2[c]	50	0.655
Trehalose lipid[d]	100	0.187
Trehalose lipid[e]	100	0.110

[a] From Itoh and Suzuki (1972).
[b] The O.D. of the normal strain with no added lipid after 50 hours was 0.890.
[c] From normal strain of *P. aeruginosa*.
[d] From *Nocardia*.
[e] From *Corynebacteria*.

Corynebacterium caused some growth stimulation but these were much less effective than the rhamnolipids.

C. SOPHOROSE LIPIDS

Species of the yeast *Torulopsis* produce glycolipids that have some similarity to the bacterial rhamnolipids (Gorin et al., 1961; Tulloch et al., 1967). There has been no study made of the potential surfactant properties of these lipids. However, it has been suggested that they have emulsifying capability useful to the organism for hydrocarbon fermentation (Gutierrez and Erickson, 1977).

Gorin et al. (1961) were the first to isolate these sophorose lipids from *Torulopsis magnoliae*. Like the R-1 lipid from *P. aeruginosa*, these contain a disaccharide (i.e., sophorose) attached glycosidically to the hydroxyl function of a hydroxycarboxylic acid (Fig. 5). Unlike R-1, there is only one fatty acid and there are also one or two acetate groups attached to the sophorose by acyl bounds (site of attachment not determined). Furthermore, the hydroxyl groups are not on the β carbons of the fatty acids but on the penultimate carbon of the alkyl chain. Therefore, the major acid found is 17-L-hydroxyoctadecanoic acid. A modified sophorose lipid has been isolated from *Torulopsis gropengiesseri* in which the hydroxyl group and the carboxyl functions of the fatty acids are both attached to the sophorose, generating a macrocyclic lactose ring (Jones, 1967; Tulloch et al., 1967). This lipid was diacylated and the sites of attachment were determined (Fig. 6).

An interesting feature of the production of the sophorose lipids is that their structure and yield can be influenced by the addition of secondary substrates (Tulloch et al., 1962; Jones and Howe, 1968; Jones, 1967). Just as the saccharide portion of the trehalose lipids was altered by growing coryneform bacteria on different carbohydrates (Suzuki et al., 1974; Itoh and Suzuki, 1974), the nature of the fatty acid in the sophorose lipids was influenced by the addition of carboxylic acids or methyl esters, hydrocarbons or glycerides to the culture medium of *Torulopsis*. For example, Table III

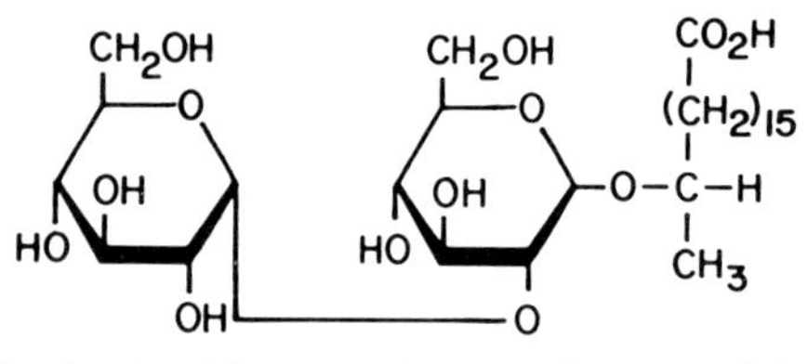

FIG. 5. A sophorose lipid isolated from *Torulopsis*. At least one of the carbohydrate hydroxyl functions was acylated but the site of attachment was not determined. Adapted from Cooper et al., 1979c, by permission of the U.S. Environmental Protection Agency.

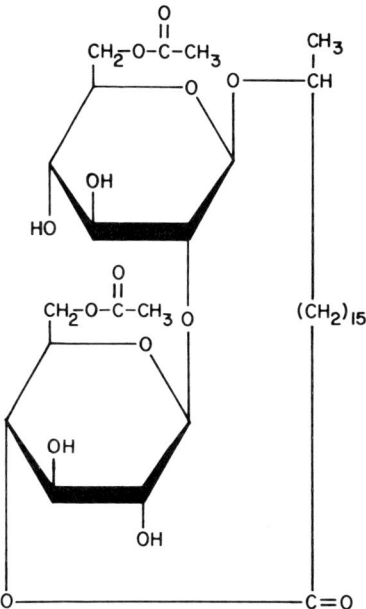

FIG. 6. The lactone structure of a sophorose lipid from *Torulopsis*. Adapted from Cooper et al., 1979c, by permission of the U.S. Environmental Protection Agency.

TABLE III
THE EFFECT OF ADDED METHYL CARBOXYLATES ON THE DISTRIBUTION OF HYDROXYCARBOXYLIC ACIDS IN THE SOPHOROSE LIPIDS[a,b]

Carbons in acid	15-OH C_{1e}	16-OH C_{1e}	16-OH C_{1e}	17-OH C_{17}	17-OH C_{18}	17-OH Oleic[c]	18-OH Oleic[c]	18-OH C_{19}	Others	Yield[d] (gm)
16	40	40			5	14	1			0.73
17			87	13						0.64
18[e]	2	2		80	9				8	0.68
19	3	5	24	9	6			53		0.63
21	4	9	54	17	5			8	3[f]	0.60
Oleic[c]	2	2			6	77	13			0.85
Linoleic[c]	1	2			4	37	56			0.60
Erucic[g]	2	4			5	65	12		12	0.51

[a] From Tulloch et al. (1962).
[b] Expressed as percentage of total.
[c] Unsaturated acids containing 18 carbon atoms.
[d] One gram of ester added.
[e] Ethyl ester.
[f] 20-Hydroxyhenicosanoic acid.
[g] Unsaturated acid containing 22 carbon atoms.

lists the fatty acids isolated from the sophorose lipids of *Torulopsis magnoliae* grown in the presence of various fatty acid esters (Tulloch *et al.*, 1962). In each case, the yield of glycolipid is greatly increased by the addition of an ester. The distribution of hydroxy fatty acids is also altered by the esters added. When the ester fatty acid had 16, 17, 18, or 19 carbons the major hydroxy fatty acid found appeared to be derived directly from the ester with a hydroxyl function added to either the penultimate or terminal carbon. The longer fatty acids appeared to be shortened to 18 carbon atoms. These sophorose lipids are a potentially interesting family of biosurfactants, which can be altered by growth conditions and by the species of *Torulopsis* used to produce them. Furthermore, by the addition of cosubstrates, it is possible to get yields as high as 17 gm/liter (Tulloch *et al.*, 1962).

D. Diglycosyl Diglycerides

Diglycosyl diglycerides are the most common type of glycolipids found in microorganisms (Shaw, 1970, 1974). Figure 7 shows the structure of one of the five common types, α-diglucosyldiglyceride. The other four are β-diglucosyl-, dimannosyl-, digalactosyl-, and galactosylglucosyldiglycerides. Much smaller amounts of the mono-, tri-, and tetraglycosyldiglycerides have also been isolated.

The potential surfactant properties of the glycosyldiglycerides have not been tested. However, Brundish *et al.* (1967) commented on the surfactant-like structure of this class of molecules with a polar, water-soluble head and two lipophilic alkyl tails. Wicken and Knox (1970) reported micelle formation by glycosyldiglycerides isolated from *Lactobacillus fermenti*.

E. Polysaccharide–Lipid Complexes

Kaeppeli and Fiechter (1976, 1977) found a polysaccharide–lipid complex that "bound" hydrocarbon substrates of the yeast *Candida tropicalis*. This complex, isolated from the cell wall, was found to emulsify hexadecane and water (Kaeppeli and Fiechter, 1977). The lipid content of the polysaccharide from cells grown of hydrocarbon substrate was 2.5% of the total weight but only 0.1% of the total weight of the complex isolated from cells growing on glucose (Kaeppeli and Fiechter, 1976, 1977). The lipids were a mixture of saturated and unsaturated fatty acids, most of which had 14, 16, or 18 carbon atoms (Kaeppeli and Fiechter, 1977).

A surface-active polymer has been reported from an *Arthrobacter* species (Rosenberg *et al.*, 1979a,b; Zuckerberg *et al.*, 1979). This polymer had a molecular weight of about 9.76×10^5. It contained D-galactosamine (20–30%), an unidentified amino uronic acid (33%), D-glucose (5%), and fatty

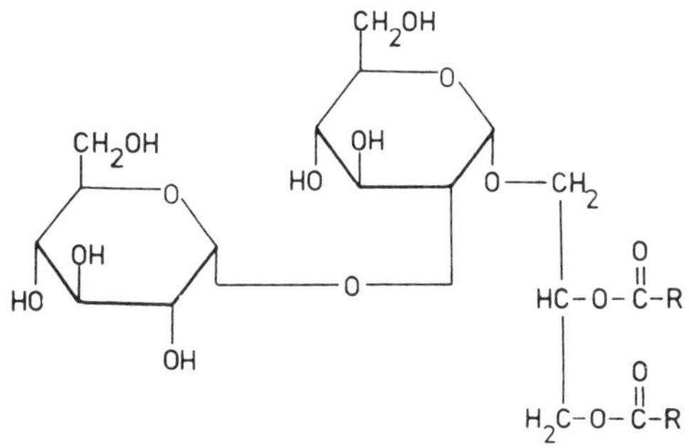

FIG. 7. A generalized structure of α-diglucosyldiglyceride; R represents an alkyl substituent.

acids (15%). The authors found that a wide number of oils could be emulsified with water using this polymer. However, there was also a large variation in the stability of the various emulsions. The most stable were obtained when the oil contained both aliphatic straight chain components and either aliphatic or aromatic cyclic components.

III. Amino Acid-Containing Surfactants

A. LIPOPEPTIDES

Lipopeptides have been isolated from a wide variety of bacteria and yeasts, but only a few have been thoroughly characterized and even fewer have been tested for surfactant properties. However, a lipopeptide produced by *Bacillus subtilis* is the most effective biosurfactant reported in the literature. This compound has been given two trivial names, surfactin (Arima *et al.*, 1968) and subtilysin (Bernheimer and Avigad, 1970). The structure (determined by Kakinuma *et al.*, 1969a,b,c) is shown in Fig. 8. There is a chain of seven amino acids covalently bonded at one end to the carboxyl function and at the other end to the hydroxyl function of a β-hydroxy fatty acid. Two of the amino acid residues, glutamic acid and aspartic acid, contain free carboxylic acid functions; the rest have small alkyl groups. Surfactin is an extremely effective surfactant. As little as 0.005% by weight lowers the surface tension of 0.1 M $NaHCO_3$ from 71.6 dyn/cm to 27.9 dyn/cm (Arima *et al.*, 1968). A striking property of this compound is its ability to lyse mammalian erythrocytes and bacterial spheroplasts (Bernheimer and Avigad, 1970; Arima *et al.*, 1968).

Iguchi *et al.* (1969) isolated a hydrocarbon-emulsifying factor from *Candida petrophilum* growing on alkanes. This factor was found to contain a peptide and one or more fatty acids. The peptide was composed of aspartic

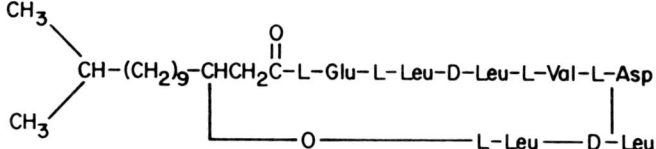

Fig. 8. The structure of surfactin or subtilysin, a lipopeptide isolated from *Bacillus subtilis*. Adapted from Cooper *et al.*, 1979c, by permission of the U.S. Environmental Protection Agency.

acid, glutamic acid, alanine, and leucine, which is not dissimilar from the peptide fraction of surfactin.

Many other lipopeptides have been isolated from bacteria, but in general, despite their amphipathic structures, there have been very few measurements of their surfactant abilities. *Bacillus subtilis* produces at least two lipopeptides other than surfactin (Takahara *et al.*, 1976; Peypoux *et al.*, 1976; Besson *et al.*, 1977). These are similar to surfactin with a cyclic peptide incorporating a fatty acid (e.g., Fig. 9). Unlike surfactin, the acids in these compounds have a β-amino function instead of a β-hydroxyl function. These lipopeptides have also been shown to have lytic and antimicrobial properties (Takahara *et al.*, 1976; Walton and Woodruff, 1949).

Streptomyces canus produces a lipopeptide that is reported to be highly surface active although no details are given (Heineman *et al.*, 1953).

Corynebacterium lepus produces a lipopeptide that reduces the surface tension of distilled water to 52 dyn/cm (Cooper *et al.*, 1979a). The lipopeptide is 35% by weight protein and the remainder carboxylic acids. These are 25% saturated fatty acids and 75% corynomycolic acids. Both types have been shown to include a wide range of isomers attributed to growth on a substrate containing a mixture of alkanes (Cooper *et al.*, 1979b).

A variety of other small lipopeptides have been isolated, including the following examples from Asselineau (1966, p. 190). Esperine isolated from *Bacillus mesentericus* contained a β-hydroxycarboxylic acid, L-Asp, L-Glu, L-Val, L-Leu, and D-Leu. Peptidolipid "Na" from *Nocardia asteroides* con-

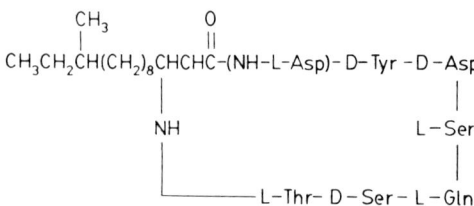

Fig. 9. One of the surfactin analogs isolated from *Bacillus subtilis* containing a β-amino fatty acid instead of a β-hydroxy fatty acid.

tained β-hydroxyeicosanoic acid, two L-Thr, L-Val, L-Pro, L-Ala, D-Ala, and D-*allo*-Ile. Peptidolipid "J," from *Mycobacterium paratuberculosis*, contained simple fatty acids, eiconanoic acid, L-Phe, D-Phe, L-Ala, L-Leu, and L-Ile. Fortuitine, from *Mycobacterium fortuitum*, also contained simple carboxylic acids, mainly eicosanoic and docosanoic acids, three L-Val, two L-Thr, one L-Ala, and one L-Pro and two MeLeu. At least one of these lipids, peptidolipid Na, had a cyclic structure similar to surfactin.

B. Ornithine Lipids

Wilkinson (1972) has isolated a lipid from *Pseudomonas rubescens* containing only one amino acid, ornithine, which caused emulsification. The compound (Fig. 10) contained both a β-hydroxycarboxylic acid and a simple carboxylic acid esterified to the hydroxyl function of the first acid. The β-hydroxy acid was amide linked to the α-amino group of ornithine. The lipid was a zwitterion, having both a free carboxyl group and a free amine group. A similar lipid with different carboxylic acids was isolated from *Thiobacillus thiooxidans* (Knoche and Shiveley, 1972). *Agrobacterium tumefaciens* produces a lipid of the same general structure, but with ornithine replaced by lysine (Tahara *et al.*, 1976a,b). From *Gluconobacter cerinus* Tahara *et al.* (1976a) isolated a compound that was a modification of the ornithine lipid; they named it cerilipin. In this analog (Fig. 11) a taurine molecule has been amide bonded to the ornithine carboxyl function, thus introducing a sulfate group into the lipid.

Another modification of the ornithine lipid has two fatty acids, but both are bonded directly to the ornithine (Gorcheim, 1968, Kawanami *et al.*, 1968; Kawanami and Otsuka, 1969; Kawanami, 1971). One acid is amide linked to the β-amino group of the ornithine. The other is attached at the carboxyl group of the ornithine by ester linkages to a bridging ethylene glycol (Kawanami, 1971).

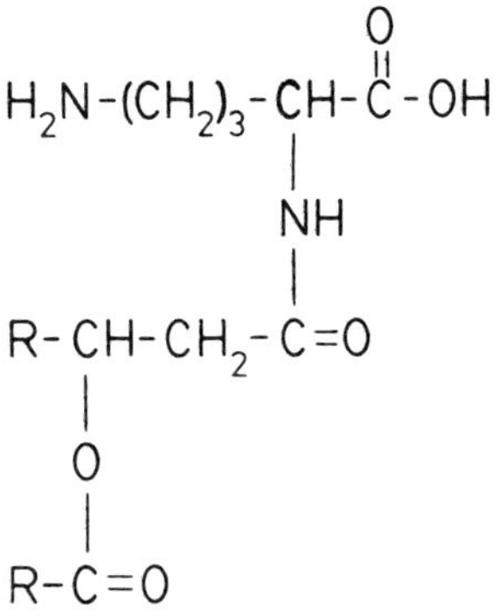

$$H_2N-(CH_2)_3-CH-\overset{O}{\underset{|}{C}}-OH$$
$$|$$
$$NH$$
$$|$$
$$R-CH-CH_2-C=O$$
$$|$$
$$O$$
$$|$$
$$R-C=O$$

FIG. 10. The structure of the ornithine-containing lipid isolated from *Pseudomonas rubescens*; R represents an alkyl substituent.

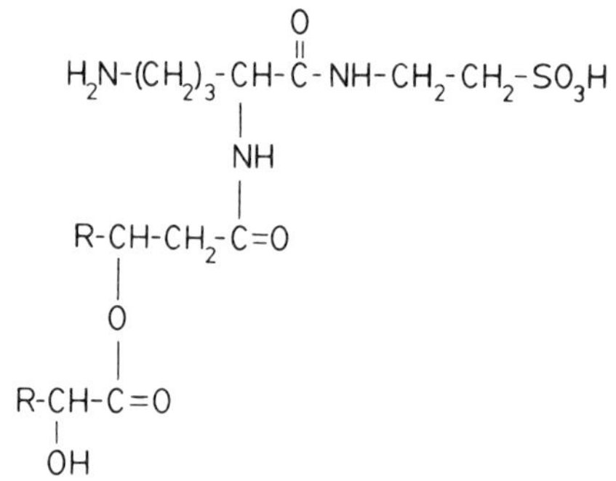

$$H_2N-(CH_2)_3-CH-\overset{O}{\underset{|}{C}}-NH-CH_2-CH_2-SO_3H$$
$$\underset{|}{NH}$$
$$R-CH-CH_2-C=O$$
$$\underset{|}{O}$$
$$R-CH-C=O$$
$$\underset{|}{OH}$$

FIG. 11. The structure of cerilipin isolated from *Gluconobacter cerinus*; R represents an alkyl substituent.

C. PROTEIN

Hisatsuka et al. (1972, 1975, 1977) isolated a "protein-like activator" for n-alkane oxidation from *P. aeruginosa*. This had a molecular weight of about 14,300 and 147 amino acid residues but was not characterized further. This activator was able to emulsify hexadecane and water, and it stimulated the growth of *P. aeruginosa* on hexadecane. If the compound was removed from the cells by treatment with ethylenediaminetetraacetic acid (EDTA) the ability of the organism to oxidize n-hexadecane was impaired unless the hexadecane was first emulsified by sonication.

IV. Phospholipids

Phospholipids are found in every microorganism (Shaw, 1974; Komura et al., 1975a,b; Asselineau, 1966). Less frequent are examples of significant production of extracellular phospholipids or the measure of surfactant properties of these lipids. Figure 12 is a generalized structure of phospholipids. All contain a glycerol unit esterified to two fatty acids and one phosphate

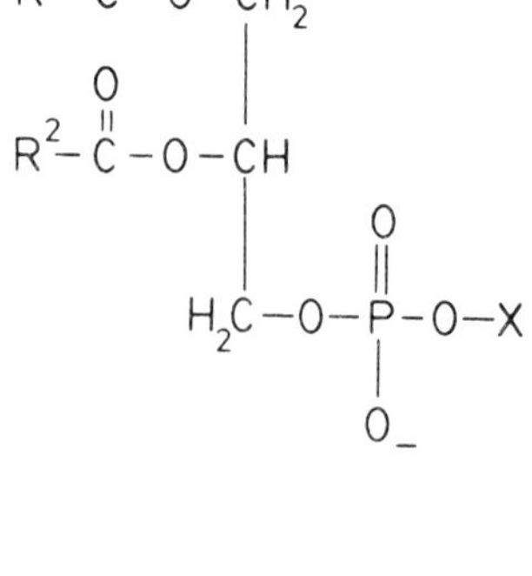

FIG. 12. General structures of some phospholipids isolated from microorganisms: R^1 and R^2 are alkyl substituents; X = H, phosphatidic acid; X = CH_2NH_2, phosphatidylethanolamine; X = $CH_2CH(NH_2)$-COOH, phosphatidylserine; X = inositol, phosphatidylinositol; X = inositol substituted with one or more mannose, phosphatidylinositol mannosides; X = $CH_2CHOHCH_2OH$, phosphatidylglycerol; X = $CH_2CHOHCH_2OPO_3$, phosphatidylglycerol phosphate. Cardiolipin contains two units of phosphatidic acid esterified to the 1 and 3 positions of glycerol.

$$R^1-\overset{O}{\underset{\|}{C}}-O-CH_2$$
$$R^2-\overset{O}{\underset{\|}{C}}-O-CH$$
$$H_2C-O-\overset{O}{\underset{\|}{P}}-O-X$$
$$\underset{|}{O_-}$$

group that usually has further substitution. Lysophospholipids have only one fatty acid ester.

Thiobacillus thiooxidans produces substances that wet elemental sulfur necessary for its growth (Jones and Starkey, 1961; Schaeffer and Umbreit, 1963, Jones and Benson, 1965; Beebe and Umbreit, 1971). During growth the surface tension of the medium fell from 72 dyn/cm to 49 dyn/cm (Jones and Starkey, 1961). The cell-free supernatant of a 7-day-old culture retained the ability to wet elemental sulfur. Various phospholipids have been isolated from the cell-free broth. Schaeffer and Umbreit (1963) decided that phosphatidylinositol was the major surfactant present. Jones and Benson (1965) isolated phosphatidylglycerol from the broth and Beebe and Umbreit (1971) isolated both phosphatidylglycerol and phosphatidic acid. Phosphatidylglycerol and phosphatidylinositol, isolated from other sources, were both able to wet elemental sulfur (Schaeffer and Umbreit, 1963; Beebe and Umbreit, 1971). The best wetting agent, of all the lipids studied, was found to be phosphatidylethanolamine. This was not isolated from *T. thiooxidans* cell-free medium but it was present in the cells (Beebe and Umbreit, 1971).

From *Corynebacterium lepus* a mixture of surface-active lipids was isolated that lowered the surface tension of distilled water to 49 dyn/cm (Cooper *et al.*, 1979a). This mixture included several phospholipids, including phosphatidylglycerol, phosphatidylinositol, phosphatidylglycerol phosphate, cardiolipin, and a phosphatidylinositol mannoside.

Although it is relatively uncommon for microorganisms to produce appreciable amounts of extracellular phospholipids, it is possible to induce the excretion of these lipids. When *Corynebacterium alkanolyticum* was treated with penicillins or cephalósporins the amount of extracellular phospholipids recovered was increased by more than an order of magnitude (Nakao *et al.*, 1973; Kikuchi *et al.*, 1973). The major lipids were cardiolipin and phosphatidylethanolamine.

The nature of the mixtures of phospholipids produced by microorganisms is influenced by the substrates. The amount of phospholipids produced by *Candida tropicalis* was much larger for cells grown on *n*-alkanes than for cells grown on glucose (Mishina *et al.*, 1977).

The relative amounts of individual phospholipids are also affected by substrate and other growth conditions. For example, Table IV contains some data for the phospholipids of *Micrococcus cerificans* (Makula and Finnerty, 1970). The amount of phosphatidylethanolamine increases from 47.1% to 61.5% when the substrate was changed from acetate at pH 7 to hexadecane at pH 7. The cardiolipin, phosphatidylglycerol phosphate, and phosphatidylglycerol content decreases. These trends were even more pronounced when the substrate was changed to hexadecane and the pH lowered

TABLE IV
EFFECT OF MEDIUM ON PHOSPHOLIPID DISTRIBUTION IN *Micrococcus cerificans*[a,b]

Phospholipid	Acetate pH 7	Hexadecane	
		pH 7	pH 2
Cardiolipin	14.5	12.9	1.0
Phosphatidylglycerol phosphate	7.7	5.7	Trace
Phosphatidylglycerol	30.6	19.9	12.2
Phosphatidylethanolamine	47.1	61.5	86.8

[a] From Makula and Finnerty (1970).
[b] Percentage of total micromoles of phosphorous per gram (dry wt.).

to 2. Under these conditions, phosphatidylethanolamine accounts for 86.8% of the phospholipids.

The fatty acids isolated from microbial phospholipids are also influenced by substrate. Table V includes a partial carboxylic acid distribution for the phospholipids of *Micrococcus cerificans* grown on nutrient broth, or one of four pure hydrocarbons (Makula and Finnerty, 1972). When pure n-alkane substrates were used, the major carboxylic acid observed had the same

TABLE V
DISTRIBUTION OF CARBOXYLIC ACIDS FROM PHOSPHOLIPIDS OF *Micrococcus cerificans* GROWN ON PURE HYDROCARBONS[a,b]

Fatty acid	NB[c]	Tetradecane	Pentadecane	Hexadecane	Heptadecane
$C_{10:0}$				0.4	
$C_{12:0}$		2.8			
$C_{13:0}$			0.5		
$C_{14:0}$	0.8	25.8	0.1	0.9	
$C_{14:1}$		9.1			
$C_{15:0}$			44.5		5.5
$C_{15:1}$			29.5	28.5	4.4
$C_{16:0}$	33.8	5.2	1.7	65.9	4.1
$C_{16:1}$	10.1	18.8	2.2		26.4
$C_{17:0}$			0.6		85.1
$C_{17:1}$			3.3		
$C_{18:0}$			1.5	1.4	8.4
$C_{18:1}$	37.8	24.3	3.4		

[a] From Makula and Finnerty (1972).
[b] Expressed as micromoles per gram (dry wt.).
[c] Nutrient broth.

number of carbon atoms as the hydrocarbon. Furthermore, both pentadecane and heptadecane resulted in an appreciable amount of unsaturated and saturated carboxylic acids with an odd number of carbon atoms.

Studies of monolayers of phospholipids of nonmicrobial origin have shown the importance of the nature of both the polar function and the carboxylic acid components on the surface properties of phospholipids (VanDeenen et al., 1962; Sacré and Tocanne, 1977; Phillips and Chapman, 1968, Chapman et al., 1966; Phillips and Hauser, 1974). Therefore, the small changes observed in the phospholipid composition of microorganisms with changes in substrates and growing conditions could have significant effects on the surfactant properties of the lipid mixtures isolated. VanDeenen et al. (1962) showed that phosphatidylcholine, phosphatidylserine, phosphatidylethanolamine, and phosphatidic acid films all had different surface properties. However, even larger differences were observed when the chain lengths of the fatty acids of each type of phospholipid were varied. Large changes were also observed when unsaturation was introduced into the alkyl chains.

V. Fatty Acids and Neutral Lipids

A. INTRODUCTION

Fatty acids and neutral lipids are found in all microbial cells and are often produced extracellularly (Shaw, 1974; Kates, 1972; Asselineau, 1966). Most of these lipids, including carboxylic acids, alcohols, esters, monoglycerides, diglycerides, and triglycerides, have been shown to have some degree of surface activity (Helenius and Simons, 1975; Phillips and Hauser, 1974; Singleton, 1960, 1968; Goddard, 1973; Lin et al., 1973). However, the neutral lipids tested have usually been from nonmicrobial sources. Because they are identical to, or very similar to, lipids that have been isolated from microbes conclusions about them are pertinent to this review.

Most of the examples of the extracellular production of neutral lipids or fatty acids by microbes have involved organisms growing on hydrocarbons. This suggests that they may be important for hydrocarbon emulsification. The following subsections include both examples of microbial production of these lipids and examples of the surfactant properties of similar lipids from other sources.

Carboxylic acids are much more important biosurfactants than the neutral lipids and are included in most of the papers cited in this section.

B. CARBOXYLIC ACIDS

Corynebacterium lepus produces a mixture of biosurfactants when grown on kerosene, which reduces the surface tension of the whole broth to 30 dyn/cm (Cooper et al., 1979a). Early in the fermentation, during the exponential growth phase, the major surfactant is a mixture of corynomycolic acids (Fig. 1). The concentration of these acids increases to a maximum and then decreases later in the fermentation to be replaced by a mixture of polar lipids in the stationary growth phase.

The surface tension of the mixture of corynomycolic acids in water (0.5 gm/liter) was about 40 dyn/cm and varied only slightly as the pH changed. The interfacial tension against hexadecane was about 10 dyn/cm and was also independent of pH (Cooper et al., 1980).

The β-hydroxy α-branched corynomycolic acids (R^1—CH(OH)—CH(R^2)—COOH) showed the influence of the substrate as both even and odd chain lengths were found (Cooper et al., 1979b). This reflects the mixture of odd- and even-chain length alkanes found in kerosene. The R^1 chain had from 16 to 25 carbon atoms and R^2 had from 8 to 13.

It is now well documented that the chain length of the hydrocarbon substrate influences the chain length of the simple fatty acids produced by the microorganism (Makula and Finnerty, 1968, 1972; Yanagawa et al., 1972; Mishina et al., 1977; Stewart and Kallio, 1959). Usually odd-number carbon hydrocarbons result in appreciable amounts of odd fatty acids. Often the major fatty acid found has the same carbon number as the hydrocarbon substrate. Similar results have been observed for dicarboxylic acids (Uchio and Shiio, 1972) and alcohols (Suzuki and Ogawa, 1972; Stewart and Kallio, 1959) grown on various pure hydrocarbons.

Odier (1976) has studied microorganisms that grow in fuel tanks. A range of organisms was reported, most of which were degrading the paraffinic fraction. Many of these, including *Pseudomonas*, *Mycococcus*, *Penicillium*, *Aspergillus*, and *Acinetobacter*, produced appreciable quantities of extracellular fatty acids. Makula and Finnerty (1972) observed that *Micrococcus cerificans* produced extracellular fatty acids when grown on hydrocarbons but not when grown on soluble substrates.

A number of yeasts produce extracellular dicarboxylic fatty acids when grown on hydrocarbons (Shiio and Uchio, 1971). The best producer discovered by these authors was *Candida cloacae* 310 (Uchio and Shiio, 1972). This yeast yielded as much as 29.3 gm/liter of dicarboxylic acid when grown on *n*-hexadecane.

C. Surface Properties of Carboxylic Acids

Fatty acids were one of the first types of compounds to be studied in relation to surface phenomena (Adam, 1941, and references therein). Measurements were made of the surface pressure and packing of films of fatty acids spread on the surface and interfacial tension effects of fatty acid so-

TABLE VI
THE SURFACE AND INTERFACIAL TENSIONS OF SATURATED
SOLUTIONS OF SODIUM CARBOXYLATES IN 0.1% NaOH AT 70°C[a]

Acid	Surface tension (dyn/cm)	Interfacial tension (dyn/cm)
Dodecanoic	36	6
Tetradecanoic	34	3
Hexadecanoic	31	2
Octadecanoic	28	

[a] From Powney and Addison (1938).

lutions. Powney and Addison (1938) measured the surface tension and interfacial tension against xylene of the sodium salts of four carboxylic acids at 70°C (Table VI). Both the surface tension and the interfacial tension decreased as the length of the fatty acid increased.

A more complete study included saturated acids from hexanoic to nonadecanoic (Cooper et al., 1980). Plots of the surface tensions and interfacial tensions against hexadecane, of the acids in water, and at various pH values, versus chain length demonstrated a marked dependence of surfactant ability on the hydrocarbon chain length (e.g., Fig. 13). Dodecanoic and tridecanoic acids represent a maximum in surfactant effectiveness for the

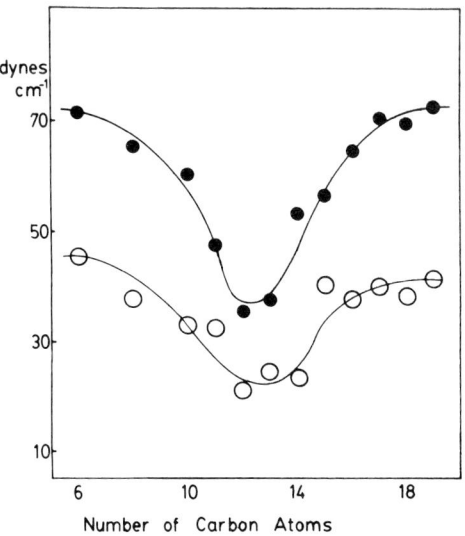

FIG. 13. Surface tension (●) and interfacial tension against hexadecane (○) of saturated solutions of carboxylic acids (0.5 mg/ml) versus the number of carbon atoms in each acid at pH 7.

data at pH 7.0. The longest and shortest acids had very little effect on either the surface or the interfacial tensions of water. At pH 9.0 the best surfactant is tetradecanoic acid and at pH 4.0 it is undecanoic acid.

Kondo and Kanai (1977) studied the mycobactericidal activity of a series of saturated fatty acids from C_2 to C_{20}. Tetradecanoic acid was the most active against mycobacteria. Both shorter and longer homologs were less effective. The toxic effects of fatty acids and other amphipathic compounds have been attributed to their surfactant capabilities (Kondo and Kanai, 1977; Miller and Barran, 1977; Newton, 1960). Therefore, this may be another example of the effect of the length of the alkyl group on the surface properties of fatty acids. It is significant that the most toxic fatty acid was tetradecanoic acid, which is similar to the maximum in surfactant ability (Cooper et al., 1980).

Corynomycolic acids and other hydroxy fatty acids have been shown to be much more effective surfactants than the simple fatty acids (Cooper et al., 1980). Octadecanoic acid and hexadecanoic acid are relatively poor surfactants, but both 2-hydroxyoctadecanoic acid (Fig. 14) and 2-hydroxyhexadecanoic acid cause significant lowering of the surface and interfacial tensions of water. Furthermore, the position of the hydroxyl function is significant for surfactant ability. Neither 12-hydroxyoctadecanoic acid (Fig. 14) nor 16-hydroxyhexadecanoic acid were as effective as the 2-hydroxy analogs.

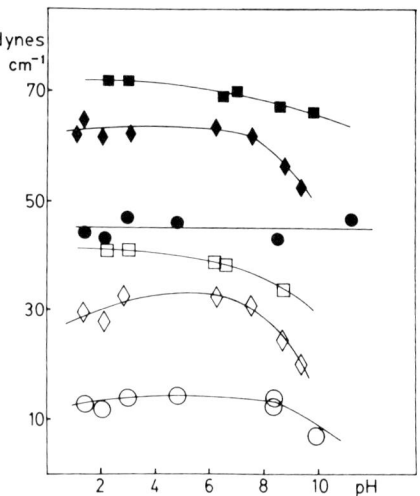

FIG. 14. Data for saturated solutions (0.5 mg/ml) of three carboxylic acids versus pH; octadecanoic acid, surface tension (■) and interfacial tension (□); 12-hydroxyoctadecanoic acid, surface tension (◆) and interfacial tension (◇); 2-hydroxyoctadecanoic acid, surface tension (●) and interfacial tension (○).

D. Neutral Lipids and Mixtures with Fatty Acids

Holdorn and Turner (1969) isolated a surface-active ether extract of *Mycobacterium rhodochrous* grown on decane. This extract, which was presumably neutral lipids, caused emulsification of water and decane and enhanced the growth of *M. rhodochrous* on decane. When added to water (0.1 gm/liter), the extract lowered the surface tension to 44 dyn/cm.

Arthrobacter parafineus produces extracellular fatty alcohols when grown on hydrocarbons (Suzuki and Ogawa, 1972). However, the maximum accumulation was not at the end of the fermentation. The maximum concentrations (ca. 0.5 mg/ml) were observed in the early log phase of growth. Fatty alcohols have also been isolated from *Mycobacterium lacticolum* growing on n-hexadecane (Mil'ko et al., 1976).

Stewart and Kallio (1959) isolated a gram-negative coccus that produced extracellular esters of long-chain alcohols and carboxylic acids when grown on hydrocarbons.

Makula et al. (1975) studied the lipids of a species of *Acinetobacter* grown on n-hexadecane. Table VII lists the amounts of various lipid classes isolated from the cells and the culture medium. This table also contains data for this bacteria grown on a nutrient broth–yeast extract medium. The amount of extracellular lipids isolated was much larger when the organism was grown on hexadecane, but the intracellular lipids were of similar concentrations for both media. The extracellular lipids isolated from the hexadecane-grown cells were mainly monoglycerides, diglycerides, and wax esters. Tri-

TABLE VII
Distribution of Neutral Lipids Isolated from a Species of *Acinetobacter* Growing on Nutrient Broth–Yeast Extract or Hexadecane[a]

Lipid	Intracellular[b]		Extracellular[c]	
	NB–YE[d]	C_{16}	NB–YE[d]	C_{16}
Triglyceride	1.78	2.5	2.4	25.6
Mono- and diglyceride	0.38	6.8	0	410.0
Free fatty acid	7.5	8.2	4	6.0
Free fatty alcohol	Trace	2.6	0	0.5
Wax ester	11.5	18.0	0	280

[a] From Makula et al. (1975).
[b] Expressed as micromoles per gram of cells (dry wt.).
[c] Expressed as micromoles per liter.
[d] Nutrient broth–yeast extract medium.

glycerides and free fatty acids were also observed as well as trace amounts of free fatty alcohols.

Thiobacillus thiooxidans produces a mixture of lipids that act as wetting agents for elemental sulfur (Jones and Starkey, 1961; Schaeffer and Umbreit, 1963; Jones and Benson, 1965; Beebe and Umbreit, 1971). These were mainly phospholipids; however, Beebe and Umbreit (1971) also identified several neutral lipids, including glycerides, fatty acids, and hydroxy fatty acids.

VI. Conclusions

Microorganisms produce many potentially useful surface-active agents. This wide range of surfactant properties provides a broad spectrum for industrial applications. The list includes both ionic and nonionic surfactants and ranges from short fatty acids to large polymers. Some of the biosurfactants are insensitive to pH changes and others can have their properties altered by adjusting the pH. Although the lipophilic portion is usually the hydrocarbon chain of a fatty acid, the polar groups include many different functional groups. These range from mildly hydrophilic alcohols and esters to ionic groups, such as phosphates and sulfates. Most of these biosurfactants are biodegradable and many are nontoxic.

There are many examples of biosurfactant modification by altering substrates. It is also possible to obtain different homologs of a class of compounds by using different microorganisms. Therefore, it is possible to isolated families of compounds with gradations of surfactant properties.

The majority of biosurfactants characterized have been lipids. However, most of the lipids known to be produced by microorganisms have not been thoroughly tested for surfactant properties. There is an excellent potential for the discovery of many more useful biosurfactants.

References

Abbott, B. J., and Gledhill, W. E. (1971). *Adv. Appl. Microbiol.* **14**, 249–388.
Adam, N. K. (1941). "The Physics and Chemistry of Surfaces," 3rd ed. Oxford Univ. Press, London and New York.
Anderson, R. J., and Newman, M. S. (1933). *J. Biol. Chem.* **101**, 499–504.
Arima, K., Kakinuma, A., and Tamura, G. (1968). *Biochem. Biophys. Res. Commun.* **31**, 488–494.
Asselineau, J. (1966). "The Bacterial Lipids." Holden-Day, San Francisco, California.
Beebe, J. L., and Umbreit, W. W. (1971). *J. Bacteriol.* **108**, 612–614.
Bernheimer, A. W., and Avigad, L. S. (1970). *J. Gen. Microbiol.* **61**, 361–369.
Besson, F., Peypoux, F., Michel, G., and Delcambe, L. (1977). *Eur. J. Biochem.* **77**, 61–68.
Brundish, D. E., Shaw, N., and Baddiley, J. (1967). *Biochem. J.* **105**, 885–889.
Chapman, D., Owens, N. F., and Walker, D. A. (1966). *Biochim. Biophys. Acta* **120**, 148–155.

Cooper, D. G., Zajic, J. E., and Gerson, D. F. (1979a). *Appl. Environ. Microbiol.* **37**, 4–10.
Cooper, D. G., Zajic, J. E., and Gracey, D. E. F. (1979b). *J. Bacteriol.* **137**, 795–801.
Cooper, D. G., Zajic, J. E., and Gerson, D. F. (1979c). Proceedings of Workshop: "Microbial Degradation of Pollutants in Marine Environments," pp. 231–240. Washington, D.C.: Environmental Protection Agency.
Cooper, D. G., Zajic, J. E., and Denis, C. (1980). *J. Am. Oil Chem. Soc.* (in press).
Cundell, A. M., and Traxler, R. W. (1973). *Mar. Pollut. Bull.* **4**, 125–127.
Edwards, J. R., and Hayashi, J. A. (1965). *Arch. Biochem. Biophys.* **111**, 415–421.
Einsele, A., and Fiechter, A. (1971). *Adv. Biochem. Eng.* **6**, 169–194.
Etémadi, A. H. (1966). *C. R. Hebd. Seances Acad. Sci. Ser. C* **263**, 1257–1259.
Etémadi, A. H., Okuda, R., and Lederer, E. (1964a). *Bull. Soc. Chim. Fr.* pp. 868–870.
Etémadi, A. H., Miquel, A., Lederer, E., and Barber, M. (1964b). *Bull. Soc. Chim. Fr.* pp. 3274–3276.
Gerson, D. F., and Zajic, J. E. (1977). *Can. Inst. Min. Metall. Spec. Vol. 17*, 705–710.
Gerson, D. F., and Zajic, J. E. (1978a). *Dev. Ind. Microbiol.* **19**, 577–599.
Gerson, D. F., and Zajic, J. E. (1978b). In "Oil Sands and Oil Shale" (O. Strauz, ed.), pp. 145–161. Verlag Chemie, Weinhelm.
Goddard, E. D. (1973). *Croat. Chem. Acta* **45**, 13–30.
Gorcheim, A. (1968). *Biochim. Biophys. Acta* **152**, 358–367.
Gorin, P. A. J., Spencer, J. F. T., and Tullock, A. P. (1961). *Can. J. Chem.* **39**, 846–855.
Gutierrez, J. R., and Erickson, L. E. (1977). *Biotechnol. Bioeng.* **19**, 1331–1350.
Hedrick, H. G., Reynolds, R. J., and Crum, M. G. (1968). *Dev. Ind. Microbiol.* **9**, 415–425, **10**, 222–233.
Heineman, B., Kaplan, M. A., Muir, R. D., and Hooper, I. R. (1953). *Antibiot. Chemother. (Washington, D.C.)* **3**, 1239–1242.
Helenius, A., and Simons, K. (1975). *Biochim. Biophys. Acta* **415**, 29–79.
Hisatsuka, K., Nakahara, T., Sano, N., and Yamada, K. (1971). *Agric. Biol. Chem.* **35**, 686–692.
Hisatsuka, K., Nakahara, T., and Yamada, K. (1972). *Agric. Biol. Chem.* **36**, 1361–1369.
Hisatsuka, K., Nakahara, T., Minoda, Y., and Yamada, K. (1975). *Agric. Biol. Chem.* **39**, 999–1005.
Hisatsuka, K., Nakahara, T., Minoda, Y., and Yamada, K. (1977). *Agric. Biol. Chem.* **41**, 445–450.
Holdorn, R. S., and Turner, A. G. (1969). *J. Appl. Bacteriol.* **34**, 448–456.
Hug, H., Blanch, H. W., and Fiechter, A. (1974). *Biotechnol. Bioeng.* **16**, 965–985.
Humphrey, A. E. (1967). *Biotechnol. Bioeng.* **9**, 3–24.
Humphrey, A. E., and Erickson, L. E. (1972). *J. Appl. Chem. Biotechnol.* **4**, 125–147.
Iguchi, T., Takeda, I., and Ohsawa, H. (1969). *Agric. Biol. Chem.* **33**, 1657–1658.
Ioneda, T., Lenz, M., and Pudles, J. (1963). *Biochem. Biophys. Res. Commun.* **13**, 110–114.
Ioneda, T., Lederer, E., and Rozanis, J. (1970). *Chem. Phys. Lipids* **4**, 375–392.
Itoh, S., and Suzuki, T. (1972). *Agric. Biol. Chem.* **36**, 2233–2235.
Itoh, S., and Suzuki, T. (1974). *Agric. Biol. Chem.* **38**, 1443–1449.
Itoh, S., Honda, H., Tomita, F., and Suzuki, T. (1971). *J. Antibiot.* **24**, 855–859.
Jobson, A., Cook, F. D., and Westlake, D. W. S. (1972). *Appl. Microbiol.* **23**, 1082–1089.
Jones, D. F. (1967). *J. Chem. Soc. C* pp. 479–484.
Jones, D. F., and Howe, R. (1968). *J. Chem. Soc. C* pp. 2801–2808.
Jones, G. E., and Benson, A. A. (1965). *J. Bacteriol.* **89**, 260–261.
Jones, G. E., and Starkey, R. L. (1961). *J. Bacteriol.* **82**, 788–789.
Kaeppeli, O., and Fiechter, A. (1976). *Biotechnol. Bioeng.* **18**, 967–974.
Kaeppeli, O., and Fiechter, A. (1977). *J. Bacteriol.* **131**, 917–921.

Kakinuma, A., Hori, M., Isono, M., Tamura, G., and Arima, K. (1969a). *Agric. Biol. Chem.* **33**, 971–972.
Kakinuma, A., Sugino, H., Isono, M., Tamura, G., and Arima, K. (1969b). *Agric. Biol. Chem.* **33**, 973–976.
Kakinuma, A., Hori, M., Sugino, H., Yoshida, I., Isono, M., Tamura, G., and Arima, K. (1969c). *Agric. Biol. Chem.* **33**, 1523–1524.
Kates, M. (1972). "Techniques of Lipidology." North-Holland Publ., Amsterdam.
Kawanami, J. (1971). *Chem. Phys. Lipids* **7**, 159–172.
Kawanami, J., and Otsuka, H. (1969). *Chem. Phys. Lipids* **3**, 135–139.
Kawanami, J., Kimura, A., and Otsuka, H. (1968). *Biochim. Biophys. Acta* **152**, 808–810.
Kikuchi, M., Kanamaru, T., and Nakao, Y. (1973). *Agric. Biol. Chem.* **37**, 2405–2408.
Klug, M. H., and Markovitz, A. J. (1971). *Adv. Microb. Physiol.* **5**, 1–43.
Knettig, E., and Zajic, J. E. (1972). *Biotechnol. Bioeng.* **14**, 379–390.
Knoche, H. W., and Shiveley, J. M. (1972). *J. Biol. Chem.* **247**, 170–178.
Komura, I., Yamada, K., and Komagata, K. (1975a). *J. Gen. Appl. Microbiol.* **21**, 97–107.
Komura, I., Yamada, K., Otsuka, S., and Komagata, K. (1975b). *J. Gen. Appl. Microbiol.* **21**, 251–261.
Konda, E., and Kanai, K. (1977). *Jpn. J. Med. Sci. Biol.* **30**, 171–178.
Lederer, E. (1967). *Chem. Phys. Lipids* **1**, 294–315.
Lin, I. J., Friend, J. P., and Zimmels, Y. (1973). *J. Colloid Interface Sci.* **45**, 378–385.
McKenna, E. J., and Kallio, R. E. (1965). *Annu. Rev. Microbiol.* **19**, 183–208.
Makula, R. A., and Finnerty, W. R. (1968). *J. Bacteriol.* **95**, 2102–2107.
Makula, R. A., and Finnerty, W. R. (1970). *J. Bacteriol.* **103**, 348–355.
Makula, R. A., and Finnerty, W. R. (1972). *J. Bacteriol.* **112**, 398–407.
Makula, R. A., Lockwood, P. J., and Finnerty, W. R. (1975). *J. Bacteriol.* **121**, 250–258.
Mallee, F. M., and Blanch, H. W. (1977) *Biotechnol. Bioeng.* **19**, 1793–1816.
Margaritis, A., Zajic, J. E., and Gerson, D. F. (1979). *Biotechnol. Bioeng.* **21**, 1151–1162.
Markovits, J., Pinte, F., and Etémadi, A. H. (1966). *C. R. Hebd. Seances Acad., Sci., Ser. C* **263**, 960–922.
Matales, R. I., Baruah, J. N., and Tannenbaum, S. R. (1968). *Science* **157**, 1322–1323.
Mil'ko, E. S., Egorov, N. N., and Revenko, A. A. (1976). *Mikrobiologiya* **45**, 808–811.
Miller, R. W., and Barran, L. R. (1977). *Can. J. Microbiol.* **23**, 1373–1383.
Mishina, M., Isurugi, M., Tanaka, A., and Fukui, S. (1977). *Agric. Biol. Chem.* **41**, 635–640.
Mulkins-Phillips, G. J., and Stewart, J. E. (1974). *Appl. Microbiol.* **28**, 915–922.
Nakahara, T., Erickson, L. E., and Gutierrez, J. R. (1977). *Biotechnol. Bioeng.* **19**, 9–25.
Nakao, Y., Kanamaru, T., Kikuchi, M., and Yamatodani, S. (1973), *Agric. Biol. Chem.* **37**, 2399–2404.
Newton, B. A. (1960). *J. Appl. Bacteriol.* **23**, 345–349.
Odier, E. (1976). *Ann. Microbiol. (Paris)* **127b**, 213–225.
Peypoux, T., Michel, G., and Decambe, L. (1976). *Eur. J. Biochem.* **63**, 391–398.
Phillips, M. C., and Chapman, D. (1968). *Biochim. Biophys. Acta* **163**, 301–313.
Phillips, M. C., and Hauser, H. (1974). *J. Colloid Interface Sci.* **49**, 31–39.
Powney, J., and Addison, C. C. (1938). *Trans. Faraday Soc.* **34**, 372–377.
Rapp, P., and Wagner, F. (1976). *Abstr. Pap., Int. Ferment. Symp., 5th, 1976* p. 133.
Rosenberg, E., Zuckerberg, A., Rubinovitz, C., and Gutnick, D. L. (1979a). *Appl. Environ. Microbiol.* **37**, 402–408.
Rosenberg, E., Perry, A., Gibson, D. F., and Gutnick, D. L. (1979b). *Appl. Environ. Microbiol.* **37**, 409–413.
Sacré, M. M., and Tocanne, J. F. (1977). *Chem. Phys. Lipids* **18**, 334–354.
Schaeffer, W. I., and Umbreit, W. W. (1963). *J. Bacteriol.* **85**, 492–493.

Shaw, N. (1970). *Bacteriol. Rev.* **34**, 365-377.
Shaw, N. (1974). *Adv. Appl. Microbiol.* **17**, 63-108.
Shiio, I., and Uchio, R. (1971). *Agric. Biol. Chem.* **35**, 2033-2042.
Singleton, W. S. (1960). *In* "Fatty Acids, Their Chemistry, Properties, Production and Uses" (E. Scott Pattison, ed.), Part 1, pp. 609-633. Dekker, New York.
Singleton, W. S. (1968). *In* "Fatty Acids, Their Chemistry, Properties, Production and Uses" (E. Scott Pattison, ed.), Part 5, pp. 3445-3454. Dekker, New York.
Small, D. M. (1968). *J. Am. Oil Chem. Soc.* **45**, 108-119.
Steck, P. A., Schwartz, B. A., Rosendahl, M. S., and Gray, G. R. (1978). *J. Biol. Chem.* **253**, 5625-5629.
Stewart, J. E., and Kallio, R. E. (1959). *J. Bacteriol.* **78**, 726-730.
Suzuki, T., and Ogawa, K. (1972). *Agric. Biol. Chem.* **36**, 457-463.
Suzuki, T., Tanaka, K., Matsubara, I., and Kinoshita, S. (1969). *Agric. Biol. Chem.* **33**, 1619-1627.
Suzuki, T., Tanaka, H., and Itoh, S. (1974). *Agric. Biol. Chem.* **38**, 557-563.
Tahara, Y., Kameda, M., Yamada, Y., and Kondo, K. (1976a). *Agric. Biol. Chem.* **40**, 243-244.
Tahara, Y., Yamada, Y., and Kondo, K. (1976b). *Agric. Biol. Chem.* **40**, 1449-1450.
Takahara, Y., Hirose, Y., Yasuda, N., Mitsugi, K., and Murao, S. (1976). *Agric. Biol. Chem.* **40**, 1901-1903.
Tulloch, A. P., Spencer, J. F. T., and Gorin, P. A. J. (1962). *Can. J. Chem.* **40**, 1326-1338.
Tulloch, A. P., Hill, A., and Spencer, J. F. T. (1967). *Chem. Commun.* pp. 584-586.
Uchio, R., and Shiio, I. (1972). *Agric. Biol. Chem.* **36**, 426-433.
VanDeenen, L. L. M., Houtsmuller, U. M. T., DeHaas, G. H., and Mulder, E. (1962). *J. Pharm. Pharmacol.* **14**, 429-444.
Vilkas, E., and Rojas, A. (1964). *Bull. Soc. Chim. Biol.* **46**, 689-701.
Walker, J. D., and Colwell, R. R. (1974). *Appl. Microbiol.* **27**, 1053-1060.
Walton, R. B., and Woodruff, H. B. (1949). *J. Clin. Invest.* **28**, 924-926.
Whitworth, D. A., Moo-Young, M., and Viswanatha, T., (1973). *Biotechnol. Bioeng.* **15**, 649-675.
Wicken, A. J., and Knox, K. W. (1970). *J. Gen. Microbiol.* **60**, 293-301.
Wilkinson, S. G. (1972). *Biochim. Biophys. Acta* **270**, 1-17.
Yanagawa, S., Fujii, K., Tanaka, D., and Fukui, S. (1972). *Agric. Biol. Chem.* **36**, 2123-2128.
Zajic, J. E., and Knettig, E. (1971). *Dev. Ind. Microbiol.* **12**, 87-97.
Zajic, J. E., Gerson, D. F., and Camp, S. E. (1977a). *Can. Fed. Biol. Sci.* **20**, 33.
Zajic, J. E., Guignard, H., and Gerson, D. F. (1977b). *Biotechnol. Bioeng.* **19**, 1303-1320.
Zuckerberg, A., Diver, A., Peeri, Z., Gutnick, D. L., and Rosenberg, E. (1979). *Appl. Environ. Microbiol.* **37**, 414-420.

INDEX

A

Acid hydrolysis, of cellulosic wastes, 180–182
Alcohol, as fuel, 151–152, *see also* Ethanol
Alicyclic hydrocarbon, oxidation by bacterial extracts, 77–78
Alkane
  oxidation, by bacterial extracts, 74
    by P(40) fraction of *Methylosinus*, 53
  oxidative attack, 97–98
  as substrates, cell yield from, 107
Alkane utilizer, isolation, 90–91
Alkene
  gaseous, epoxidation of, 41–42
  oxidation by methane-grown bacteria, 50
    methane monooxygenase, 74–76
    P(40) fraction of *Methylosinus*, 53
Amino acid, in surfactants, 239–342
Anaerobic methane oxidation, in aquatic environments, 12–18
Anthrobacter globiformis, carbon assimilation pathway, 28
Aquatic environment, anaerobic methane oxidation in, 12–18
Aromatic hydrocarbon, oxidation by bacterial extracts, 77–78
Assay
  for methane-utilizing bacteria, 44–45
  of microbial methyl ketone formation, 55–56

B

Bacillus PM6, carbon assimilation pathway, 28
Bacterium 5B1, carbon assimilation pathway, 28
Biosurfactant
  amino acid-containing, 239–342
  carbohydrate containing, 231–239
  carboxylic acids and, 245–246
  cerilipin, structure, 242
  diglycosyl diglyceride containing, 238
  fatty acids and, 245
  importance, 229–230
  lipopeptide containing, 239–241
  microorganisms, 230–331
  neutral lipids and, 249–250
  ornithine lipid-containing, 241
  phospholipids and, 242–245
  polysaccharide–lipid complex containing, 238–239
  proteins, 242
  rhamnolipid containing, 234–236
  sophorose lipid-containing, 236–238
  surfactin, structure, 240
  trehalose lipid-containing, 231–234
  uses, 231
Brevibacterium, carbon assimilation pathway, 28
2-Butanol, oxidation to 2-butanone, 56–62
2-Butanone
  production, effect of metal-chelating agent, 61
    inhibition studies, 61
    pH, 58–59
    product inhibition, 60–61
    substrate concentration, 60
    substrate specificity, 62
    temperature, 58–59

C

Carbon assimilation, by facultative methylotrophs, 28
C_1-utilizing microbe
  conversion of 2-butanol to 2-butanone, 57
  epoxidation by, 41–54
Carboxylic acid
  biosurfactants and, 245–246
  interfacial tension, 247
  saturated solutions, 248
Cassava root, ethanol production, 171–174
Cellulose
  acid and enzymatic hydrolysis, comparison, 197
  ball milling, 187
  fermentation of hydrolyzate to ethanol, 194–195

INDEX

Cellulose (continued)
  enzymatic hydrolysis, 182
  ethanol production, 176-199
  γ-irradiation, 187-188
  pretreatment for ethanol production, 186-189
  sodium hydroxide treatment, 188
  sulfur dioxide treatment, 188-189
  waste, 195
Cellulosic waste, in ethanol production, 196-197
Cerilipin, structure, 242
Cheese whey, in ethanol production, 199-201
Coffee, waste, 201
Corn
  ethanol production, 165-168
    energy balance, 219
Corynomycolic acid, structure, 232

D

Diglycosyl diglyceride
  in biosurfactants, 238
  structure, 239

E

Enrichment culture, methane, 16
Epoxidation
  by C_1-utilizing microbes, 41-54
  of gaseous 1-alkenes, 41-42
  inhibition studies, 50-51
  of propylene, 45-48
    effect of concentration, 49
    effect of inhibitors, 51
    pH, 48-49
    temperature, 48-49
  substrate specificity of methylotrophs, 49-50
1,2-Epoxide, 42-43
Ethanol
  from acid hydrolysis of cellulosic wastes, 180-182
  biomass feedstock, comparison, 218
  cassava roots, 171-174
  cellulose, 176-199
    ball milling, 187
    enzymatic hydrolysis process, 189-194
    fermentation of hydrolyzate, 194-195
    γ-irradiation, 187-188
    sodium hydroxide treatment, 188
    sulfur dioxide treatment, 188-189
  cellulosic wastes, one-step process, 196-197
  cheese whey, 199-201
  coffee waste, 201
  corn, 165-168
    integration of production with wet milling, 168-171
  enzymatic hydrolysis of cellulose, 182
  fermentation, alternative liquid fuel, 147-224
  as fuel, 216-218
  net by-product credits, 217
  noval starch sources, 175-176
  pineapple waste, 202
  potato peel waste, 202
  from potatoes, 174-175
  production, cellulose vs. corn, 198-199
    comparison of processes, 216
    economic analysis, 204-218
    energy requirements, 221-224
    future developments, 197-198
    processes, 153-154
  from spent sulfite liquor, 202-204
  starch, 165
  sugarcane, productivity factors, 155
  sugars, 154-165
    "ABC" process, 162-163
    continuous fermentation, 157-161
    fermentation of pentoses, 164
    yeasts tolerant to high concentrations, 161
  Trichoderma cellulase, 182-186
  wood, 177-180
Ether, oxidation by methane monooxygenase, 76-77

F

Facultative methylotroph
  one-carbon metabolism, regulation, 30-32
  plasmids and methane oxidation, 32-35
Fatty acid
  biosurfactants and, 245
  of propane utilizer, 104-105

Formaldehyde fixation, Icl$^+$ serine pathway, 24

G

Gaseous alkane utilizer, isolation, 94–95
Glycolipid, and growth, 235
Growth
   transition from nonhydrocarbons to hydrocarbons, 102
   using hydrocarbon substrates, 106–108

H

Heterocyclic hydrocarbon, oxidation by bacterial extracts, 77–78
Hydrocarbon
   oxidation by methane monooxygenases, 71–85
   as substrates, growth yields, 106–108

I

Icl$^+$ serine pathway, of formaldehyde fixation, 24
Inhibitor
   effect on epoxidation of propylene, 51
   hydroxylation of methane, 51
Interfacial tension, of carboxylic acids, 247
Isocitrate lyase, of propane-utilizer, 104–105
Isolation
   of alkane utilizers, 90–91
   gaseous alkane-utilizing organism, 94–95
   methane-oxidizing organisms, 18–21

K

Ketone, formation by C_1-utilizing microbes, 54–68

L

Lipid, composition of utilizing organisms, 103–104

Lipopeptide, in surfactants, 239–241
Liquid fuel, ethanol production, 147–224

M

Metal-chelating agent, 2-butanone production, 61
Methane
   epoxidation of propylene, 51
   hydroxylation, inhibitors, 51
   oxidation, coded by plasmids, 32–35
      by extracts of *M. capsulatus*, 75
      inhibitors, 79
   oxidizing activities of methylotrophs, 52
   production in Lake Mendota, Wisconsin, 7–12
Methane-grown bacteria, oxidation of gaseous alkenes and methane, 50
Methane monooxygenase
   electron transfer, mechanism, 84
   of *M. trichosporium*, 78–80
   oxidation of alkanes, 74–75
      alkenes, 75–76
      ethers, 76–77
      hydrocarbons, 71–85
      propene, 85–86
   purification from *M. capsulatus* Bath, 80–84
   specificity of, 96–97
   substrates, 73–78
Methane oxidation, by yeasts, 26–27
Methane-oxidizing bacteria, *see also* Methylotroph
   in aquatic environments, 7–12
   diversity of, 21–26
   isolation of, 18–21
   occurrence, 4
Methane-utilizing microbe
   assay method, 44–45
   characteristics, 91–92
   strains, 43–44
Methanol
   and bacteria, 27–30
   and fungi, 27–30
Methyl ketone
   formation by microorganisms, 54–68
   microbial formation, assay for, 55–56
Methylobacter, methane oxidation, 23

Methylobacterium organophilum, carbon assimilation, pathway, 28
Methylobacterium organophilum XX, DNA homology, 35
Methylococcus, methane oxidation, 23
Methylococcus capsulatus Bath, methane monooxygenase, purification, 81–84
Methylocystis, methane oxidation, 23
Methylomonas, methane oxidation, 23
Methylomonas methanica, oxidation of hydrocarbons, 85
Methylosinus
  methane oxidation, 23
  oxidation of *n*-alkenes and *n*-alkanes, 53
Methylosinus trichosporium, methane monooxygenase, 78–80
Methylotroph
  cell fractions, 52
  conversion of 2-butanol to 2-butanone, 63
  detection, techniques for, 4–6
  ecology, 3–35
  epoxidation, inhibition studies, 50–51
  epoxidation, of propylene, 46
    and substrate specificity, 49–50
  facultative, carbon assimilation by, 28
  propylene oxide production, time course, 47
Microwave irradiation
  biologic systems, effect on, 137–138
  microorganisms, effect on, 129–143
    survival, 135–136
  Staphylococcus aureus, effect on enzymatic activity, 141–142
  thermal vs. nonthermal, 138–143
Mycobacterium 10, carbon assimilation pathway, 28
Mycobacterium vaccae, growth on gaseous alkanes, 106

N

Neutral lipid
  from *Acinetobacter*, 249
  biosurfactants and, 249–250

O

Organism JB_1, carbon assimilation pathway, 28
Ornithine lipid, in biosurfactant, 241

P

Pentose, ethanol production, 164
Phospholipid
  biosurfactants, 242–245
  structure, 242
Pichia, secondary alcohol dehydrogenase, purification, 65
Pineapple waste, ethanol production, 202
Plasmid, methane oxidation, coding for, 32–35
Propane
  cooxidations, 110–111
  metabolism, 99
  as substrate, growth yields, 106–107
  toxicity, 111
  utilization, 89–112
Propane-oxidizing bacteria
  fatty acid production, 109–110
  products from, 108–110
  single-cell proteins, production, 109
  water-soluble vitamin content, 109
Propane-oxidizing system
  induction of, 95–98
  molecular oxygen involvement, 95–96
Propane utilizer
  characteristics, 92–93
  lipid composition of, 103–104
  prospecting with, 111–112
  three carbon substrates, metabolism, 104–106
Propene, oxidation by methane monooxygenase, 85–86
Propylene
  denitrifying, characteristics, 119
  epoxidation, effect of inhibitors, 51
    methane, 51
  metabolism, 101–102
  microbial epoxidation of, 45–48
  oxidation of, 98
  secondary alcohol dehydrogenase, purification, 64
  whole cell oxidation, 86
Pseudomonas aminovorans, carbon assimilation pathway, 28
Pseudomonas AM1, carbon assimilation pathway, 28
Pseudomonas butanovora
  and *n*-butane, production of protein from, 117–126
  cell growth, 121–123

culture conditions, 121–122
extracellular protein, accumulation, 123–124
grown on n-butane, cell composition, 121
substrate utilization, 120
taxonomic properties, 118–119
Pseudomonas C, carbon assimilation pathway, 28
Pseudomonas MA, carbon assimilation pathway, 28
Pseudomonas MS, carbon assimilation pathway, 28
Pseudomonas oleovorans, carbon assimilation pathway, 28

Sophorose lipid
methyl carboxylates, 237
from *Torulopsis*, 236, 237
Spent sulfite liquor, ethanol production, 202–204
Staphylococcus aureus
microwave irradiation, effect on, 141
thermonuclease activity, 142
Starch
ethanol production, 165
novel sources, ethanol production, 175–176
Streptomyces, carbon assimilation pathway, 28
Surfactin, analogs, structure, 240

S

Secondary alcohol, oxidation of, by C_1 utilizers grown on methanol, 62
Secondary alcohol dehydrogenase
inhibition studies, 67
metal content, 67
properties, 65–68
purification, 62–67
substrate specificity, 66–67

T

Threhalose lipid
in biosurfactants, 231–234
effect on growth of *Nocardia*, 233
Trichoderma cellulose, ethanol production, 182–186

CONTENTS OF PREVIOUS VOLUMES

Volume 1

Protected Fermentation
 Miloš Herold and Jan Nečásek

The Mechanism of Penicillin Biosynthesis
 Arnold L. Demain

Preservation of Foods and Drugs by Ionizing Radiations
 W. Dexter Bellamy

The State of Antibiotics in Plant Disease Control
 David Pramer

Microbial Synthesis of Cobamides
 D. Perlman

Factors Affecting the Antimicrobial Activity of Phenols
 E. O. Bennett

Germfree Animal Techniques and Their Applications
 Arthur W. Phillips and James E. Smith

Insect Microbiology
 S. R. Dutky

The Production of Amino Acids by Fermentation Processes
 Shukuo Kinoshita

Continuous Industrial Fermentations
 Philip Gerhardt and M. C. Bartlett

The Large-Scale Growth of Higher Fungi
 Radcliffe F. Robinson and R. S. Davidson

AUTHOR INDEX—SUBJECT INDEX

Volume 2

Newer Aspects of Waste Treatment
 Nandor Porges

Aerosol Samplers
 Harold W. Batchelor

A Commentary on Microbiological Assaying
 F. Kavanagh

Application of Membrane Filters
 Richard Ehrlich

Microbial Control Methods in the Brewery
 Gerhard J. Hass

Newer Development in Vinegar Manufactures
 Rudolph J. Allgeier and Frank M. Hildebrandt

The Microbiological Transformation of Steroids
 T. H. Stoudt

Biological Transformation of Solar Energy
 William J. Oswald and Clarence G. Golueke

SYMPOSIUM ON ENGINEERING ADVANCES IN FERMENTATION PRACTICE

Rheological Properties of Fermentation Broths
 Fred H. Deindoerfer and John M. West

Fluid Mixing in Fermentation Process
 J. Y. Oldshue

Scale-Up of Submerged Fermentations
 W. H. Bartholemew

Air Sterilization
  Arthur E. Humphrey

Sterilization of Media for Biochemical Processes
  Lloyd L. Kempe

Fermentation Kinetics and Model Processes
  Fred H. Deindoerfer

Continuous Fermentation
  W. D. Maxon

Control Applications in Fermentation
  George J. Fuld

AUTHOR INDEX—SUBJECT INDEX

Volume 3

Preservation of Bacteria by Lyophilization
  Robert J. Heckly

Sphaerotilus, Its Nature and Economic Significance
  Norman C. Dondero

Large-Scale Use of Animal Cell Cultures
  Donald J. Merchant and C. Richard Eidam

Protection against Infection in the Microbiological Laboratory: Devices and Procedures
  Mark A. Chatigny

Oxidation of Aromatic Compounds by Bacteria
  Martin H. Rogoff

Screening for the Biological Characterizations of Antitumor Agents Using Microorganisms
  Frank M. Schabel, Jr., and Robert F. Pittillo

The Classification of Actinomycetes in Relation to Their Antibiotic Activity
  Elio Baldacci

The Metabolism of Cardiac Lactones by Microorganisms
  Elwood Titus

Intermediary Metabolism and Antibiotic Synthesis
  J. D. Bu'Lock

Methods for the Determination of Organic Acids
  A. C. Hulme

AUTHOR INDEX—SUBJECT INDEX

Volume 4

Induced Mutagenesis in the Selection of Microorganisms
  S. I. Alikhanian

The Importance of Bacterial Viruses in Industrial Processes, Especially in the Dairy Industry
  F. J. Babel

Applied Microbiology in Animal Nutrition
  Harlow H. Hall

Biological Aspects of Continuous Cultivation of Microorganisms
  T. Holme

Maintenance and Loss in Tissue Culture of Specific Cell Characteristics
  Charles C. Morris

Submerged Growth of Plant Cells
  L. G. Nickell

AUTHOR INDEX—SUBJECT INDEX

Volume 5

Correlations between Microbiological Morphology and the Chemistry of Biocides
  Adrian Albert

Generations of Electricity by Microbial Action
  J. B. David

Microorganisms and the Molecular Biology of Cancer
G. F. Gause

Rapid Microbiological Determinations with Radioisotopes
Gilbert V. Levin

The Present Status of the 2,3-Butylene Glycol Fermentation
Sterling K. Long and Roger Patrick

Aeration in the Laboratory
W. R. Lockhart and R. W. Squires

Stability and Degeneration of Microbial Cultures on Repeated Transfer
Fritz Reusser

Microbiology of Paint Films
Richard T. Ross

The Actinomycetes and Their Antibiotics
Selman A. Waksman

Fusel Oil
A. Dinsmoor Webb and John L. Ingraham

AUTHOR INDEX—SUBJECT INDEX

Volume 6

Global Impacts of Applied Microbiology: An Appraisal
Carl-Göran Hedén and Mortimer P. Starr

Microbial Processes for Preparation of Radioactive Compounds
D. Perlman, Aris P. Bayan, and Nancy A. Giuffre

Secondary Factors in Fermentation Processes
P. Margalith

Nonmedical Uses of Antibiotics
Herbert S. Goldberg

Microbial Aspects of Water Pollution Control
K. Wuhrmann

Microbial Formation and Degradation of Minerals
Melvin P. Silverman and Henry L. Ehrlich

Enzymes and Their Applications
Irwin W. Sizer

A Discussion of the Training of Applied Microbiologists
B. W. Koft and Wayne W. Umbreit

AUTHOR INDEX—SUBJECT INDEX

Volume 7

Microbial Carotenogenesis
Alex Ciegler

Biodegradation: Problems of Molecular Recalcitrance and Microbial Fallibility
M. Alexander

Cold Sterilization Techniques
John B. Opfell and Curtis E. Miller

Microbial Production of Metal-Organic Compounds and Complexes
D. Perlman

Development of Coding Schemes for Microbial Taxonomy
S. T. Cowan

Effects of Microbes on Germfree Animals
Thomas D. Luckey

Uses and Products of Yeasts and Yeast-Like Fungi
Walter J. Nickerson and Robert G. Brown

Microbial Amylases
Walter W. Windish and Nagesh S. Mhatre

The Microbiology of Freeze-Dried Foods
Gerald J. Silverman and Samuel A. Goldblith

Low-Temperature Microbiology
Judith Farrell and A. H. Rose

AUTHOR INDEX—SUBJECT INDEX

Volume 8

Industrial Fermentations and Their Relations to Regulatory Mechanisms
Arnold L. Demain

Genetics in Applied Microbiology
S. G. Bradley

Microbial Ecology and Applied Microbiology
Thomas D. Brock

The Ecological Approach to the Study of Activated Sludge
Wesley O. Pipes

Control of Bacteria in Nondomestic Water Supplies
Cecil W. Chambers and Norman A. Clarke

The Presence of Human Enteric Viruses in Sewage and Their Removal by Conventional Sewage Treatment Methods
Stephen Alan Kollins

Oral Microbiology
Heiner Hoffman

Media and Methods for Isolation and Enumeration of the Enterococci
Paul A. Hartman, George W. Reinbold, and Devi S. Saraswat

Crystal-Forming Bacteria as Insect Pathogens
Martin H. Rogoff

Mycotoxins in Feeds and Foods
Emanuel Borker, Nino F. Insalata, Colette P. Levi, and John S. Witzeman

AUTHOR INDEX—SUBJECT INDEX

Volume 9

The Inclusion of Antimicrobial Agents in Pharmaceutical Products
A. D. Russell, June Jenkins, and I. H. Harrison

Antiserum Production in Experimental Animals
Richard H. Hyde

Microbial Models of Tumor Metabolism
G. F. Gause

Cellulose and Cellulolysis
Brigitta Norkrans

Microbiological Aspects of the Formation and Degradation of Cellulose Fibers
L. Jurášek, J. Ross Colvin, and D. R. Whitaker

The Biotransformation of Lignin to Humus—Facts and Postulates
R. T. Oglesby, R. F. Christman, and C. H. Driver

Bulking of Activated Sludge
Wesley O. Pipes

Malo-Lactic Fermentation
Ralph E. Kunkee

AUTHOR INDEX—SUBJECT INDEX

Volume 10

Detection of Life in Soil on Earth and Other Planets, Introductory Remarks
Robert L. Starkey

For What Shall We Search?
Allan H. Brown

Relevance of Soil Microbiology to Search for Life on Other Planets
G. Stotzky

Experiments and Instrumentation for Extraterrestrial Life Detection
Gilbert V. Levin

Halophilic Bacteria
D. J. Kushner

Applied Significance of Polyvalent Bacteriophages
S. G. Bradley

Proteins and Enzymes as Taxonomic Tools
  Edward D. Garber and John W. Rippon

Mycotoxins
  Alex Ciegler and Eivind B. Lillehoj

Transformation of Organic Compounds by Fungal Spores
  Claude Vézina, S. N. Sehgal, and Kamar Singh

Microbial Interactions in Continuous Culture
  Henry R. Bungay, III and Mary Lou Bungay

Chemical Sterilizers (Chemosterilizers)
  Paul M. Borick

Antibiotics in the Control of Plant Pathogens
  M. J. Thirumalachar

AUTHOR INDEX—SUBJECT INDEX

CUMULATIVE AUTHOR INDEX—CUMULATIVE TITLE INDEX

Volume 11

Successes and Failures in the Search for Antibiotics
  Selman A. Waksman

Structure-Activity Relationships of Semisynthetic Penicillins
  K. E. Price

Resistance to Antimicrobial Agents
  J. S. Kiser, G. O. Gale, and G. A. Kemp

Micromonospora Taxonomy
  George Luedemann

Dental Caries and Periodontal Disease Considered as Infectious Diseases
  William Gold

The Recovery and Purification of Biochemicals
  Victor H. Edwards

Ergot Alkaloid Fermentations
  William J. Kelleher

The Microbiology of the Hen's Egg
  R. G. Board

Training for the Biochemical Industries
  I. L. Hepner

AUTHOR INDEX—SUBJECT INDEX

Volume 12

History of the Development of a School of Biochemistry in the Faculty of Technology, University of Manchester
  Thomas Kennedy Walker

Fermentation Processes Employed in Vitamin C Synthesis
  Miloš Kulhánek

Flavor and Microorganisms
  P. Margalith and Y. Schwartz

Mechanisms of Thermal Injury in Nonsporulating Bacteria
  M. C. Allwood and A. D. Russell

Collection of Microbial Cells
  Daniel I. C. Wang and Anthony J. Sinskey

Fermentor Design
  R. Steel and T. L. Miller

The Occurrence, Chemistry and Toxicology of the Microbial Peptide-Lactones
  A. Taylor

Microbial Metabolites as Potentially Useful Pharmacologically Active Agents
  D. Perlman and G. P. Peruzzotti

AUTHOR INDEX—SUBJECT INDEX

Volume 13

Chemotaxonomic Relationships Among the Basidiomycetes
  Robert G. Benedict

Proton Magnetic Resonance Spectroscopy—An Aid in Identification and Chemotaxonomy of Yeasts
P. A. J. Gorin and J. F. T. Spencer

Large-Scale Cultivation of Mammalian Cells
R. C. Telling and P. J. Radlett

Large-Scale Bacteriophage Production
K. Sargent

Microorganisms as Potential Sources of Food
Jnanendra K. Bhattacharjee

Structure–Activity Relationships among Semisynthetic Cephalosporins
M. L. Sassiver and Arthur Lewis

Structure–Activity Relationships in the Tetracycline Series
Robert K. Blackwood and Arthur R. English

Microbial Production of Phenazines
J. M. Ingram and A. C. Blackwood

The Gibberellin Fermentation
E. G. Jeffreys

Metabolism of Acylanilide Herbicides
Richard Bartha and David Pramer

Therapeutic Dentrifrices
J. K. Peterson

Some Contributions of the U.S. Department of Agriculture to the Fermentation Industry
George E. Ward

Microbiological Patents in International Litigation
John V. Whittenburg

Industrial Applications of Continuous Culture: Pharmaceutical Products and Other Products and Processes
R. C. Righelato and R. Elsworth

Mathematical Models for Fermentation Processes
A. G. Frederickson, R. D. Megee, III, and H. M. Tsuchija

AUTHOR INDEX—SUBJECT INDEX

Volume 14

Development of the Fermentation Industries in Great Britain
John J. H. Hastings

Chemical Composition as a Criterion in the Classification of Actinomycetes
H. A. Lechevalier, Mary P. Lechevalier, and Nancy N. Gerber

Prevalence and Distribution of Antibiotic-Producing Actinomycetes
John N. Porter

Biochemical Activities of Nocardia
R. L. Raymond and V. W. Jamison

Microbial Transformations of Antibiotics
Oldrich K. Sebek and D. Perlman

In Vivo Evaluation of Antibacterial Chemotherapeutic Substances
A. Kathrine Miller

Modification of Lincomycin
Barney J. Magerlein

Fermentation Equipment
G. L. Solomons

The Extracellular Accumulation of Metabolic Products by Hydrocarbon-Degrading Microorganisms
Bernard J. Abbott and William E. Gledhill

AUTHOR INDEX—SUBJECT INDEX

Volume 15

Medical Applications of Microbial Enzymes
Irwin W. Sizer

Immobilized Enzymes
  K. L. Smiley and G. W. Strandberg

Microbial Rennets
  Joseph L. Sardinas

Volatile Aroma Components of Wines and Other Fermented Beverages
  A. Dinsmoor Webb and Carlos J. Muller

Correlative Microbiological Assays
  Ladislav J. Haňka

Insect Tissue Culture
  W. F. Hink

Metabolites from Animal and Plant Cell Culture
  Irving S. Johnson and George B. Boder

Structure-Activity Relationships in Coumermycins
  John C. Godfrey and Kenneth E. Price

Chloramphenicol
  Vedpal S. Malik

Microbial Utilization of Methanol
  Charles L. Cooney and David W. Levine

Modeling of Growth Processes with Two Liquid Phases: A Review of Drop Phenomena, Mixing and Growth
  P. S. Shah, L. T. Fan, I. C. Kao, and L. R. Erickson

Microbiology and Fermentations in the Prairie Regional Laboratory of the National Research Council of Canada 1946–1971
  R. H. Haskins

AUTHOR INDEX—SUBJECT INDEX

Volume 16

Public Health Significance of Feeding Low Levels of Antibiotics to Animals
  Thomas H. Jukes

Intestinal Microbial Flora of the Pig
  R. Kenworthy

Antimycin A., a Piscicidal Antibiotic
  Robert E. Lennon and Claude Vézina

Ochratoxins
  Kenneth L. Applegate and John R. Chipley

Cultivation of Animal Cells in Chemically Defined Media, A Review
  Kiyoshi Higuchi

Genetic and Phenetic Classification of Bacteria
  R. R. Colwell

Mutation and the Production of Secondary Metabolites
  Arnold L. Demain

Structure-Activity Relationships in the Actinomycins
  Johannes Meienhofer and Eric Atherton

Development of Applied Microbiology at the University of Wisconsin
  William B. Sarles

AUTHOR INDEX—SUBJECT INDEX

Volume 17

Education and Training in Applied Microbiology
  Wayne W. Umbreit

Antimetabolites from Microorganisms
  David L. Pruess and James P. Scannell

Lipid Composition as a Guide to the Classification of Bacteria
  Norman Shaw

Fungal Sterols and the Mode of Action of the Polyene Antibiotics
  J. M. T. Hamilton-Miller

Methods of Numerical Taxonomy for Various Genera of Yeasts
I. Campbell

Microbiology and Biochemistry of Soy Sauce Fermentation
F. M. Young and B. J. B. Wood

Contemporary Thoughts on Aspects of Applied Microbiology
P. S. S. Dawson and K. L. Phillips

Some Thoughts on the Microbiological Aspects of Brewing and Other Industries Utilizing Yeast
G. G. Stewart

Linear Alkylbenzene Sulfonate: Biodegradation and Aquatic Interactions
William E. Gledhill

The Story of the American Type Culture Collection—Its History and Development (1899–1973)
William A. Clark and Dorothy H. Geary

Microbial Penicillin Acylases
E. J. Vandamme and J. P. Voets

SUBJECT INDEX

Volume 18

Microbial Foundation of Environmental Pollutants
Martin Alexander

Microbial Transformation of Pesticides
Jean-Marc Bollag

Taxonomic Criteria for Mycobacteria and Nocardiae
S. G. Bradley and J. S. Bond

Effect of Structural Modifications on the Biological Properties of Aminoglycoside Antibiotics Containing 2-Deoxystreptamine
Kenneth E. Price, John C. Godfrey, and Hiroshi Kawaguchi

Recent Developments of Antibiotic Research and Classification of Antibiotics According to Chemical Structure
János Bérdy

SUBJECT INDEX

Volume 19

Culture Collections and Patent Depositions
T. G. Pridham and C. W. Hesseltine

Production of the Same Antibiotics by Members of Different Genera of Microorganisms
Hubert A. Lechevalier

Antibiotic-Producing Fungi: Current Status of Nomenclature
C. W. Hesseltine and J. J. Ellis

Significance of Nucleic Acid Hybridization to Systematics of Actinomycetes
S. G. Bradley

Current Status of Nomenclature of Antibiotic-Producing Bacteria
Erwin F. Lessel

Microorganisms in Patent Disclosures
Irving Marcus

Microbiological Control of Plant Pathogens
Y. Henis and I. Chet

Microbiology of Municipal Solid Waste Composting
Melvin S. Finstein and Merry L. Morris

Nitrification and Dentrification Processes Related to Waste Water Treatment
D. D. Focht and A. C. Chang

The Fermentation Pilot Plant and Its Aims
D. J. D. Hockenhull

The Microbial Production of Nucleic Acid-Related Compounds
Koichi Ogata

Synthesis of L-Tyrosine-Related Amino Acids by β-Tyrosinase
Hideaki Yamada and Hidehiko Kumagai

Effects of Toxicants on the Morphology and Fine Structure of Fungi
Donald V. Richmond

SUBJECT INDEX

Volume 20

The Current Status of Pertussis Vaccine: An Overview
Charles R. Manclark

Biologically Active Components and Properties of Bordetella pertussis
Stephen I. Morse

Role of the Genetics and Physiology of Bordetella pertussis in the Production of Vaccine and the Study of Host–Party Relationships in Pertussis
Charlotte Parker

Problems Associated with the Development and Clinical Testing of an Improved Pertussis Vaccine
George R. Anderson

Problems Associated with the Control Testing of Pertussis Vaccine
Jack Cameron

Vinegar: Its History and Development
Hubert A. Conner and Rudolph J. Allgeier

Microbial Rennets
M. Sternberg

Biosynthesis of Cephalosporins
Toshihiko Kanzaki and Yukio Fujisawa

Preparation of Pharmaceutical Compounds by Immobilized Enzymes and Cells
Bernard J. Abbott

Cytotoxic and Antitumor Antibiotics Produced by Microorganisms
J. Fuska and B. Proksa

SUBJECT INDEX

Volume 21

Production of Polyene Macrolide Antibiotics
Juan F. Martin and Lloyd E. McDaniel

Use of Antibiotics in Agriculture
Tomomasa Misato, Keido Ko, and Isamu Yamaguchi

Enzymes Involved in β-Lactam Antibiotic Biosynthesis
E. J. Vandamme

Information Control in Fermentation Development
D. J. D. Hockenhull

Single-Cell Protein Production by Photosynthetic Bacteria
R. H. Shipman, L. T. Fan, and I. C. Kao

Environmental Transformation of Alkylated and Inorganic Forms of Certain Metals
Jitendra Saxena and Philip H. Howard

Bacterial Neuraminidase and Altered Immunological Behavior of Treated Mammalian Cells
Prasanta K. Ray

Pharmacologically Active Compounds from Microbial Origin
Hewitt W. Matthews and Barbara Fritche Wade

SUBJECT INDEX

Volume 22

Transformation of Organic Compounds by Immobilized Microbial Cells
Ichiro Chibata and Tetsuya Tosa

Microbial Cleavage of Sterol Side Chains
Christoph K. A. Martin

Zearalenone and Some Derivatives: Production and Biological Activities
P. H. Hidy, R. S. Baldwin, R. L. Greasham, C. L. Keith, and J. R. McMullen

Mode of Action of Mycotoxins and Related Compounds
F. S. Chu

Some Aspects of the Microbial Production of Biotin
Yoshikazu Izumi and Koichi Ogata

Polyether Antibiotics: Versatile Carboxylic Acid Ionophores Produced by Streptomyces
J. W. Westley

The Microbiology of Aquatic Oil Spills
R. Bartha and R. M. Atlas

Comparative Technical and Economic Aspects of Single-Cell Protein Processes
John H. Litchfield

SUBJECT INDEX

Volume 23

Biology of *Bacillus popilliae*
Lee A. Bulla, Jr., Ralph N. Costilow, and Eugene S. Sharpe

Production of Microbial Polysaccharides
M. E. Slodki and M. C. Cadmus

Effects of Cadmium on the Biota: Influence of Environmental Factors
H. Babich and G. Stotzky

Microbial Utilization of Straw (A Review)
Youn W. Han

The Slow-Growing Pigmented Water Bacteria: Problems and Sources
Lloyd G. Herman

The Biodegration of Polyethylene Glycols
Donald P. Cox

Introduction to Injury and Repair of Microbial Cells
F. F. Busta

Injury and Recovery of Yeasts and Mold
K. E. Stevenson and T. R. Graumlich

Injury and Repair of Gram-Negative Bacteria, with Special Consideration of the Involvement of the Cytoplasmic Membrane
L. R. Beuchat

Heat Injury of Bacterial Spores
Daniel M. Adams

The Involvement of Nucleic Acids in Bacterial Injury
M. D. Pierson, R. F. Gomez, and S. E. Martin

SUBJECT INDEX

Volume 24

Preservation of Microorganisms
Robert J. Heckly

Streptococcus mutans Dextransucrase: A Review
Thomas J. Montville, Charles L. Cooney and Anthony J. Sinskey

Microbiology of Activated Sludge Bulking
Wesley O. Pipes

Mixed Cultures in Industrial Fermentation Processes
David E. F. Harrison

Utilization of Methanol by Yeasts
Yoshiki Tani, Nobuo Kato, and Hideaki Yamada

Recent Chemical Studies on Peptide Antibiotics
Jun'ichi Shoji

The CBS Fungus Collection
J. A. Von Arx and M. A. A. Schipper

Microbiology and Biochemistry of Oil-Palm Wine
Nduka Okafor

Bacterial-Amylases
M. B. Ingle and R. J. Erickson

SUBJECT INDEX

Volume 25

Introduction to Extracellular Enzymes: From the Ribosome to the Market Place
Rudy J. Wodzinski

Applications of Microbial Enzymes in Food Systems and in Biotechnology
Matthew J. Taylor and Tom Richardson

Molecular Biology of Extracellular Enzymes
Robert F. Ramaley

Increasing Yields of Extracellular Enzymes
Douglas E. Eveleigh and Bland S. Montenecourt

Regulation of Chorismate-Derived Antibiotic Production
Vedpal S. Malik

Structure–Activity Relationships in Fusidic Acid-Type Antibiotics
W. von Daehne, W. O. Godtfredsen, and P. R. Rasmussen

Antibiotic Tolerance in Producer Organisms
Leo C. Vining

Microbial Models for Drug Metabolism
John P. Rosazza and Robert V. Smith

Plant Cell Cultures, a Potential Source of Pharmaceuticals
W. G. W. Kurz and F. Constabel

Bacteriophages of the Genus *Clostridium*
Seiya Ogata and Motoyoshi Hongo

SUBJECT INDEX